T0215951

Lecture Notes in Mathematics 2216

More information about this series at http://www.springer.com/series/304

Chris Wendl

Holomorphic Curves in Low Dimensions

From Symplectic Ruled Surfaces to Planar Contact Manifolds

 Springer

Chris Wendl
Institut für Mathematik
Humboldt-Universität zu Berlin
Berlin, Germany

ISSN 0075-8434   ISSN 1617-9692   (electronic)
Lecture Notes in Mathematics
ISBN 978-3-319-91369-8   ISBN 978-3-319-91371-1   (eBook)
https://doi.org/10.1007/978-3-319-91371-1

Library of Congress Control Number: 2018943453

Mathematics Subject Classification (2010): 57R17, 32Q65

Printed on acid-free paper

This Springer imprint is published by the registered company Springer International Publishing AG part of Springer Nature.
The registered company address is: Gewerbestrasse 11, 6330 Cham, Switzerland

To Clara, whose entry into the world may have slightly delayed the completion of this book.

Preface

This book is an expanded version of a set of lecture notes for a minicourse I gave at IRMA Strasbourg in October 2012 as part of the *Master Classes on Holomorphic Curves and Applications to Enumerative Geometry, Symplectic and Contact Topology*. The focus of the minicourse was on certain specifically low-dimensional aspects of the theory of pseudoholomorphic curves, which lend a distinctive flavor to the study of symplectic and contact manifolds in dimensions four and three, respectively. While most of these topics are covered to some extent elsewhere in the literature (notably in [MS12], the standard reference in the field), they usually do not take center stage, and I have sometimes encountered experts in symplectic topology who seem only vaguely aware of why holomorphic curve methods are so much more powerful in dimension four than in higher dimensions.

As a convincing demonstration of this power, I chose to explain the main results of McDuff's classic paper [McD90] characterizing rational and ruled symplectic 4-manifolds. First proved around the end of the 1980s, these results are now considered fundamental in the study of symplectic 4-manifolds, and their proofs are quite beautiful and natural and, from a modern perspective, not conceptually difficult. A beginner, however, might find them unfairly intimidating if attempting to read the original papers on the subject, which were written before many of what we would now call the "standard" techniques had been fully developed. My goal therefore was to present these proofs in the most elegant way that I could, using modern techniques that I regard as essential for researchers in the field to learn. Since I had limited time and did not want to get bogged down with analysis, most of the necessary analytical background on holomorphic curves was stated without proofs, though I have endeavored in this book at least to give precise statements of all required results and brief informal explanations of why they are true, with references to other sources where the details may be found. In contrast to other available treatments of this subject (e.g., in [MS12, LM96b]), I have placed considerable emphasis on the natural role played by Lefschetz pencils and fibrations, a distinctly topological (rather than analytical) topic which has exerted a similarly large influence on symplectic topology since the 1990s.

A second objective of the original minicourse and of this book is addressed in the last two chapters, which discuss contact topology. My own motivation to understand McDuff's rational/ruled paper came largely from this direction, as it had become clear through the work of Hofer-Wysocki-Zehnder and others that the Gromov-McDuff technique of foliating symplectic 4-manifolds by holomorphic curves also had many deep implications for contact 3-manifolds. In my own research, these implications have been most apparent through the relationship between holomorphic curves and Lefschetz fibrations on symplectic fillings, which connects naturally with the study of open book decompositions on contact manifolds. The so-called *planar* contact manifolds—those which are supported by planar open books—have turned out to play an analogous role in the three-dimensional contact world to the one played by rational and ruled surfaces in dimension four. The last chapter is an attempt to illustrate this analogy, focusing in particular on two problems of fundamental importance in the field: the existence of closed Reeb orbits (i.e., the Weinstein conjecture) and the classification of symplectic fillings. I have included also in Chap. 8 some general discussion of the context for these problems and the historical development of the techniques used to study them, reaching from Conley and Zehnder's 1982 solution of the Arnol'd conjecture on the torus to the introduction in 2000 and (as of this writing not yet complete) subsequent development of *symplectic field theory*. These last two chapters are meant as a survey, so they allow themselves the luxury of not presenting complete proofs, but cover correspondingly more ground.

Chapter 7 is a later addition that was not part of the original minicourse, but fits in thematically with the material of the first six chapters. The subject here is the relationship between McDuff's characterization of rational/ruled symplectic 4-manifolds and the Gromov-Witten invariants, in particular the beautiful theorem that a symplectic 4-manifold is *symplectically uniruled* if and only if it is a blowup of a rational or ruled surface. Several results of fundamental importance can be understood as consequences of this theorem, e.g., that the class of (blowups of) symplectic rational or ruled surfaces is invariant under birational equivalence and that the minimal blowdown of a symplectic 4-manifold is unique unless it is rational or ruled. One good reason to write this chapter was that while the theorem "uniruled $\Rightarrow$ rational/ruled" has evidently been known to experts for at least 20 years, I am not aware of any previous source in the literature that both contains the statement and explains why it is true. More seriously, McDuff's paper [McD92] on immersed spheres, which carries out the hard part of the proof, seems not to have penetrated the public consciousness nearly as much as its predecessor [McD90]. One reason for this is surely that the main proof in [McD92] is significantly more intricate than anything in [McD90], and another is that the result was superseded a short time later by developments from Seiberg-Witten theory. Since I had never planned to discuss Seiberg-Witten theory in my minicourse, it therefore seemed natural after writing the notes that I should try to supplement them with a readable account of the contents of [McD92], and while I cannot say with any certainty whether I have succeeded, the outcome of that effort is Chap. 7 of this book. Since it was relevant, I took the opportunity to add a gentle introduction to Gromov-Witten theory, and the restriction to dimension four allowed me to do this in a way that some readers may find less intimidating than the standard presentation in [MS12]. It should be added that my explanation of [McD92] would have been completely impossible without some extremely valuable input from McDuff herself, who became enthusiastic about this effort before I had quite understood what I was getting myself into. The proof I've written up in Sect. 7.3 is essentially one that she explained to me after she sat down to reconsider the original argument of [McD92].

There are several topics that might have seemed natural to include but have been glaringly omitted: foremost among these is the substantial contribution made by Seiberg-Witten theory to the classification of symplectic structures on rational and ruled surfaces, including work of Taubes [Tau95, Tau00], Lalonde-McDuff [LM96a], and Li and Liu [LL95, Li99, Liu96]. I will mention a few such results in Sects. 1.2 and 7.3.1, mainly for the sake of cultural knowledge, but without any serious attempt to explain why they are true. It would have been even more unrealistic to attempt a nontrivial discussion of Seiberg-Witten theory in contact geometry; thus, my exposition says almost nothing about Taubes's solution to the Weinstein conjecture in dimension three [Tau07], nor its connections to the SFT-like invariant known as embedded contact homology (see [Hut10]) and its many impressive applications. The interesting topic of finite energy foliations (see, e.g., [HWZ03, Wen08, FS]) will be alluded to briefly but then forgotten, and my discussion of Lefschetz fibrations on symplectic fillings will necessarily omit many additional applications for which they can profitably be used, e.g., in the study of Stein manifolds [ÖS04a] and Lagrangian intersection theory [Sei08b]. I have tried at least to supply suitable references wherever possible, so the reader should never assume that what I have to say on any given topic is all that can be said.

The target reader for this book is assumed to have at least a solid background in basic differential geometry and algebraic topology (including homological intersection numbers and the first Chern class), as well as some basic literacy concerning symplectic manifolds (Darboux's theorem, Moser's stability theorem, the Lagrangian neighborhood theorem, etc.) as found e.g. in the early chapters of [MS17]. I have tried to avoid explicitly requiring prior knowledge of holomorphic curves—hence the technical overview in Chap. 2—but readers who already have such knowledge will probably find it helpful (and some of those will be content to skip most of Chap. 2).

Acknowledgments

I would like to thank Emmanuel Opshtein, IRMA Strasbourg, and the CNRS for bringing about the workshop that gave rise to the lecture notes on which this book is based. Many thanks also to Patrick Massot and Paolo Ghiggini for their careful reading and helpful comments on various preliminary versions, Janko Latschev for providing the proof of Proposition 8.10, and especially to Dusa McDuff for her invaluable explanations of the paper [McD92] and for many constructive comments on the first draft of Chap. 7.

Much of the writing of this book was carried out at University College London, where I was supported in part by a Royal Society University Research Fellowship.

Berlin, Germany   Chris Wendl

Contents

Chapter 1
Introduction

The main subject of this book is a set of theorems that were among the earliest major applications of pseudoholomorphic curves in symplectic topology, and which illustrate the power of holomorphic curves to turn seemingly *local* information into *global* results. The term "local information" here can mean various things: in the results of Gromov and McDuff that will be our main topic, it refers to the existence of a symplectic submanifold with certain properties. In Chap. 9, we will also sketch some more recent results of this nature in contact topology, for instance classifying the symplectic fillings of a given contact manifold. The "local" information in this case is the boundary of a symplectic manifold, which sometimes completely determines the interior. Such phenomenona are consequences of the rigid analytical properties of pseudoholomorphic curves in symplectic settings.

1.1 Some Examples of Symplectic 4-Manifolds and Submanifolds

If (M, ω) is a symplectic manifold, we say that a submanifold $S \subset M$ is **symplectically embedded** (and we thus call it a **symplectic submanifold**) if $\omega|_{TS}$ defines a symplectic form on S, i.e. the restriction of ω to S is nondegenerate. Our focus will be on situations where $\dim M = 4$ and $\dim S = 2$, in which case we can take advantage of the homological intersection product

$$H_2(M) \times H_2(M) \to \mathbb{Z}$$
$$(A, B) \mapsto A \cdot B,$$

defined by counting (with signs) the intersections of any two transversely intersecting immersed submanifolds that represent A and B (see for example [Bre93]). The work of Gromov [Gro85] and McDuff [McD90] revealed that in the world of

© Springer International Publishing AG, part of Springer Nature 2018
C. Wendl, *Holomorphic Curves in Low Dimensions*, Lecture Notes in Mathematics 2216, https://doi.org/10.1007/978-3-319-91371-1_1

symplectic 4-manifolds, a special role is played by those which happen to contain a symplectically embedded 2-sphere $S \subset (M, \omega)$ with

$$[S] \cdot [S] \geqslant 0.$$

We shall state some of the important results about these in Sect. 1.2 below, but first, let us take a brief look at some specific examples. The upshot of the results we will discuss is that these are in fact the *only* examples of closed symplectic 4-manifolds containing such an object.

Example 1.1. Let σ_1 and σ_2 denote two area forms on S^2. Then

$$(S^2 \times S^2, \sigma_1 \oplus \sigma_2)$$

is a symplectic manifold, carrying what we call a **split** symplectic structure. For any $z \in S^2$, the submanifolds $S_1 := S^2 \times \{z\}$ and $S_2 := \{z\} \times S^2$ are each symplectically embedded and have self-intersection number 0. Examples with positive self-intersection may be found as follows: if we identify S^2 with the extended complex plane $\mathbb{C} \cup \{\infty\}$, then any complex submanifold of $S^2 \times S^2$ is also a symplectic submanifold of $(S^2 \times S^2, \sigma_1 \oplus \sigma_2)$. Now choose a holomorphic map $f : S^2 \to S^2$ of degree $d > 0$ (i.e. a rational function), and consider the graph

$$\Sigma_f := \{(z, f(z)) \mid z \in S^2\} \subset S^2 \times S^2.$$

This is a symplectic submanifold, and since $[\Sigma_f] = [S_1] + d[S_2] \in H_2(S^2 \times S^2)$ and $[S_1] \cdot [S_2] = 1$, we have

$$[\Sigma_f] \cdot [\Sigma_f] = [S_1] \cdot [S_1] + 2d[S_1] \cdot [S_2] + d^2[S_2] \cdot [S_2] = 2d > 0.$$

Exercise 1.2. For each $d > 0$, find explicit examples of holomorphic functions $f, g : S^2 \to S^2$ of degree d such that the homologous symplectic submanifolds $\Sigma_f, \Sigma_g \subset S^2 \times S^2$ defined as in Example 1.1 have exactly $2d$ intersections with each other, all transverse and positive.

Example 1.3. Suppose $\pi : M \to \Sigma$ is a smooth fiber bundle whose base and fibers are each closed, connected and oriented surfaces. We say that a symplectic structure ω on M is compatible with this fibration if it is nondegenerate on all the fibers; this makes (M, ω) into the total space of a **symplectic fibration** (see [MS17, Chap. 6]), and each fiber is then a symplectic submanifold with self-intersection number 0. If the fiber has genus 0, we call (M, ω) a **symplectic ruled surface**. Observe that Example 1.1 above is the simplest special case of this. By a well-known theorem of Thurston [Thu76], every smooth oriented S^2-bundle over a closed oriented surface admits a unique deformation class of symplectic structures for which it becomes a symplectic ruled surface. We will prove a generalization of this theorem in Chap. 3; see Theorem 3.33.

Example 1.4. The complex projective space $\mathbb{CP}^n$ is a complex n-dimensional manifold that also has a natural symplectic structure. It is defined as the space of all complex lines in $\mathbb{C}^{n+1}$, which we can express in two equivalent ways as follows:

$$\mathbb{CP}^n = (\mathbb{C}^{n+1}\backslash\{0\})/\mathbb{C}^* = S^{2n+1}/S^1.$$

In the first case, we divide out the natural free action (by scalar multiplication) of the multiplicative group $\mathbb{C}^* := \mathbb{C}\backslash\{0\}$ on $\mathbb{C}^{n+1}\backslash\{0\}$, and the second case is the same thing but restricting to the unit sphere $S^{2n+1} \subset \mathbb{C}^{n+1} = \mathbb{R}^{2n+2}$ and unit circle $S^1 \subset \mathbb{C} = \mathbb{R}^2$. One denotes the equivalence class in $\mathbb{CP}^n$ represented by a point $(z_0, \ldots, z_n) \in \mathbb{C}^{n+1}\backslash\{0\}$ by

$$[z_0 : \ldots : z_n] \in \mathbb{CP}^n.$$

To see the complex manifold structure of $\mathbb{CP}^n$, notice that for each $k = 0, \ldots, n$, there is an embedding

$$\iota_k : \mathbb{C}^n \hookrightarrow \mathbb{CP}^n : (z_1, \ldots, z_n) \mapsto [z_1 : \ldots, z_{k-1} : 1 : z_k : \ldots : z_n], \qquad (1.1)$$

whose image is the complement of the subset

$$\mathbb{CP}^{n-1} \cong \left\{ [z_1 : \ldots : z_{k-1} : 0 : z_k : \ldots : z_n] \in \mathbb{CP}^n \mid (z_1, \ldots, z_n) \in \mathbb{C}^n\backslash\{0\} \right\}.$$

It is not hard to show that if the maps ι_k^{-1} are thought of as complex coordinate charts on open subsets of $\mathbb{CP}^n$, then the transition maps $\iota_k^{-1} \circ \iota_j$ are all holomorphic. It follows that $\mathbb{CP}^n$ naturally carries the structure of a complex manifold such that the embeddings $\iota_k : \mathbb{C}^n \to \mathbb{CP}^n$ are holomorphic. Each of these embeddings also defines a decomposition of $\mathbb{CP}^n$ into $\mathbb{C}^n \cup \mathbb{CP}^{n-1}$, where $\mathbb{CP}^{n-1}$ is a complex submanifold of (complex) codimension one. For the case $n = 1$, this decomposition becomes $\mathbb{CP}^1 = \mathbb{C} \cup \{\text{point}\} \cong S^2$, so this is simply the Riemann sphere with its natural complex structure, where the "point at infinity" is $\mathbb{CP}^0$. In the case $n = 2$, we have $\mathbb{CP}^2 \cong \mathbb{C}^2 \cup \mathbb{CP}^1$, and one sometimes refers to the complex submanifold $\mathbb{CP}^1 \subset \mathbb{CP}^2$ as the "sphere at infinity".

The standard symplectic form on $\mathbb{CP}^n$ is defined in terms of the standard symplectic form on $\mathbb{C}^{n+1}$. The latter takes the form

$$\omega_{\text{st}} = \sum_{j=1}^{n+1} dp_j \wedge dq_j,$$

where we write the natural coordinates $(z_1, \ldots, z_{n+1}) \in \mathbb{C}^{n+1}$ as $z_j = p_j + iq_j$ for $j = 1, \ldots, n + 1$. If $\langle \, , \, \rangle$ denotes the standard Hermitian inner product on $\mathbb{C}^{n+1}$, the above can be rewritten as

$$\omega_{\text{st}}(X, Y) = \text{Im}\langle X, Y \rangle.$$

We claim that the restriction of ω_{st} to the unit sphere $S^{2n+1} \subset \mathbb{C}^{n+1}$ descends to a well-defined 2-form on the quotient $S^{2n+1}/S^1 = \mathbb{CP}^n$. Indeed, the expression above is clearly invariant under the S^1-action on S^{2n+1}, and the kernel of $\omega_{st}|_{T S^{2n+1}}$ spans the fibers of the orbits of this S^1-action (i.e. the fibers of the Hopf fibration), hence for any $p \in S^{2n+1}$ with vectors $X, X', Y \in T_p S^{2n+1}$ such that X and X' project to the same vector in $T(\mathbb{CP}^n)$, we have $\omega_{st}(X, Y) = \omega_{st}(X', Y)$. The resulting 2-form on $\mathbb{CP}^n$ will be denoted by ω_{FS}, and it is characterized by the condition

$$\mathrm{pr}^* \, \omega_{FS} = \omega_{st}|_{T S^{2n+1}}, \tag{1.2}$$

where pr denotes the quotient projection $S^{2n+1} \to S^{2n+1}/S^1 = \mathbb{CP}^n$. This expression shows that ω_{FS} is closed. The nondegeneracy of ω_{FS} follows from the observation that for the natural complex structure $i : T\mathbb{CP}^n \to T\mathbb{CP}^n$,

$$\omega_{FS}(X, iX) > 0 \text{ for every nontrivial } X \in T(\mathbb{CP}^n), \tag{1.3}$$

implying not only that ω_{FS} is symplectic, but also that every complex submanifold of $\mathbb{CP}^n$ is a symplectic submanifold. In fact, ω_{FS} is also **compatible** with the complex structure of $\mathbb{CP}^n$ in the sense that the pairing $g_{FS}(X, Y) := \omega_{FS}(X, iY)$ defines a Riemannian metric, making ω_{FS} a **Kähler form**. The metric g_{FS} is the one induced from the round metric of $S^{2n+1} \subset \mathbb{C}^{n+1}$ on the quotient $\mathbb{CP}^n = S^{2n+1}/S^1$; it is known as the **Fubini-Study** metric.

Restricting to $n = 2$, the sphere at infinity $\mathbb{CP}^1 \subset \mathbb{CP}^2$ is a complex and therefore also symplectic submanifold, and we claim that its homology class in $H_2(\mathbb{CP}^2)$ satisfies

$$[\mathbb{CP}^1] \cdot [\mathbb{CP}^1] = 1. \tag{1.4}$$

This is a well known fact about the homology of $\mathbb{CP}^2$, and can also be viewed as an example of the basic principle of projective geometry that "any two lines intersect in one point". One can see it explicitly from the following decomposition, which will be relevant to the main results below. Observe that for any $\zeta \in \mathbb{C}$, the holomorphic embedding

$$u_\zeta : \mathbb{C} \to \mathbb{C}^2 : z \mapsto (z, \zeta)$$

extends naturally to a holomorphic embedding of $\mathbb{CP}^1$ in $\mathbb{CP}^2$. Indeed, using ι_2 to include $\mathbb{C}^2$ in $\mathbb{CP}^2$, $u_\zeta(z)$ becomes the point $[z : \zeta : 1] = [1 : \zeta/z : 1/z]$, and as $z \to \infty$, this converges to the point

$$x_0 := [1 : 0 : 0]$$

in the sphere at infinity. One can check using alternate charts that this extension is indeed a holomorphic map. Together with the sphere at infinity, the collection

Fig. 1.1 $\mathbb{CP}^2\backslash\{x_0\}$ is
foliated by holomorphic
spheres that all intersect at x_0

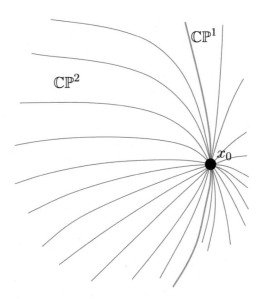

of embeddings $u_\zeta : \mathbb{CP}^1 \to \mathbb{CP}^2$ for all $\zeta \in \mathbb{C}$ thus gives a smooth family of complex submanifolds that foliate the region $\mathbb{CP}^2\backslash\{x_0\}$, but all intersect each other transversely at x_0 (see Fig. 1.1). Since they are all homologous to $\mathbb{CP}^1 \subset \mathbb{CP}^2$ and transverse intersections of complex submanifolds always count positively, (1.4) follows. From a different perspective, the spheres parametrized by u_ζ are precisely the fibers of the map

$$\pi : \mathbb{CP}^2\backslash\{[1 : 0 : 0]\} \to \mathbb{CP}^1 : [z_1 : z_2 : z_3] \mapsto [z_2 : z_3], \tag{1.5}$$

thus slightly generalizing the notion of a *symplectic fibration* discussed in Example 1.3. This is our first example of a *Lefschetz pencil*, a notion that will be examined in detail in Chap. 3.

Exercise 1.5. Generalizing the sphere at infinity, $\mathbb{CP}^n$ contains holomorphically embedded copies of $\mathbb{CP}^k$ for each $k \leq n$, defined as the set of all points $[z_0 : \ldots : z_n]$ with $n - k$ chosen coordinates set to zero.

(a) Show that for every submanifold of this form, the inclusion $\iota : \mathbb{CP}^k \hookrightarrow \mathbb{CP}^n$ satisfies $\iota^*\omega_{FS} = \omega_{FS}$, so in particular these submanifolds are all symplectic.

(b) Show that $\int_{\mathbb{CP}^1} \omega_{FS} = \pi$. *Hint: find an embedding $\varphi : \mathbb{C} \hookrightarrow S^3$ such that for the projection* $\mathrm{pr} : S^3 \to \mathbb{CP}^1 = S^3/S^1$, $\mathrm{pr} \circ \varphi$ *is a diffeomorphism of $\mathbb{C}$ to the complement of a point in $\mathbb{CP}^1$. Then use (1.2) to integrate* $(\mathrm{pr} \circ \varphi)^*\omega_{FS}$ *over $\mathbb{C}$.*

Example 1.6. The *symplectic blowup* operation provides an easy way of locally modifying any symplectic manifold to a new one with slightly more complicated topology. We will review the details of this construction in Chap. 3, but topologically, one can picture the blowup of a smooth oriented 4-manifold M as a 4-manifold obtained by picking a point $p \in M$ and an integrable complex structure near p, and replacing p with the space of complex lines in $T_p M$, i.e. with a copy of $\mathbb{CP}^1 \cong S^2$. The resulting oriented manifold $\tilde{M}$ turns out to be diffeomorphic (see Exercise 3.3) to $M \# \overline{\mathbb{CP}}^2$, where the bar over $\mathbb{CP}^2$ indicates a reversal of its usual orientation. In the symplectic category (see Sect. 3.2), blowing up can more accurately be understood as replacing a closed Darboux ball $\overline{B}_R^4 \subset (M, \omega)$ of some radius $R > 0$ with a symplectically embedded sphere $E \subset (\tilde{M}, \tilde{\omega})$, which has symplectic area πR^2 and satisfies

$$[E] \cdot [E] = -1.$$

Symplectically embedded spheres with self-intersection -1 are referred to as **exceptional spheres**. One says more generally that $(\tilde{M}, \tilde{\omega})$ is a **blowup** of (M, ω) if it can be obtained from (M, ω) by a finite sequence of symplectic blowup operations. The inverse operation, called the *symplectic blowdown*, can be defined by removing neighborhoods of exceptional spheres and replacing them with Darboux balls of appropriate size.

Observe now that if we take any of our previous examples where (M, ω) contains a symplectic sphere S with $[S] \cdot [S] \geq 0$ and blow them up along a Darboux ball disjoint from S, the resulting blowup still contains S and its self-intersection number is unchanged.

Definition 1.7. We will refer to a symplectic 4-manifold as a **blown-up symplectic ruled surface** if it is either a symplectic ruled surface or is obtained from one by a sequence of symplectic blowup operations.

Definition 1.8. A **symplectic rational surface** is a symplectic 4-manifold that is obtained from $(\mathbb{CP}^2, \omega_{\text{FS}})$ by a finite sequence of symplectic blowup and blowdown operations and symplectic deformations.

Definition 1.9. A symplectic 4-manifold (M, ω) is called **minimal** if it cannot be obtained from any other symplectic 4-manifold by blowing up, or equivalently, if it contains no exceptional spheres.

Exercise 1.10. Suppose (M, ω) is a closed symplectic 4-manifold and $E_1, \ldots,$ $E_k \subset (M, \omega)$ is a collection of exceptional spheres that are all pairwise disjoint. Prove $k \leq \dim H_2(M; \mathbb{Q})$.

Example 1.11. The following construction combines all three of the examples discussed above. Let $\mathbb{CP}^2 \# \overline{\mathbb{CP}}^2$ denote the complex blowup of $\mathbb{CP}^2$ at the point $x_0 = [1 : 0 : 0]$, i.e. at the singular point of the "fibration" (1.5). As we will review in Sect. 3.1, the complex blowup operation makes $\mathbb{CP}^2 \# \overline{\mathbb{CP}}^2$ naturally a complex manifold such that the resulting exceptional sphere $E \subset \mathbb{CP}^2 \# \overline{\mathbb{CP}}^2$ is a complex submanifold, and there is a natural identification

$$\beta : (\mathbb{CP}^2 \# \overline{\mathbb{CP}}^2) \backslash E \xrightarrow{\cong} \mathbb{CP}^2 \backslash \{x_0\}$$

which extends to a holomorphic map $\beta : \mathbb{CP}^2 \# \overline{\mathbb{CP}}^2 \to \mathbb{CP}^2$ collapsing E to the point x_0. Taking π to be the map in (1.5), one then finds that $\tilde{\pi} := \pi \circ \beta : (\mathbb{CP}^2 \# \overline{\mathbb{CP}}^2) \backslash E \to \mathbb{CP}^1$ extends over E to define a smooth and holomorphic fiber bundle

$$\tilde{\pi} : \mathbb{CP}^2 \# \overline{\mathbb{CP}}^2 \to \mathbb{CP}^1.$$

Put another way, we have replaced the point $x_0 \in \mathbb{CP}^2$, where all the fibers of π intersect, with a sphere E that intersects all the fibers of $\tilde{\pi}$ at separate points, so that $\tilde{\pi}$ is an honest S^2-bundle which has E as a section. We will see in Chap. 3 that the symplectic version of this blowup operation can be arranged to produce a symplectic structure on $\mathbb{CP}^2 \# \overline{\mathbb{CP}}^2$ for which the fibers of $\tilde{\pi}$ are symplectic submanifolds (see Theorem 3.13). This shows that the symplectic blowup of $(\mathbb{CP}^2, \omega_{FS})$ is a symplectic ruled surface, and it follows that everything one can construct from $(\mathbb{CP}^2, \omega_{FS})$ by a finite sequence of blowups is a blown-up ruled surface. As explained in Remark 1.13 below, there are exactly two oriented S^2-bundles over S^2 up to diffeomorphism, thus every one is either the trivial bundle $S^2 \times S^2 \to S^2$ or $\tilde{\pi} : \mathbb{CP}^2 \# \overline{\mathbb{CP}}^2 \to \mathbb{CP}^1$.

Remark 1.12. The example above illustrates that a symplectic 4-manifold can be both a rational surface and a (blown-up) ruled surface. We will see in Theorem 7.6 that the symplectic rational surfaces are precisely those symplectic 4-manifolds that admit genus zero symplectic Lefschetz pencils, where the use of the word "pencil" implies a fibration (with isolated singularities) over the base $\mathbb{CP}^1 \cong S^2$. The classification scheme described below thus implies that up to symplectic deformation equivalence, there are exactly two symplectic ruled surfaces that are also rational, namely the trivial and unique nontrivial S^2-bundles over S^2, which are topologically $S^2 \times S^2$ and $\mathbb{CP}^2 \# \overline{\mathbb{CP}}^2$ respectively. The rest of the ruled surfaces are sometimes called **irrational** ruled surfaces.

Remark 1.13. It is not hard to denumerate the topological types of all smooth oriented S^2-bundles $\pi : M \to \Sigma$ over closed oriented surfaces Σ. The structure group of such a bundle is $\mathrm{Diff}_+(S^2)$, the group of orientation-preserving diffeomorphisms of S^2, so the main thing one needs to understand is the homotopy type of $\mathrm{Diff}_+(S^2)$, which was computed in 1959 by Smale [Sma59]. Viewing $SO(3)$ as the group of orientation-preserving isometries of the round sphere $S^2 \subset \mathbb{R}^3$ with respect to the Euclidean metric, Smale proved that the inclusion

$$SO(3) \hookrightarrow \mathrm{Diff}_+(S^2)$$

is a homotopy equivalence. Most importantly for our purposes, this inclusion induces isomorphisms $\pi_k(SO(3)) \to \pi_k(\text{Diff}_+(S^2))$ for $k = 0, 1$; see Remark 1.16 below for a sketch of the proof. Given this, the connectedness of $SO(3)$ implies that every oriented S^2-bundle $\pi : M \to \Sigma$ can be trivialized over the 1-skeleton of Σ, and thus everywhere outside the interior of some disk $\mathbb{D}^2 \subset \Sigma$. The ability to extend the trivialization from $\partial\mathbb{D}^2$ over the rest of $\mathbb{D}^2$ then depends on the homotopy class of a transition map $\partial\mathbb{D}^2 \to \text{Diff}_+(S^2)$, i.e. an element of $\pi_1(\text{Diff}_+(S^2)) = \pi_1(SO(3)) = \mathbb{Z}_2$. This means that aside from the trivial bundle $\Sigma \times S^2 \to \Sigma$, there is exactly one nontrivial oriented smooth S^2-bundle over Σ, corresponding to the unique nontrivial element in $\pi_1(\text{Diff}_+(S^2)) = \pi_1(SO(3))$. We shall denote this nontrivial bundle by

$$\Sigma \,\widetilde{\times}\, S^2 \to \Sigma,$$

so for example, Example 1.11 shows

$$S^2 \,\widetilde{\times}\, S^2 \cong \mathbb{CP}^2 \# \overline{\mathbb{CP}}^2.$$

Note that two oriented S^2-bundles over bases of different genus can never be homeomorphic, as the homotopy exact sequence of $S^2 \hookrightarrow M \to \Sigma$ implies $\pi_1(M) \cong \pi_1(\Sigma)$ (cf. Proposition 7.62).

Exercise 1.14. Fix a closed oriented surface Σ and oriented 2-sphere bundle $\pi : M \to \Sigma$.

(a) Show that $\pi : M \to \Sigma$ admits a section. *Hint: construct it inductively over the skeleta of Σ, using the fact that fibers are simply connected.*

(b) Show that for any section $S \subset M$ of $\pi : M \to \Sigma$, the self-intersection number $[S] \cdot [S]$ is even if the bundle is trivial and odd if it is nontrivial. *Hint: over the 1-skeleton of Σ, it is easy to construct both a section S and a small perturbation S' that does not intersect S, thus the interesting part happens when you try to extend both of these sections from $\partial\mathbb{D}^2$ over a disk $\mathbb{D}^2$. If the bundle is nontrivial, then the relationship between S and S' over $\partial\mathbb{D}^2$ can be described in terms of a loop in $SO(3)$ that generates $\pi_1(SO(3)) = \pi_1(\text{Diff}_+(S^2))$.*

Exercise 1.15. In contrast to Exercise 1.14, find an example of an oriented $\mathbb{T}^2$-bundle over a closed oriented surface that does not admit a section. *Hint: see Example 3.35.*

Remark 1.16. If you enjoy Serre fibrations and homotopy exact sequences (see [Hat02, §4.2]), then you might like the following proof that the inclusion $SO(3) \hookrightarrow \text{Diff}_+(S^2)$ induces isomorphisms $\pi_k(SO(3)) \to \pi_k(\text{Diff}_+(S^2))$ for every k. We use the group

$$G := \text{Conf}_+(S^2) \subset \text{Diff}_+(S^2)$$

of orientation-preserving conformal transformations as an intermediary—here "conformal" is defined with respect to the Euclidean metric on the round sphere $S^2 \subset \mathbb{R}^3$, thus G contains $\mathrm{SO}(3)$, and the goal is to prove that both of the inclusions $G \hookrightarrow \mathrm{Diff}_+(S^2)$ and $\mathrm{SO}(3) \hookrightarrow G$ induce isomorphisms on homotopy groups. For the first inclusion, identify S^2 with the extended complex plane $\mathbb{C} \cup \{\infty\}$ so that the conformal structure of the round sphere corresponds to the standard complex structure i, and let $\mathcal{J}(S^2)$ denote the space of all smooth complex structures on S^2 that are compatible with its orientation. The uniformization theorem then implies that the map

$$\mathrm{Diff}_+(S^2) \to \mathcal{J}(S^2) : \varphi \mapsto \varphi^* i$$

is surjective, and in fact, this map is a Serre fibration—this fact is somewhat nontrivial, but we will outline a proof using holomorphic curve methods in Chap. 2, see Remark 2.48. The fiber of this fibration over $i \in \mathcal{J}(S^2)$ is G, so we obtain a long exact sequence

$$\ldots \to \pi_{k+1}(\mathcal{J}(S^2)) \to \pi_k(G) \to \pi_k(\mathrm{Diff}_+(S^2)) \to \pi_k(\mathcal{J}(S^2)) \to \ldots$$

But $\mathcal{J}(S^2)$ can also be viewed as the space of almost complex structures on S^2 compatible with a fixed symplectic form, and is thus contractible (cf. Proposition 2.1), so this exact sequence implies that the maps $\pi_k(G) \to \pi_k(\mathrm{Diff}_+(S^2))$ are isomorphisms for every k.

For the inclusion $\mathrm{SO}(3) \hookrightarrow G$, we can fix a base point $p_0 \in S^2$ and use the action of $\mathrm{SO}(3)$ or G on the base point to define a pair of Serre fibrations,

$$\begin{array}{ccccc}
\mathrm{SO}(2) & \hookrightarrow & \mathrm{SO}(3) & \longrightarrow & S^2 \\
\downarrow{\scriptstyle\Phi} & & \downarrow{\scriptstyle\Psi} & & \downarrow{\scriptstyle\mathrm{Id}} \\
G_0 & \hookrightarrow & G & \longrightarrow & S^2
\end{array}$$

where $G_0 \subset G$ denotes the group of conformal transformations that fix the base point, $\mathrm{SO}(2)$ is identified with the analogous subgroup of $\mathrm{SO}(3)$, the maps Φ and Ψ are the natural inclusions, and the diagram commutes. Identifying S^2 with $\mathbb{C} \cup \{\infty\}$ and choosing ∞ as the base point, G_0 becomes the group of affine transformations $z \mapsto az + b$ on $\mathbb{C}$ with $a \neq 0$, and the subgroup $\mathrm{SO}(2) \subset G_0$ then consists of the transformations of the form $z \mapsto e^{i\theta} z$. It follows that $\mathrm{SO}(2)$ is a deformation retract of G_0, so Φ is a homotopy equivalence. Forming the homotopy exact sequence of both fibrations along with the induced maps from one to the other now produces the commutative diagram

$$\begin{array}{ccccccccc}
\pi_{k+1}(S^2) & \longrightarrow & \pi_k(\mathrm{SO}(2)) & \longrightarrow & \pi_k(\mathrm{SO}(3)) & \longrightarrow & \pi_k(S^2) & \longrightarrow & \pi_{k-1}(\mathrm{SO}(2)) \\
\downarrow{\scriptstyle\mathrm{Id}} & & \downarrow{\scriptstyle\Phi_*} & & \downarrow{\scriptstyle\Psi_*} & & \downarrow{\scriptstyle\mathrm{Id}} & & \downarrow{\scriptstyle\Phi_*} \\
\pi_{k+1}(S^2) & \longrightarrow & \pi_k(G_0) & \longrightarrow & \pi_k(G) & \longrightarrow & \pi_k(S^2) & \longrightarrow & \pi_{k-1}(G_0),
\end{array}$$

and since both instances of Φ_* are isomorphisms, the five-lemma implies that Ψ_* is as well.

Remark 1.17. The terms "ruled surface" and "rational surface" both originate in algebraic geometry. The former traditionally describes a surface that is fibered by lines, i.e. in the setting of complex projective varieties, this means a complex surface fibered by complex submanifolds biholomorphic to $\mathbb{CP}^1$. Similarly, a surface is called rational if it is birationally equivalent to the projective plane: again in the complex context, this means a complex surface that can be related to $\mathbb{CP}^2$ by a finite sequence of complex blowup and blowdown operations.

1.2   Results About Symplectically Embedded Spheres

In Sect. 1.1 we saw two fundamental examples of closed symplectic 4-manifolds (M, ω) containing symplectically embedded spheres $S \subset (M, \omega)$ with $[S] \cdot [S] \geq 0$:

(1) Symplectic ruled surfaces, whose fibers S have $[S] \cdot [S] = 0$,
(2) $(\mathbb{CP}^2, \omega_{FS})$, which contains the sphere at infinity $\mathbb{CP}^1 \subset \mathbb{CP}^2$, with $[\mathbb{CP}^1] \cdot [\mathbb{CP}^1] = 1$.

The following theorem says that except for trivial modifications such as rescaling and blowing up, these examples are the only ones. The result was first hinted at in Gromov's seminal paper [Gro85, §2.4.B_2–2.4.B_3'], and was then proved in full by McDuff [McD90, McD92] (see also [LM96b] and [MS12, §9.4]).

Theorem A. *Suppose (M, ω) is a closed and connected symplectic 4-manifold containing a symplectically embedded 2-sphere $S \subset M$ with*

$$[S] \cdot [S] \geq 0.$$

Then (M, ω) is either $(\mathbb{CP}^2, c\omega_{FS})$ for some constant $c > 0$ or it is a blown-up symplectic ruled surface (see Definition 1.7).

This result is often summarized by saying that every symplectic 4-manifold containing a nonnegative symplectic sphere is "rational or ruled". Observe that in all of the examples we discussed in Sect. 1.1, the sphere in question actually satisfies

$$[S] \cdot [S] \in \{0, 1\}, \tag{1.6}$$

thus a consequence of Theorem A is the fact that if we are given a symplectic sphere with $[S] \cdot [S] \geq 0$, we can always find *another* one for which (1.6) is satisfied. This corollary actually will be proved separately, as a step in the proof of Theorem A.

Remark 1.18. All of the symplectic manifolds occurring in Theorem A have fundamental groups isomorphic to that of a closed oriented surface (cf. Proposition 7.62). By contrast, Gompf [Gom95] has shown that every finitely-presented group can be the fundamental group of a closed symplectic 4-manifold, so the manifolds that appear in Theorem A form a rather restrictive class.

The next two theorems regarding the symplectic blowup are actually preliminary results in the background of Theorem A, and were also first proved in [McD90]. The first has the consequence that minimality is invariant under symplectic deformation equivalence. Recall that two symplectic manifolds (M_0, ω_0) and (M_1, ω_1) are called **symplectically deformation equivalent** if there exists a diffeomorphism $\varphi : M_0 \to M_1$ such that ω_0 and $\varphi^*\omega_1$ are homotopic via a smooth 1-parameter family of symplectic forms.

Theorem B. *Suppose M is a closed connected 4-manifold with a smooth 1-parameter family of symplectic structures $\{\omega_s\}_{s\in[0,1]}$, and $E_1, \ldots, E_k \subset M$ is a collection of pairwise disjoint exceptional spheres in (M, ω_0). Then there are smooth 1-parameter families of embedded spheres $E_1^s, \ldots, E_k^s \subset M$ for $s \in [0, 1]$ such that*

- *$E_i^0 = E_i$ for $i = 1, \ldots, k$;*
- *For every $s \in [0, 1]$, $E_i^s \cap E_j^s = \varnothing$ for $i \neq j$;*
- *For every $s \in [0, 1]$ and $i = 1, \ldots, k$, E_i^s is symplectically embedded in (M, ω_s).*

In particular, (M, ω_0) is minimal if and only if (M, ω_1) is minimal.

Observe that by Exercise 1.10, any maximal collection of pairwise disjoint exceptional spheres in a closed symplectic 4-manifold is finite. The next result will turn out to be an easy consequence of this fact in combination with Theorem B.

Theorem C. *Suppose (M, ω) is a closed symplectic 4-manifold and $E_1, \ldots, E_k \subset M$ is a maximal collection of pairwise disjoint exceptional spheres. Then the manifold (M_0, ω_0) obtained by blowing down (M, ω) at all of these spheres is minimal.*

For this reason, many questions about symplectic 4-manifolds in general can be reduced to questions about the minimal case, including Theorem A. In fact, for the minimal case one also has the following somewhat stronger formulation:

Theorem D. *Suppose (M, ω) is a closed, connected and minimal symplectic 4-manifold that contains a symplectically embedded 2-sphere $S \subset (M, \omega)$ with $[S] \cdot [S] \geq 0$. One then has the following possibilities:*

- *(1) If $[S] \cdot [S] = 0$, then (M, ω) admits a symplectomorphism to a symplectic ruled surface such that S is identified with a fiber.*
- *(2) If $[S] \cdot [S] = 1$, then (M, ω) admits a symplectomorphism to $(\mathbb{CP}^2, c\omega_{FS})$ for some constant $c > 0$, such that S is identified with the sphere at infinity $\mathbb{CP}^1 \subset \mathbb{CP}^2$.*

(3) *If* $[S] \cdot [S] > 1$, *then* (M, ω) *is symplectomorphic to one of the following:*

(a) $(\mathbb{CP}^2, c\omega_{FS})$ *for some constant* $c > 0$,
(b) $(S^2 \times S^2, \sigma_1 \oplus \sigma_2)$ *for some pair of area forms* σ_1, σ_2 *on* S^2.

The appearance of the specific rational ruled surface $(S^2 \times S^2, \sigma_1 \oplus \sigma_2)$ in this result comes about due to the following stronger result of Gromov [Gro85] and McDuff [McD90]:

Theorem E. *Suppose* (M, ω) *is a closed, connected and minimal symplectic 4-manifold containing a pair of symplectically embedded spheres* $S_1, S_2 \subset (M, \omega)$ *that satisfy* $[S_1] \cdot [S_1] = [S_2] \cdot [S_2] = 0$ *and have exactly one intersection with each other, which is transverse and positive. Then* (M, ω) *admits a symplectomorphism to* $(S^2 \times S^2, \sigma_1 \oplus \sigma_2)$ *identifying* S_1 *with* $\{S^2\} \times \{0\}$ *and* S_2 *with* $\{0\} \times S^2$, *where* σ_1, σ_2 *are any two area forms on* S^2 *such that*

$$\int_{S^2} \sigma_i = \int_{S_i} \omega \qquad for\ i = 1, 2.$$

Finally, here is a closely related result that was not stated explicitly in the work of Gromov or McDuff but follows by similar arguments and implies Theorems A and D above. We will discuss in Chap. 3 the notions of Lefschetz pencils and Lefschetz fibrations, which are something like symplectic fibrations but with isolated singular points. These singular points come in two types: (1) *Lefschetz critical points*, at which two smooth pieces of a single fiber (called a *singular fiber*) have a positive transverse intersection, and (2) *pencil singularities*, also known as *base points*, at which all fibers come together and intersect each other positively and transversely. A fibration that includes singularities of the second type is called a *Lefschetz pencil*, and the base[1] of such a fibration is necessarily $\mathbb{CP}^1$. The term *Lefschetz fibration* is reserved for the case where pencil singularities do not appear (but Lefschetz critical points are allowed), in which case the base can be any oriented surface. Our proof of Theorems A and D will also yield a proof of the following generalization of Theorem A.

Theorem F. *Suppose* (M, ω) *is a closed and connected symplectic 4-manifold that contains a symplectically embedded 2-sphere* $S \subset (M, \omega)$ *with*

$$m := [S] \cdot [S] \geqslant 0.$$

Then for any choice of pairwise distinct points $p_1, \ldots, p_m \in S$, (M, ω) *admits a symplectic Lefschetz pencil with base points* $p_1, \ldots, p_m$ *(or a symplectic Lefschetz*

---

[1] Be aware that the standard terminology for Lefschetz pencils employs the word "base" with two distinct meanings that may occasionally appear in the same sentence: the notion of "base points" is completely unrelated to the "base of the fibration".

fibration if $m = 0$), in which S is a smooth fiber and no singular fiber contains
more than one critical point. Moreover, the set of singular fibers of this pencil (or
fibration) is empty if and only if $m \in \{0, 1\}$ and $(M \backslash S, \omega)$ is minimal.

For the sake of completeness, let us now state a few related results that come
from Seiberg-Witten theory. While their proofs are beyond the scope of this book,
it is important to be aware of them since they frequently appear in applications
as sufficient conditions to establish the hypotheses of the theorems above. (For
more detailed accounts of Taubes-Seiberg-Witten theory and its applications to
symplectic 4-manifolds, see [MS17, §13.3] or the earlier survey papers [LM96b,
MS96, HT99].)

As preparation, recall that every symplectic manifold (M, ω) has a well-defined
first Chern class $c_1(M, \omega) \in H^2(M)$, defined as the first Chern class of the
complex vector bundle (TM, J) for any choice of almost complex structure J
compatible with ω (see e.g. [MS17, §2.7]). For any $A \in H_2(M)$, we shall abbreviate
the evaluation of $c_1(M, \omega)$ on A by

$$c_1(A) := \langle c_1(M, \omega), A \rangle.$$

If S is a closed oriented surface and $E \to S$ is a complex vector bundle, we also
often use the abbreviated notation

$$c_1(E) := \langle c_1(E), [S] \rangle \in \mathbb{Z}$$

for the **first Chern number** of E. Now if S is a closed symplectically embedded
surface in (M, ω), Proposition 2.2 below constructs a compatible almost complex
structure J such that $J(TS) = TS$, in which case TS is a complex subbundle of
$(TM|_S, J)$ and has a complex normal bundle $N_S \subset (TM|_S, J)$ satisfying $TS \oplus
N_S = TM|_S$. The first Chern number of TS is just the Euler characteristic $\chi(S)$,
while $c_1(N_S)$ can be expressed as a signed count of zeroes of a generic section of
N_S used to push S to a small perturbation of itself and count intersections, giving
the relation $c_1(N_S) = [S] \cdot [S]$. This proves

$$c_1([S]) = \chi(S) + [S] \cdot [S],$$

so in particular every exceptional sphere $E \subset (M, \omega)$ satisfies $c_1([E]) = 1$. Note
that the fact that E is *symplectically* embedded fixes an orientation of E, and the
definition of the homology class $[E] \in H_2(M)$ depends on this orientation—if E
were only a smooth (but not symplectic) submanifold, we would have to make an
additional choice of an orientation for E before defining the class $[E]$.

Theorem 1.19 (Taubes [Tau95] and Li and Liu [LL99]). *In a closed symplectic
4-manifold (M, ω), any smoothly embedded oriented 2-sphere $E \subset M$ satisfying
$[E] \cdot [E] = -1$ and $c_1([E]) = 1$ is homologous to a symplectic exceptional sphere.
Moreover, if M is connected with $b_2^+(M) \geqslant 2$, then the condition $c_1([E]) = 1$ is
always satisfied after possibly reversing the orientation of E.* □

Here, the topological invariant $b_2^+(M)$ is defined as the maximal dimension of a subspace of $H^2(M; \mathbb{R})$ on which the cup product pairing is positive-definite (see Sect. 7.3.6); note that the existence of a symplectic form implies $b_2^+(M) \geqslant 1$ since $\langle [\omega] \cup [\omega], [M] \rangle > 0$. In the case $b_2^+(M) \geqslant 2$, the theorem implies that symplectic minimality in dimension four is not actually a *symplectic* condition at all, but depends only on the *smooth* topology of M. When $b_2^+(M) = 1$, we have the additional condition involving the first Chern class of ω, but this is a relatively weak symplectic invariant. In the language of Gromov's h-principle (see [EM02]), $c_1(M, \omega)$ depends only on the **formal homotopy class** of ω, meaning its homotopy class as a nondegenerate (but not necessarily closed) 2-form, or equivalently, the homotopy class of almost complex structures compatible with ω. Gromov famously proved that symplectic forms on open manifolds are determined up to symplectic deformation by their formal homotopy classes, but this is known to be false in the closed case [Rua94, IP99].

Corollary 1.20. *A closed symplectic 4-manifold (M, ω) with $b_2^+(M) \geqslant 2$ is minimal if and only there is no closed oriented smooth 4-manifold M' for which M is diffeomorphic to $M' \# \overline{\mathbb{CP}}^2$. For $b_2^+(M) = 1$, (M, ω) is minimal if and only if every symplectic form formally homotopic to ω is minimal.* $\qquad\square$

The next result is often used for establishing the hypothesis of Theorem A.

Theorem 1.21 (Liu [Liu96]). *A closed and connected symplectic 4-manifold (M, ω) admits a symplectically embedded 2-sphere of nonnegative self-intersection number whenever either of the following conditions holds:*

(1) $\langle c_1(M, \omega) \cup [\omega], [M] \rangle > 0$;
(2) (M, ω) is minimal and $\langle c_1(M, \omega) \cup c_1(M, \omega), [M] \rangle < 0$.

$\qquad\square$

We will prove in Sect. 7.3.6 that both statements have relatively easy converses in light of Theorem A; see Proposition 7.67 and Exercises 7.69 and 7.70. These imply:

Corollary 1.22. *A closed connected symplectic 4-manifold (M, ω) is symplectically deformation equivalent to one satisfying $\langle c_1(M, \omega) \cup [\omega], [M] \rangle > 0$ if and only if it is a symplectic rational surface or blown-up ruled surface.*

Corollary 1.23 ("Gompf's Conjecture"). *A closed, connected, minimal symplectic 4-manifold (M, ω) satisfies $\langle c_1(M, \omega) \cup c_1(M, \omega), [M] \rangle < 0$ if and only if it is a symplectic ruled surface with base of genus at least 2.*

Remark 1.24. It is common in the literature to express the conditions in Theorem 1.21 in terms of the **canonical class** $K := -c_1(M, \omega) \in H^2(M)$ of a symplectic 4-manifold. This is, by definition, the first Chern class of the **canonical line bundle**

$$T^{2,0}M \to M,$$

where the latter denotes the bundle of complex-bilinear alternating forms on M with respect to any almost complex structure compatible with ω. Using "·" to denote the product on $H^*(M; \mathbb{R})$ and identifying $H^4(M; \mathbb{R})$ with $\mathbb{R}$ via integration over the fundamental class, the two conditions in the theorem then take the simple form

$$K \cdot [\omega] < 0 \quad \text{and} \quad K \cdot K < 0.$$

In Chap. 7, we will prove a further corollary of these results which generalizes Theorem A by allowing higher-genus symplectically embedded surfaces with positive first Chern number; see Theorem 7.36. It should be emphasized however that results of this type do not supersede Theorem A, they merely weaken the hypotheses needed to apply it.

1.3 Summary of the Proofs

The results stated in Sect. 1.2 are based on the powerful theory of pseudoholomorphic curves, first introduced by Gromov in [Gro85]. The technical details can be quite intricate—depending how deeply one wants to delve into them—nonetheless it is not so hard to give intuitive explanations for why most of these statements are true, and we shall do this in the next several paragraphs. The proofs we will explain in this book are in spirit the same as what was originally explained by McDuff, but they will differ in several details. The main reason for this is that the most "natural" way to prove these results requires certain technical ingredients that were not yet developed at the time [McD90] was written. As a consequence, several steps that required very clever arguments in [McD90] can now be replaced by more straightforward applications of machinery that has meanwhile become standard in the field. Other, similarly modern treatments can be found in [LM96b] and [MS12, §9.4], but ours will also differ from theirs in a few places—in particular, we will make use of the topological notion of Lefschetz pencils, thus relating McDuff's results to another strain of ideas that has become quite important in symplectic topology since the 1990s.

For the following discussion we assume that the reader has at least some basic familiarity with holomorphic curves. The essential technical background will be summarized in Chap. 2.

The starting point for all of the above results is the following easy but fundamental lemma (see also Proposition 2.2 for a sketch of the proof):

Lemma 1.25. *Suppose (M, ω) is a symplectic manifold and $S \subset M$ is a smooth 2-dimensional submanifold. Then S is a symplectic submanifold if and only if there exists an ω-tame almost complex structure J preserving TS.*

1.3.1   Exceptional Spheres

By definition, an exceptional sphere $E \subset (M, \omega)$ is embedded symplectically, hence by the lemma above, one can choose an ω-tame almost complex structure J so that E becomes the image of an embedded J-holomorphic curve. The technical work underlying Theorem B is to show that this curve is remarkably stable under changes in the data: for a generic homotopy of tame almost complex structures, one can find a corresponding isotopy of pseudoholomorphic exceptional spheres. Theorem B will thus be essentially a consequence of the following technical result, which is important enough to deserve special mention in this summary (see Theorem 5.1 for a more precise statement):

Proposition. *If (M, ω) is a closed symplectic 4-manifold, then for generic ω-tame almost complex structures J, every homology class $A \in H_2(M)$ for which there exists an exceptional sphere $E \subset (M, \omega)$ with $[E] = A$ can be represented by a unique (up to parametrization) embedded J-holomorphic sphere. Moreover, such J-holomorphic exceptional spheres deform smoothly under generic deformations of J.*

1.3.2   The Case $[S] \cdot [S] = 0$

Consider now the case of a closed and connected symplectic 4-manifold (M, ω) containing a symplectic sphere $S \subset (M, \omega)$ with $[S] \cdot [S] = 0$. By the fundamental lemma stated above, one can choose an ω-tame almost complex structure J such that S is the image of an embedded J-holomorphic sphere. This implies that a certain connected component of the moduli space of (unparametrized) J-holomorphic spheres is nonempty: call this component $\mathcal{M}_S(J)$. The hard part is then to use the analytical properties of J-holomorphic curves to show the following:

Lemma 1.26 (cf. Proposition 2.53).   *After a generic perturbation of J, the component $\mathcal{M}_S(J)$ is a nonempty, smooth, oriented 2-dimensional manifold whose elements are each embedded J-holomorphic spheres with pairwise disjoint images, foliating an open subset of M.*

Lemma 1.27 (cf. Theorem 4.6).   *If $(M \backslash S, \omega)$ is minimal, then $\mathcal{M}_S(J)$ is also compact.*

We claim that these two results together imply (M, ω) is a symplectic ruled surface if it is minimal. To see this, let $\mathcal{U} \subset M$ denote the subset consisting of every point that lies in the image of some curve in $\mathcal{M}_S(J)$. By Lemma 1.26, $\mathcal{U}$ is open, and Lemma 1.27 implies that it is also closed if $(M \backslash S, \omega)$ is minimal. Since M is connected, it follows that $\mathcal{U} = M$, so the images of the curves in $\mathcal{M}_S(J)$ form a smooth foliation of (M, ω). We can then define a smooth map

$$\pi : M \to \mathcal{M}_S(J) : x \mapsto u_x, \tag{1.7}$$

where u_x denotes the unique curve in $\mathcal{M}_S(J)$ that has x in its image. This map is a fibration, and its fibers are embedded spheres which are J-holomorphic, and therefore also symplectic.

We can say a bit more if we are given not just one but *two* symplectic spheres $S_1, S_2 \subset (M, \omega)$ with zero self-intersection which have one positive and transverse intersection with each other. In this case, if (M, ω) is minimal, the above argument gives two transverse fibrations for which the fiber of one can be identified with the base of the other. Then (1.7) is a trivial sphere-fibration over S^2, and we will be able to prove Theorem E by using a Moser deformation argument to identify (M, ω) with $(S^2 \times S^2, \sigma_1 \oplus \sigma_2)$.

A brief word on what happens when $(M \backslash S, \omega)$ is not minimal: in this case Lemma 1.27 fails, $\mathcal{M}_S(J)$ is not compact. It does however have a very nice compactification $\overline{\mathcal{M}}_S(J)$, which is obtained from $\mathcal{M}_S(J)$ by adding finitely many nodal curves, each consisting of two embedded J-holomorphic spheres that have self-intersection -1 and intersect each other once transversely. In topological terms, these nodal curves look exactly like Lefschetz singular fibers, with the result that (1.7) becomes a Lefschetz fibration.

1.3.3 The Case $[S] \cdot [S] > 0$

If $S \subset (M, \omega)$ has self-intersection 1, then defining a suitable J as above, the resulting moduli space $\mathcal{M}_S(J)$ is no longer a surface, but is 4-dimensional. This is too many dimensions to define a foliation of M, but we can bring the dimension back down to 2 by imposing a constraint: pick any point $p \in S$ and consider the moduli space $\mathcal{M}_S(J; p)$ consisting of curves in $\mathcal{M}_S(J)$ with a marked point that is constrained to pass through p. Now $\mathcal{M}_S(J; p)$ is again 2-dimensional, and just as in the $[S] \cdot [S] = 0$ case, if $(M \backslash S, \omega)$ is minimal then the curves in $\mathcal{M}_S(J; p)$ give a foliation of (M, ω) by symplectically embedded spheres, except that they all intersect each other at p. In topological language, the fibration (1.7) is now replaced by a *Lefschetz pencil*

$$\pi : M \backslash \{p\} \to \mathcal{M}_S(J; p) : x \mapsto u_x,$$

where the structure of the singularity at p dictates that its base $\mathcal{M}_S(J; p)$ must be diffeomorphic to $\mathbb{CP}^1$. This therefore gives a decomposition of (M, ω) matching the decomposition of $(\mathbb{CP}^2, \omega_{FS})$ explained in Example 1.4. We can then use a Moser deformation argument to show that (M, ω) is symplectomorphic to $(\mathbb{CP}^2, c\omega_{FS})$ for a suitable constant $c > 0$.

The case $[S] \cdot [S] > 1$ follows the same general idea, but now a topological coincidence kicks in to simplify matters. Writing $k := [S] \cdot [S] \geqslant 2$, we can define a suitable 2-dimensional moduli space $\mathcal{M}_S(J; p_1, \ldots, p_k)$ by picking distinct points $p_1, \ldots, p_k \subset S$ and defining our curves to have k marked points constrained to pass through the points $p_1, \ldots, p_k$. There are now two possibilities for (M, ω):

(1) It does not contain any symplectically embedded sphere S' with $0 \leqslant [S'] \cdot [S'] < k$.
(2) It does.

In the second case, we can go back to the beginning of the argument using S', and repeat if possible until the situation is reduced to $[S] \cdot [S] \in \{0, 1\}$, which we already understand. In the first case, the usual analytical arguments applied to $\mathcal{M}_S(J; p_1, \ldots, p_k)$ give us a Lefschetz pencil

$$\pi : M \backslash \{p_1, \ldots, p_k\} \to \mathcal{M}_S(J; p_1, \ldots, p_k) : x \mapsto u_x, \tag{1.8}$$

where again the structure of the singularities dictates $\mathcal{M}_S(J; p_1, \ldots, p_k) \cong \mathbb{CP}^1$. This Lefschetz pencil has k base points but no singular fibers. As it turns out, *this can never happen*:

Lemma (cf. Proposition 3.31). *On any closed oriented 4-manifold, a Lefschetz pencil with fibers diffeomorphic to S^2 and at least two base points always has at least one singular fiber.*

The reader who already has a bit of intuition about Lefschetz pencils will find it easy to understand why this is true: if we have such a pencil with $k \geqslant 2$ base points, then blowing up $k - 1$ of them produces a pencil with one base point and no singular fibers—this can only be $\mathbb{CP}^2$. But $\mathbb{CP}^2$ is not the blowup of anything: it has no homology class $A \in H_2(\mathbb{CP}^2)$ with $A \cdot A = -1$, and thus no exceptional spheres.

The upshot of this discussion is that for $[S] \cdot [S] = k \geqslant 1$, one can always reduce the case $k > 1$ to the case $k \in \{0, 1\}$. Another way to say it is that the Lefschetz pencil (1.8) does exist in general, but it must always have some singular fibers, the components of which are symplectically embedded spheres S' with $[S'] \cdot [S'] < k$. Analyzing the possible singular fibers a bit more closely, one finds in fact that one of the following most hold if $k > 1$:

(1) There exists a symplectically embedded sphere $S' \subset (M, \omega)$ with $[S'] \cdot [S'] = 1$.
(2) There exist two symplectically embedded spheres $S_1, S_2 \subset (M, \omega)$ with $[S_1] \cdot [S_1] = [S_2] \cdot [S_2] = 0$ and one intersection which is positive and transverse.

In the first case, we saw above that (M, ω) must be $\mathbb{CP}^2$, while in the second, Theorem E says that it must be $S^2 \times S^2$ with a split symplectic form.

The methods we have just sketched also produce a slightly more technical result that is sometimes useful in applications. We state it here as an extension of Theorem F; the proof will be sketched as an exercise in Chap. 6.

Theorem G (cf. Exercise 6.3). *Suppose (M, ω) is a closed and connected symplectic 4-manifold that contains a symplectically embedded 2-sphere $S \subset (M, \omega)$*

with $[S] \cdot [S] =: m \geqslant 0$, and $p_1, \ldots, p_m \in S$ is any fixed set of pairwise distinct points. Then:

    (1) For generic ω-tame almost complex structures J, the Lefschetz pencil or fibration of Theorem F with base points $p_1, \ldots, p_m$ is isotopic (with fixed base points) to one for which the irreducible components of all fibers are embedded J-holomorphic spheres. Moreover, every J-holomorphic curve that is homologous to the fiber and passes through all the base points is one of these.

    (2) Given ω and J as above and a smooth 1-parameter family of symplectic forms $\{\omega_s\}_{s \in [0,1]}$ with $\omega_0 = \omega$, for generic smooth 1-parameter families $\{J_s\}_{s \in [0,1]}$ of ω_s-tame almost complex structures with $J_0 = J$, there exists a smooth isotopy of Lefschetz pencils or fibrations with fixed base points $p_1, \ldots, p_m$ and J_s-holomorphic fibers for $s \in [0, 1]$.

Notice that by the second statement in this result, the rigid symplectic Lefschetz pencil or fibration structure on a rational or ruled surface cannot be destroyed by deforming the symplectic form. In particular, the class of symplectic 4-manifolds containing symplectic spheres with nonnegative self-intersection is invariant under symplectic deformation equivalence.

One subtlety in the statement of Theorem G is that the definition of the word "generic" depends on the choice of the points $p_1, \ldots, p_m \in S$, cf. Remark 2.24. For the case $(M, \omega) = (\mathbb{CP}^2, \omega_{FS})$ however, which is minimal and contains a symplectic sphere with $[S] \cdot [S] = 1$, it turns out that one can remove the genericity assumption in Theorem G altogether. One obtains from this a proof of the following fundamental result of Gromov [Gro85, 0.2.B], usually summarized with the statement that $\mathbb{CP}^2$ contains a unique "J-holomorphic line" through any two points:

Corollary (cf. Corollary 6.5). *For any tame almost complex structure J on $(\mathbb{CP}^2, \omega_{FS})$ and any two distinct points $p_1, p_2 \in \mathbb{CP}^2$, there is a unique J-holomorphic sphere homologous to $[\mathbb{CP}^1] \in H_2(\mathbb{CP}^2)$ passing through p_1 and p_2, and it is embedded.*

1.4   Outline of the Remaining Chapters

The rest of the book will be organized as follows. In Chap. 2, we will explain the necessary technical background on closed holomorphic curves, omitting most of the proofs but supplying sketches in a few cases where the relevant results might not be considered "standard" knowledge. In Chap. 3 we will discuss the symplectic blowup and the basic theory of symplectic Lefschetz pencils, including the proof of an important result of Gompf saying that a Lefschetz pencil up to isotopy determines a symplectic structure up to deformation. We then begin the serious analytical work in Chap. 4 by proving some compactness results that will be needed in the proofs of all the major theorems. These results are in some sense easy consequences of

Gromov's compactness theorem, but they also depend crucially on the intersection theory of holomorphic curves and are thus unique to dimension four. In Chap. 5 we will prove the main results on exceptional spheres, notably Theorems B and C. The proofs of Theorems A, D, E and F will then be completed in Chap. 6, with the proof of Theorem G sketched as an exercise. The remaining chapters cover a pair of topics that we have not discussed in this introduction: first, Chap. 7 gives a brief outline of the Gromov-Witten invariants and discusses McDuff's generalization [McD92] of Theorem A to a statement about symplectic 4-manifolds containing certain *immersed* symplectic spheres. In modern terms, the result is a complete characterization of the symplectic 4-manifolds that are *symplectically uniruled*. Chapters 8 and 9 then give an overview of some applications of similar ideas outside the realm of closed symplectic manifolds, namely in 3-dimensional contact topology. Appendix A outlines a proof of the folk theorem, important for the proof of Theorem F and everything that depends on it, that 2-dimensional moduli spaces of embedded J-holomorphic curves look like Lefschetz fibrations near nodal singularities.

Chapter 2
Background on Closed Pseudoholomorphic Curves

This chapter is preparatory: readers who are already sufficiently familiar with pseudoholomorphic curves may prefer to skip to Chap. 3 and later consult this chapter for reference as necessary. Its purpose is to fix definitions and notation and give a quick summary, mostly without proofs, of the technical ingredients that are required to prove the main results. More details on most of the material in Sect. 2.1 can be found in the lecture notes [Wenc], and much of the rest can also be found in [MS12]. We will give additional references to the literature as needed.

2.1 Holomorphic Curves in General

In this section we summarize the essential facts about closed pseudoholomorphic curves in arbitrary symplectic manifolds.

2.1.1 Symplectic and Almost Complex Structures

For any smooth real vector bundle $E \to B$ of finite even rank, a **complex structure** on E is a fiberwise linear bundle map $J : E \to E$ (i.e. a section of the vector bundle $\text{End}(E) \to B$) such that $J^2 = -\mathbb{1}$. In this book we shall only consider *smooth* complex structures, meaning that the section of $\text{End}(E)$ is assumed to be smooth. If M is a smooth manifold of dimension $2n$, a complex structure on the bundle $TM \to M$ is called an **almost complex structure** on M, and the pair (M, J) is then a smooth **almost complex manifold**. An important special case arises from complex manifolds: if M admits an atlas of coordinate charts with holomorphic transition functions, then it carries a natural almost complex structure $J : TM \to TM$ defined by choosing local holomorphic coordinates and

© Springer International Publishing AG, part of Springer Nature 2018
C. Wendl, *Holomorphic Curves in Low Dimensions*, Lecture Notes
in Mathematics 2216, https://doi.org/10.1007/978-3-319-91371-1_2

multiplying by i. Any almost complex structure that can be obtained in this way is said to be **integrable**, and one can in that case simply call it a *complex structure on M* (without the "almost"). Most almost complex structures are not integrable, except in real dimension 2: it is a nontrivial fact that every smooth almost complex structure on a surface is integrable. Almost complex manifolds of real dimension 2 are therefore equivalent to complex 1-dimensional manifolds: these are what we call **Riemann surfaces**. The most common example we will encounter is the **Riemann sphere**

$$(S^2, i) := \mathbb{C} \cup \{\infty\},$$

which is covered by the two holomorphically compatible coordinate charts $\varphi_0 = \mathrm{Id} : S^2 \backslash \{\infty\} \to \mathbb{C}$ and $\varphi_\infty : S^2 \backslash \{0\} \to \mathbb{C} : z \mapsto 1/z$. The uniformization theorem implies that every complex structure j on S^2 is **biholomorphically equivalent** to the standard complex structure i defined in this way, i.e. there exists a diffeomorphism $\varphi : S^2 \to S^2$ satisfying $\varphi^* j = i$.

If (M, ω) is a symplectic manifold of dimension $2n$, then following Gromov [Gro85], one says that J is **tamed** by ω if all the J-complex lines in TM are also symplectic subspaces carrying the same induced orientation, which means

$$\omega(X, JX) > 0 \text{ for all nonzero } X \in TM.$$

Further, one says that J is **compatible** with ω (or sometimes also *callibrated* by ω) if it is tamed and the 2-tensor defined by

$$g_J(X, Y) = \omega(X, JY)$$

is also symmetric, which means it is a Riemannian metric. We shall denote

$$\mathcal{J}(M) = \{\text{smooth almost complex structures compatible with the orientation of } M\}$$

$$\mathcal{J}_\tau(M, \omega) = \{J \in \mathcal{J}(M) \mid J \text{ is tamed by } \omega\}$$

$$\mathcal{J}(M, \omega) = \{J \in \mathcal{J}(M) \mid J \text{ is compatible with } \omega\},$$

all of which are regarded as topological spaces with the natural C^∞-topology, e.g. we consider a sequence $J_k \in \mathcal{J}(M)$ to converge in $\mathcal{J}(M)$ if all its derivatives converge uniformly on compact subsets of M.

The following foundational lemma is originally due to Gromov:

Proposition 2.1. *On any symplectic manifold (M, ω), the spaces $\mathcal{J}_\tau(M, \omega)$ and $\mathcal{J}(M, \omega)$ are both nonempty and contractible.* $\square$

Most of the theory of pseudoholomorphic curves in symplectic manifolds works equally well whether one considers tame or compatible almost complex structures, but many authors prefer to work only with compatible structures because they make the proofs of certain basic results (including Proposition 2.1) slightly simpler. For our purposes in this chapter, it will make no difference if we work with $\mathcal{J}(M, \omega)$ or $\mathcal{J}_\tau(M, \omega)$, so in order to avert the appearance of lost generality, we shall use $\mathcal{J}_\tau(M, \omega)$. Every statement made in the following remains true if $\mathcal{J}_\tau(M, \omega)$ is replaced with $\mathcal{J}(M, \omega)$, though the situation will become less clear-cut when we generalize to punctured curves in Chap. 8, cf. Remark 8.22.

We can now restate and sketch the proof of Lemma 1.25, which will be the first step in proving the results stated in the introduction.

Proposition 2.2. *Suppose (M, ω) is a symplectic manifold and $S \subset M$ is a smooth 2-dimensional submanifold. Then S is a symplectic submanifold if and only if there exists an ω-tame almost complex structure J preserving TS.*

Proof. In one direction this is immediate: if $J \in \mathcal{J}_\tau(M, \omega)$ and $J(TS) = TS$, then for any $p \in S$ and nonzero tangent vector $X \in T_pS$, (X, JX) forms a basis of T_pS and tameness implies $\omega(X, JX) > 0$, hence $\omega|_{TS}$ is nondegenerate. Conversely, if $S \subset (M, \omega)$ is symplectic, we can find a neighborhood $\mathcal{U} \subset M$ of S and a splitting

$$TM|_{\mathcal{U}} = \tau \oplus \nu$$

such that τ and ν are everywhere symplectic orthogonal complements and $\tau|_S = TS$. Using Exercise 2.3 below, choose complex structures j_τ on τ and j_ν on ν, so that an almost complex structure on $\mathcal{U}$ can be defined by $J_S := j_\tau \oplus j_\nu$. It is straightforward to check that this is ω-tame, in fact, it is ω-compatible. Now choose any other ω-tame almost complex structure J' on a neighborhood of $M \setminus \mathcal{U}$ whose closure does not intersect S. By Exercise 2.4 below, one can then find another tame almost complex structure on M that matches J_S near S and J' outside of $\mathcal{U}$.   □

The following two exercises both depend on the fact that certain spaces of complex structures are nonempty and contractible.

Exercise 2.3. Show that if $E \to M$ is any smooth orientable vector bundle of real rank 2 over a smooth finite-dimensional manifold M, then E admits a complex structure.

Exercise 2.4. Suppose (M, ω) is a symplectic manifold, $A \subset M$ is a closed subset and J_A is an ω-compatible (or ω-tame) almost complex structure defined on a neighborhood of A. Show that M then admits an ω-compatible (or ω-tame) almost complex structure J that matches J_A on a neighborhood of A. *Hint: choose any $J' \in \mathcal{J}(M, \omega)$ and a smooth homotopy between J' and J_A on the latter's domain of definition, then use a cutoff function.*

2.1.2 Simple Holomorphic Curves and Multiple Covers

If (Σ, j) is a Riemann surface and (M, J) is an almost complex manifold of real dimension $2n$, then a C^1-smooth map $u : \Sigma \to M$ is called a **pseudoholomorphic** (or J-**holomorphic**) **curve** if it satisfies the **nonlinear Cauchy-Riemann equation**

$$Tu \circ j = J \circ Tu. \tag{2.1}$$

Note that if J is integrable, this reduces to the usual Cauchy-Riemann equation and just means that u is a holomorphic map between complex manifolds: in particular, it is therefore always smooth. In the nonintegrable case, the latter is still true but is much harder to prove: one can exploit the fact that (2.1) is a first-order elliptic PDE and use elliptic regularity theory to show that every C^1-map satisfying (2.1) is actually smooth.[1]

If (Σ, j) and (Σ', j') are two closed and connected Riemann surfaces, then any holomorphic map

$$\varphi : (\Sigma, j) \to (\Sigma', j')$$

has a well-defined mapping degree $\deg(\varphi) \in \mathbb{Z}$, see e.g. [Mil97]. Moreover, the fact that φ is holomorphic then implies the following:

Proposition 2.5. *If* $\varphi : (\Sigma, j) \to (\Sigma', j')$ *is a holomorphic map between two closed and connected Riemann surfaces, then one of the following is true:*

- $\deg(\varphi) = 0$ *and* φ *is constant;*
- $\deg(\varphi) = 1$ *and* φ *is* **biholomorphic** *i.e.* φ *is a diffeomorphism with a holomorphic inverse;*
- $\deg(\varphi) \geqslant 2$ *and* φ *is a* **branched cover***, i.e. it is a covering map outside of finitely many branch points at which it takes the form* $\varphi(z) = z^k$ *in suitable local holomorphic coordinates, for integers* $k \in \{2, \ldots, \deg(\varphi)\}$.

$\square$

Notice that if $\varphi : (\Sigma, j) \to (\Sigma', j')$ is holomorphic and $u' : (\Sigma', j') \to (M, J)$ is a J-holomorphic curve, then one can compose them to define another J-holomorphic curve

$$u = u' \circ \varphi : (\Sigma, j) \to (M, J). \tag{2.2}$$

If $\deg(\varphi) = 1$, then Proposition 2.5 implies that u is merely a reparametrization of u'. If however $k := \deg(\varphi) \geqslant 2$, then we say u is a k-**fold multiple cover** of u'. Any curve u which is not a multiple cover of any other curve is called **simple**.

A smooth map $u : \Sigma \to M$ is said to be **somewhere injective** if there is a point $z \in \Sigma$ at which $du(z) : T_z\Sigma \to T_{u(z)}M$ is injective and $u^{-1}(u(z)) = \{z\}$; in this

---

[1]In fact it suffices to assume $u : \Sigma \to M$ is of Sobolev class $W^{1,p}$ for any $p > 2$. This is useful in setting up the Fredholm theory for J-holomorphic curves.

case we call z an **injective point**. We will refer to any point $z \in \Sigma$ at which $du(z)$ fails to be injective as a **non-immersed point**. Note that if u is a J-holomorphic curve, then $z \in \Sigma$ is a non-immersed point if and only if $du(z) = 0$.[2] Observe moreover that multiply covered J-holomorphic curves are obviously not somewhere injective. The following less obvious result says that the converse is also true.

Proposition 2.6. *If* (Σ, j) *is a closed connected Riemann surface and* $u : (\Sigma, j) \to$ (M, J) *is a J-holomorphic curve, then the following conditions are equivalent:*

(1) u is somewhere injective;
(2) u is simple;
(3) u has at most finitely many self-intersections and non-immersed points.

Moreover, if u is not simple, then it is either constant or it is a k-fold multiple cover of a simple curve for some integer $k \geqslant 2$. □

In many applications it is useful to note that if u is a multiply covered holomorphic sphere, then its underlying simple curve must be a sphere as well:

Proposition 2.7. *Suppose* $\varphi : (\Sigma, j) \to (\Sigma', j')$ *is a nonconstant holomorphic map between two closed and connected Riemann surfaces. Then if Σ has genus 0, so does Σ'.*

Proof. Suppose $\Sigma \cong S^2$ and Σ' has positive genus. Then the universal cover of Σ' is contractible, thus $\pi_2(\Sigma') = 0$, so $\varphi_*[\Sigma] = 0 \in H_2(\Sigma')$ and thus $\deg(\varphi) = 0$, implying φ is constant. □

The constant J-holomorphic curves can be characterized by their homology whenever J is tamed by a symplectic form ω. This follows from the observation that any J-holomorphic curve $u : \Sigma \to M$ must then satisfy

$$\int u^*\omega \geqslant 0,$$

with equality if and only if u is constant. As a consequence:

Proposition 2.8. *If (M, ω) is a symplectic manifold, $J \in \mathcal{J}_\tau(M, \omega)$ and (Σ, j) is a closed and connected Riemann surface, then a J-holomorphic curve $u : \Sigma \to M$ is constant if and only if it is homologous to zero, i.e.* $[u] := u_*[\Sigma] = 0 \in H_2(M)$.
□

---

[2]Non-immersed points of J-holomorphic curves are also sometimes called **critical points**, though we will avoid using the term in this way since its standard meaning in differential topology is different—strictly speaking, if $\dim M \geqslant 4$, then *every* point of a J-holomorphic curve $u : \Sigma \to M$ is critical since $du(z) : T_z\Sigma \to T_{u(u)}M$ can never be surjective.

2.1.3 Smoothness and Dimension of the Moduli Space

Assume (M, J) is an almost complex manifold of dimension $2n$. Fix integers $m \geqslant 0$ and $g \geqslant 0$, and a homology class $A \in H_2(M)$. The **moduli space of unparametrized J-holomorphic curves** in M homologous to A, with genus g and m **marked points** is defined as

$$\mathcal{M}_{g,m}(A; J) = \{(\Sigma, j, u, (\zeta_1, \ldots, \zeta_m))\} \Big/ \sim$$

where

- (Σ, j) is a closed connected Riemann surface of genus g;
- $u : (\Sigma, j) \to (M, J)$ is a pseudoholomorphic curve representing the homology class A, i.e. $[u] := u_*[\Sigma] = A \in H_2(M)$;
- $(\zeta_1, \ldots, \zeta_m)$ is an ordered set of m distinct points in Σ;
- $(\Sigma, j, u, (\zeta_1, \ldots, \zeta_m))$ and $(\Sigma', j', u', (\zeta'_1, \ldots, \zeta'_m))$ are defined to be equivalent if and only if there exists a biholomorphic map $\varphi : (\Sigma, j) \to (\Sigma', j')$ such that $u = u' \circ \varphi$ and $\varphi(\zeta_i) = \zeta'_i$ for all $i = 1, \ldots, m$.

For situations where the homology class is not specified, we shall write

$$\mathcal{M}_{g,m}(J) = \bigcup_{A \in H_2(M)} \mathcal{M}_{g,m}(A; J),$$

and for the case with no marked points we will sometimes abbreviate

$$\mathcal{M}_g(A; J) := \mathcal{M}_{g,0}(A; J), \qquad \mathcal{M}_g(J) := \mathcal{M}_{g,0}(J).$$

The space $\mathcal{M}_{g,m}(J)$ admits a metrizable topology with the following notion of convergence: we have

$$[(\Sigma_k, j_k, u_k, (\zeta_1^k, \ldots, \zeta_m^k))] \to [(\Sigma, j, u, (\zeta_1, \ldots, \zeta_m))] \in \mathcal{M}_{g,m}(J)$$

if and only if one can choose representatives of the form

$$\left(\Sigma, j'_k, u'_k, (\zeta_1, \ldots, \zeta_m)\right) \sim \left(\Sigma_k, j_k, u_k, (\zeta_1^k, \ldots, \zeta_m^k)\right)$$

such that $u'_k \to u$ and $j'_k \to j$ in C^∞. We shall sometimes abuse notation and abbreviate an element $[(\Sigma, j, u, (\zeta_1, \ldots, \zeta_m))] \in \mathcal{M}_{g,m}(J)$ simply as $u \in \mathcal{M}_{g,m}(J)$ when the rest is clear from context.

Remark 2.9. Observe that the above definition of convergence in $\mathcal{M}_{g,m}(J)$ does not depend on J, so in particular one can similarly define convergence of a sequence $u_k \in \mathcal{M}_{g,m}(J_k)$ to an element $u \in \mathcal{M}_{g,m}(J)$, where the almost complex structures J_k and J need not all be the same.

We shall denote by

$$\mathcal{M}^*_{g,m}(A; J) \subset \mathcal{M}_{g,m}(A; J), \qquad \mathcal{M}^*_{g,m}(J) \subset \mathcal{M}_{g,m}(J)$$

the subsets consisting of all curves $[(\Sigma, j, u, (\zeta_1, \ldots, \zeta_m))]$ for which the map $u : \Sigma \to M$ is somewhere injective. Observe that this is always an open subset.

Remark 2.10. While the marked points $(\zeta_1, \ldots, \zeta_m)$ may appear superfluous at this stage, their importance lies in the fact that there is a well-defined and continuous **evaluation map**

$$\mathrm{ev} = (\mathrm{ev}_1, \ldots, \mathrm{ev}_m) : \mathcal{M}_{g,m}(J) \to M^m :$$

$$[(\Sigma, j, u, (\zeta_1, \ldots, \zeta_m))] \mapsto (u(\zeta_1), \ldots, u(\zeta_m)). \qquad (2.3)$$

This can be used for instance to define subspaces of $\mathcal{M}_{g,m}(J)$ consisting of curves whose marked points are mapped to particular submanifolds; see Sects. 2.1.4 and 2.2.3 for more on this.

For reasons that will become clear in Theorem 2.11 below, we define the **virtual dimension** of the moduli space $\mathcal{M}_{g,m}(A; J)$ to be the integer

$$\text{vir-dim}\, \mathcal{M}_{g,m}(A; J) := (n-3)(2-2g) + 2c_1(A) + 2m, \qquad (2.4)$$

where we use the abbreviation

$$c_1(A) := \langle c_1(TM, J), A \rangle.$$

One can equivalently define $c_1(A)$ as the first Chern number of the complex vector bundle $(u^*TM, J) \to \Sigma$ if u is any representative of a curve in $\mathcal{M}_{g,m}(A; J)$. Such a map $u : \Sigma \to M$ also naturally represents an element of $\mathcal{M}_g(A; J)$, the moduli space with no marked points, and the virtual dimension of this space is also sometimes called the **index** of u:

$$\mathrm{ind}(u) := (n-3)\chi(\Sigma) + 2c_1([u]). \qquad (2.5)$$

Another way to write (2.4) is thus

$$\text{vir-dim}\, \mathcal{M}_{g,m}(A; J) = \mathrm{ind}(u) + 2m.$$

We would now like to state a result describing the local structure of $\mathcal{M}_{g,m}(A; J)$. This will require an explanation of the term "Fredholm regular," the proper definition of which is rather technical—trying to state it here precisely would take us too far afield, but we will summarize the idea. Due to Theorem 2.12 below, one usually does not need to know the precise definition in order to make use of it.

As preparation, imagine that instead of a moduli space of holomorphic curves, we want to study the set of solutions to an equation of the form $s(x) = 0$, where $s : B \to E$ is a smooth section of a finite-dimensional real vector bundle $E \to B$. At any point $x \in s^{-1}(0)$, the section has a well-defined *linearization* $Ds(x) :$ $T_x B \to E_x$, and the implicit function theorem tells us that if $Ds(x)$ is surjective, then a neighborhood of x in $s^{-1}(0)$ can be identified with a neighborhood of 0 in $\ker Ds(x)$, hence it is a smooth manifold of dimension $\dim \ker Ds(x) = \dim B -$ rank E. When this surjectivity condition holds, we say that the solution $x \in s^{-1}(0)$ is *regular*. The implicit function theorem then implies that the (necessarily open) subset of regular points in $s^{-1}(0)$ is a smooth manifold.

One can study the space of holomorphic curves in the same manner, but now the vector bundle and its base both become infinite dimensional, and one needs to apply the implicit function theorem in a Banach space setting (see e.g. [Lan93, Chap. XIV] and [Lan99]). The base will now be a smooth Banach manifold consisting of suitable pairs (j, u) where $u : \Sigma \to M$ is a map (in some Sobolev or Hölder regularity class) and j is a complex structure on Σ. The corresponding vector bundle is an infinite-dimensional Banach space bundle $\mathcal{E} \to \mathcal{B}$ whose fiber over (j, u) is a Banach space of sections (again in a suitable regularity class) of the bundle

$$\overline{\text{Hom}}_{\mathbb{C}}((T\Sigma, j), (u^*TM, J)) \to \Sigma,$$

consisting of complex antilinear bundle maps $(T\Sigma, j) \to (u^*TM, J)$. The nonlinear Cauchy-Riemann equation can then be expressed in terms of a smooth section of this bundle,

$$\bar{\partial}_J : \mathcal{B} \to \mathcal{E} : (j, u) \mapsto Tu + J \circ Tu \circ j,$$

whose zero set can be identified with the moduli space of J-holomorphic curves. At a solution $(j, u) \in \bar{\partial}_J^{-1}(0)$, there is now a linearization

$$D\bar{\partial}_J(j, u) : T_{(j,u)}\mathcal{B} \to \mathcal{E}_{(j,u)},$$

which is a bounded linear operator between two Banach spaces. Most importantly—and this is another deep consequence of elliptic regularity theory—the linearization in this case is a *Fredholm operator*, i.e. the dimension of its kernel and codimension of its image are both finite, and its **Fredholm index**

$$\text{ind } D\bar{\partial}_J(j, u) := \dim \ker D\bar{\partial}_J(j, u) - \text{codim im } D\bar{\partial}_J(j, u) \in \mathbb{Z}$$

is constant under small perturbations of the setup. The Fredholm index of $D\bar{\partial}_J(j, u)$ can be derived from the Riemann-Roch formula: up to adjustments accounting for symmetries of the domain and marked points, it is precisely what we called $\text{ind}(u)$ in (2.5) above, i.e. it is the virtual dimension of the moduli space. Since kernels of Fredholm operators always have closed complements, the Fredholm property makes it possible to apply the implicit function theorem in much the same way as in the finite-dimensional example: if $D\bar{\partial}_J(j,u)$ is surjective, then a neighborhood of u in the moduli space can be identified with a neighborhood of 0 in $\ker D\bar{\partial}_J(j,u)$, whose dimension is precisely the Fredholm index. The miracle of Fredholm theory is that this dimension is finite even though both $\dim \mathcal{B}$ and $\operatorname{rank} \mathcal{E}$ are infinite.

With the preceding understood, we shall say that a curve $u \in \mathcal{M}_{g,m}(A; J)$ is **Fredholm regular** if the linearization $D\bar{\partial}_J(j,u) : T_{(j,u)}\mathcal{B} \to \mathcal{E}_{(j,u)}$ outlined above is a surjective operator. This definition may sound ad hoc at first, as the functional analytic setup underlying the section $\bar{\partial}_J : \mathcal{B} \to \mathcal{E}$ depends on a number of choices (e.g. which Banach spaces to use as local models for $\mathcal{B}$ and $\mathcal{E}$), but one can show using the theory of elliptic operators that the notion of Fredholm regularity does not depend on these choices.

There is one caveat that must be added to the above sketch: the local identification between $\mathcal{M}_{g,m}(A; J)$ and $\bar{\partial}_J^{-1}(0)$ is only correct up to a finite ambiguity which depends on symmetries. It depends in particular on the **automorphism group**

$$\operatorname{Aut}(\Sigma, j, u, (\zeta_1, \ldots, \zeta_m)),$$

which is defined to be the group of all biholomorphic maps $\varphi : (\Sigma, j) \to (\Sigma, j)$ such that $u = u \circ \varphi$ and $\varphi(\zeta_i) = \zeta_i$ for all $i = 1, \ldots, m$. This group is always finite unless u is constant, and it is trivial whenever u is somewhere injective.

Theorem 2.11. *Let $\mathcal{M}_{g,m}^{\mathrm{reg}}(A; J) \subset \mathcal{M}_{g,m}(A; J)$ denote the (necessarily open) subset consisting of all curves in $\mathcal{M}_{g,m}(A; J)$ which are Fredholm regular and have trivial automorphism group.*[3] *Then $\mathcal{M}_{g,m}^{\mathrm{reg}}(A; J)$ naturally admits the structure of a smooth oriented finite-dimensional manifold with*

$$\dim \mathcal{M}_{g,m}^{\mathrm{reg}}(A; J) = \operatorname{vir-dim} \mathcal{M}_{g,m}(A; J).$$

$\square$

On its own, Theorem 2.11 is quite difficult to apply since one needs a way of checking which curves are Fredholm regular. In practice however, one can usually appeal to more general results instead of checking explicitly. For intuition on this, consider again the example of a finite-dimensional vector bundle $E \to B$ with a smooth section $s : B \to E$. The zero-set $s^{-1}(0) \subset B$ is geometrically the intersection of two submanifolds of the total space, namely $s(B) \subset E$ and the zero-section $B \subset E$. Another way of expressing the implicit function theorem is then the statement that this intersection is a smooth submanifold whenever $s(B)$ and B intersect *transversely*; this is in fact equivalent to the requirement that the

---

[3] A version of Theorem 2.11 holds without the assumption on the automorphism group, but in general $\mathcal{M}_{g,m}^{\mathrm{reg}}(A; J)$ will then be a smooth *orbifold* rather than a manifold. More precisely, every Fredholm regular curve $u \in \mathcal{M}_{g,m}(A; J)$ has a neighborhood that looks like the quotient of a vector space with dimension $\operatorname{ind}(u) + 2m$ by an action of its automorphism group. This is explained in detail in [Wenc].

linearization $Ds(x) : T_x B \to E_x$ be surjective for all $x \in s^{-1}(0)$. Transversality, of course, is a "generic" property: it's something that can typically be achieved by making small perturbations, and one expects it to hold outside of exceptional ("non-generic") cases, see for example [Hir94]. Thus in the finite-dimensional example, one can make a generic perturbation of the section s so that every solution $x \in s^{-1}(0)$ is regular.

Going back to holomorphic curves, the intuition from the finite-dimensional example suggests that if we perturb the Cauchy-Riemann equation in some sufficiently generic way, we may be able to ensure after this perturbation that every curve in our moduli space is Fredholm regular. The most obvious piece of data to try and perturb is the almost complex structure. It turns out that this is not generic enough to achieve transversality in general, but it does suffice for making all *somewhere injective* curves regular, which is good enough for many important applications.

To set up the result in its most useful form, we shall suppose (M, ω) is a $2n$-dimensional symplectic manifold, $\mathcal{U} \subset M$ is an open subset with compact closure and $J_0 \in \mathcal{J}_\tau(M, \omega)$ is a fixed ω-tame almost complex structure. Define the subset

$$\mathcal{J}_\tau(M, \omega; \mathcal{U}, J_0) = \{J \in \mathcal{J}_\tau(M, \omega) \mid J \equiv J_0 \text{ on } M \backslash \mathcal{U}\} \subset \mathcal{J}_\tau(M, \omega).$$

Note that in some applications, if M is compact, it suffices to take $\mathcal{U} = M$ so that J_0 is irrelevant and $\mathcal{J}_\tau(M, \omega; \mathcal{U}, J_0) = \mathcal{J}_\tau(M, \omega)$, but this more general definition gives us the useful option of perturbing J only within a fixed open subset. Now for any fixed integers $g, m \geq 0$ and a homology class $A \in H_2(M)$, define

$$\mathcal{J}_\tau^{\text{reg}}(M, \omega; \mathcal{U}, J_0) = \left\{ J \in \mathcal{J}_\tau(M, \omega; \mathcal{U}, J_0) \; \middle| \; \begin{array}{l} \text{every } u \in \mathcal{M}_{g,m}(A; J) \text{ with an injective point} \\ \text{mapped into } \mathcal{U} \text{ is Fredholm regular} \end{array} \right\}.$$

Note that by Proposition 2.6, every somewhere injective curve that passes through $\mathcal{U}$ necessarily maps some injective point into $\mathcal{U}$. In the special case $M = \mathcal{U}$, we simply write

$$\mathcal{J}_\tau^{\text{reg}}(M, \omega) = \mathcal{J}_\tau^{\text{reg}}(M, \omega; M, J_0),$$

where J_0 is in this case arbitrary.

Recall that a subset Y of a topological space X is called **meager** if it is a countable union of nowhere dense subsets. Equivalently, a set is meager if its complement contains a countable intersection of open dense sets, and this complement is then called **comeager**.[4] Any countable intersection of comeager subsets is also comeager, and by the Baire category theorem, if X is a complete metric space then every comeager subset is also dense. In the following, the role of X will be played by $\mathcal{J}_\tau(M, \omega; \mathcal{U}, J_0)$, which is a complete metrizable space

---

[4]It has become conventional among symplectic topologists to refer to comeager sets as "Baire sets" or "sets of second category," though this seems to be at odds with the standard usage in other fields.

whenever $\mathcal{U} \subset M$ is precompact. The meager sets play a similar role in infinite-dimensional settings to the sets of measure zero in finite dimensions: in particular, any statement (such as the theorem below) that is true for all choices of J belonging to some comeager subset of all the allowed almost complex structures is said to hold for **generic** choices.

Theorem 2.12. *Fix any symplectic manifold (M, ω), an open subset $\mathcal{U} \subset M$ with compact closure, integers $g, m \geqslant 0$, $A \in H_2(M)$, and $J_0 \in \mathcal{J}_\tau(M, \omega)$. Then the set $\mathcal{J}_\tau^{\text{reg}}(M, \omega; \mathcal{U}, J_0)$ as defined above is a comeager subset of $\mathcal{J}_\tau(M, \omega; \mathcal{U}, J_0)$. In particular, every $J \in \mathcal{J}_\tau(M, \omega)$ matching J_0 outside $\mathcal{U}$ admits a C^∞-small ω-tame perturbation in $\mathcal{U}$ for which every somewhere injective curve in $\mathcal{M}_{g,m}(A; J)$ passing through $\mathcal{U}$ is Fredholm regular.* $\square$

Since every somewhere injective curve has trivial automorphism group, Theorems 2.11 and 2.12 combine to endow the space of somewhere injective curves $\mathcal{M}_{g,m}^*(A; J)$ with the structure of a smooth finite-dimensional manifold for generic J. For reasons that we will discuss in Sect. 2.1.8, this manifold always carries a canonical orientation, so the result can be summarized as follows:

Corollary 2.13. *If (M, ω) is a closed symplectic manifold, then for generic ω-tame almost complex structures J, the space $\mathcal{M}_{g,m}^*(A; J)$ is a smooth oriented finite-dimensional manifold with its dimension given by (2.4).* $\square$

Remark 2.14. Our definition of $\mathcal{J}_\tau^{\text{reg}}(M, \omega; \mathcal{U}, J_0)$ above depends on the choices $g, m \geqslant 0$ and $A \in H_2(M)$, though we have suppressed this in the notation. In practice it makes no difference: since the set of all possible triples (g, m, A) is countable, the *intersection* of all the spaces $\mathcal{J}_\tau^{\text{reg}}(M, \omega; \mathcal{U}, J_0)$ for all possible choices of (g, m, A) is still a comeager subset, thus one can still say that for a generic $J \in \mathcal{J}_\tau(M, \omega)$, *all* the spaces $\mathcal{M}_{g,m}^*(J)$ are smooth.

There is a similar result for 1-parameter families of data. Fix a smooth 1-parameter family of symplectic structures $\{\omega_s\}_{s \in [0,1]}$ on M and define $\mathcal{J}_\tau(M, \{\omega_s\})$ to be the space of all smooth 1-parameter families of almost complex structures $\{J_s\}_{s \in [0,1]}$ such that $J_s \in \mathcal{J}_\tau(M, \omega_s)$ for all $s \in [0, 1]$. The space $\mathcal{J}_\tau(M, \{\omega_s\})$ carries a natural C^∞-topology and is again nonempty and contractible as a consequence of Proposition 2.1. Fixing any $J \in \mathcal{J}_\tau(M, \omega_0)$ and $J' \in \mathcal{J}_\tau(M, \omega_1)$, the same is true of the subset

$$\mathcal{J}_\tau(M, \{\omega_s\}; J, J') := \left\{ \{J_s\} \in \mathcal{J}_\tau(M, \{\omega_s\}) \mid J_0 \equiv J \text{ and } J_1 \equiv J' \right\}.$$

Given a family $\{J_s\} \in \mathcal{J}_\tau(M, \{\omega_s\})$, we define the **parametric** moduli space

$$\mathcal{M}_{g,m}(A; \{J_s\}) = \{(s, u) \mid s \in [0, 1] \text{ and } u \in \mathcal{M}_{g,m}(A; J_s)\},$$

and write

$$\mathcal{M}_{g,m}(\{J_s\}) = \bigcup_{A \in H_2(M)} \mathcal{M}_{g,m}(A; \{J_s\}).$$

This has a natural topology for which a sequence (s_k, u_k) converges to (s, u) if and only if $s_k \to s$ and $u_k \to u$, with the latter defined in the same way as convergence in $\mathcal{M}_{g,m}(J)$ (see Remark 2.9). Define also the open subsets

$$\mathcal{M}_{g,m}^*(A; \{J_s\}) \subset \mathcal{M}_{g,m}(A; \{J_s\}), \qquad \mathcal{M}_{g,m}^*(\{J_s\}) \subset \mathcal{M}_{g,m}(\{J_s\}),$$

consisting of pairs (s, u) for which u is somewhere injective.

Theorem 2.15. *Given* $\{J_s\} \subset \mathcal{J}_\tau(M, \{\omega_s\})$, *let*

$$\mathcal{M}_{g,m}^{reg}(A; \{J_s\}) \subset \mathcal{M}_{g,m}(A; \{J_s\})$$

denote the (necessarily open) subset consisting of all pairs (s, u) for which $u \in \mathcal{M}_{g,m}^{reg}(A; J_s)$, i.e. u is Fredholm regular and has trivial automorphism group. Then $\mathcal{M}_{g,m}^{reg}(A; \{J_s\})$ naturally admits the structure of a smooth oriented manifold with boundary

$$\partial \mathcal{M}_{g,m}^{reg}(A; \{J_s\}) = -\left(\{0\} \times \mathcal{M}_{g,m}^{reg}(A; J_0)\right) \cup \left(\{1\} \times \mathcal{M}_{g,m}^{reg}(A; J_1)\right),$$

and the natural projection

$$\mathcal{M}_{g,m}^{reg}(A; \{J_s\}) \to [0, 1] : (s, u) \mapsto s$$

is a submersion.   □

Theorem 2.16. *Suppose M is a closed manifold with a smooth 1-parameter family of symplectic structures $\{\omega_s\}_{s\in[0,1]}$, $J \in \mathcal{J}_\tau^{reg}(M, \omega_0)$ and $J' \in \mathcal{J}_\tau^{reg}(M, \omega_1)$. Then for every $g, m \geq 0$ and $A \in H_2(M)$, $\mathcal{J}_\tau(M, \{\omega_s\}; J, J')$ contains a comeager subset*

$$\mathcal{J}_\tau^{reg}(M, \{\omega_s\}; J, J') \subset \mathcal{J}_\tau(M, \{\omega_s\}; J, J')$$

such that for all $\{J_s\} \in \mathcal{J}_\tau^{reg}(M, \{\omega_s\}; J, J')$, $\mathcal{M}_{g,m}^*(A; \{J_s\})$ naturally admits the structure of a smooth oriented manifold with boundary

$$\partial \mathcal{M}_{g,m}^*(A; \{J_s\}) = -\left(\{0\} \times \mathcal{M}_{g,m}^*(A; J_0)\right) \cup \left(\{1\} \times \mathcal{M}_{g,m}^*(A; J_1)\right),$$

and all critical values of the projection

$$\mathcal{M}_{g,m}^*(A; \{J_s\}) \to [0, 1] : (s, u) \mapsto s$$

lie in the interior $(0, 1)$.   □

It should be emphasized that in contrast to Theorem 2.15, the map $\mathcal{M}_{g,m}^*(A; \{J_s\}) \to [0, 1] : (s, u) \mapsto s$ in Theorem 2.16 need not be a submersion,

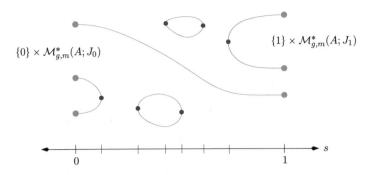

Fig. 2.1 Possible structure of the parametric moduli space $\mathcal{M}^*_{g,m}(A; \{J_s\})$ in a case where vir-dim $\mathcal{M}^*_{g,m}(A; J_s) = 0$, with the map $\mathcal{M}^*_{g,m}(A; \{J_s\}) \to [0, 1] : (s, u) \mapsto s$ shown as having six critical values in the interior $(0, 1)$

as there may exist pairs $(s, u) \in \mathcal{M}^*_{g,m}(A; \{J_s\})$ for which the curve u is not Fredholm regular. Such points in $\mathcal{M}^*_{g,m}(A; \{J_s\})$ can give rise e.g. to birth-death bifurcations in the family of moduli spaces $\mathcal{M}^*_{g,m}(A; J_s)$; see Fig. 2.1.

Remark 2.17. As in the discussion of the space $\mathcal{J}^{\mathrm{reg}}_\tau(M, \omega; \mathcal{U}, J_0)$ above, Theorem 2.16 and all the genericity results discussed in the next two subsections admit generalizations to allow for perturbations that match fixed data outside of some chosen precompact open subset $\mathcal{U} \subset M$. The caveat is always that such results only apply to the open set of J-holomorphic curves $u : \Sigma \to M$ that have an injective point $z \in \Sigma$ with $u(z) \in \mathcal{U}$. The crucial step in the proofs of such theorems is typically a lemma about the smoothness of a "universal" moduli space consisting of pairs (u, J), where J belongs to the space of admissible perturbed data, and u is in the moduli space determined by J. This space is at best an infinite-dimensional Banach manifold since the space of perturbations of J is quite large, but if it is smooth, then one can apply Smale's infinite-dimensional version of Sard's theorem [Sma65] to the projection $(u, J) \mapsto J$, giving a comeager set of regular values J for which the corresponding moduli space is smooth. The standard arguments for proving smoothness of the universal moduli space only require the data to be perturbable in an arbitrarily small neighborhood of the point $u(z)$, so long as z is an injective point. (For the technical details, see [MS12, Prop. 3.2.1] or [Wenc, Prop. 4.55], and also [Wene, Lemma A.3] for the corresponding result needed for Sect. 2.1.5 below.)

This level of generality is useful for at least two reasons. First, it will sometimes be convenient in our proofs of the main theorems from Chap. 1 to make perturbations that leave unchanged some region where the holomorphic curves are already sufficiently well understood. A second reason involves the generalization to punctured holomorphic curves in Chaps. 8 and 9, where the target spaces will always be noncompact, thus it is necessary for technical reasons to choose a compact region in which the perturbations take place (cf. the discussion preceding Theorem 8.31).

Note that for closed curves $u : \Sigma \to M$, Proposition 2.6 implies that having an injective point $z \in \Sigma$ with $u(z) \in \mathcal{U}$ is equivalent to u being a somewhere injective curve that intersects $\mathcal{U}$. We will see in Sect. 8.3 that the same is true for punctured curves, but the reader should beware that it is *not* true in general for curves with boundary and totally real boundary conditions, see [Laz00].

Let us point out explicitly what these genericity results mean in cases where the virtual dimension of the moduli space is *negative*. By definition, any smooth manifold with negative dimension is empty. One sees from the example of the finite-dimensional vector bundle $E \to B$ that this is the right convention: if rank $E > \dim B$, then $Ds(x) : T_x B \to E_x$ can never be surjective, so the fact that generic perturbations make $s(B)$ and B transverse actually means that after a generic perturbation, $s^{-1}(0)$ will be empty. This is also what happens in the infinite-dimensional setting if the Fredholm index is negative, hence:

Corollary 2.18. *Suppose (M, ω) is a symplectic manifold, $\mathcal{U} \subset M$ is an open subset with compact closure and $J \in \mathcal{J}_\tau(M, \omega)$. Then after a generic ω-tame perturbation of J on $\mathcal{U}$, every J-holomorphic curve u with an injective point mapped into $\mathcal{U}$ satisfies $\mathrm{ind}(u) \geqslant 0$.* $\square$

Since $\dim \mathcal{M}^*_{g,m}(A; \{J_s\}) = \dim \mathcal{M}^*_{g,m}(A; J_0) + 1$ in general, the transversality theory also implies the following result, which can be strengthened a bit further in light of Remark 2.20 below.

Corollary 2.19. *If M is a closed manifold with a smooth 1-parameter family of symplectic structures $\{\omega_s\}_{s \in [0,1]}$, then for generic families $\{J_s\} \in \mathcal{J}_\tau(M, \{\omega_s\})$, every somewhere injective J_s-holomorphic curve u for every $s \in [0, 1]$ satisfies $\mathrm{ind}(u) \geqslant -1$.* $\square$

Remark 2.20. It is sometimes useful to observe that since the index (2.5) is always even, Corollary 2.19 actually implies $\mathrm{ind}(u) \geqslant 0$. As we'll see in Chap. 4, this plays an important role in the proof of Theorem B. It is not true however in more general settings, e.g. curves with totally real boundary conditions or finite-energy punctured holomorphic curves can have odd index in general, cf. Remark 8.32.

It is not generally true that the entire moduli space $\mathcal{M}_{g,m}(J)$ can be made smooth or that all curves with negative index can be eliminated just by perturbing J, unless one can somehow rule out multiply covered curves. This fact causes enormous headaches throughout the field of symplectic topology, but for our applications in dimension four it will not pose a problem.

2.1.4 Moduli Spaces with Marked Point Constraints

A straightforward extension of the above results is to consider moduli spaces of holomorphic curves with marked points satisfying constraints on their images in the target. Such constraints can be defined using the evaluation map $\mathrm{ev} : \mathcal{M}_{g,m}(J) \to M^m$, see (2.3).

Assume M is a $2n$-dimensional manifold with either a fixed symplectic structure ω or a fixed smooth 1-parameter family of symplectic structures $\{\omega_s\}_{s\in[0,1]}$. For this subsection we will assume M is closed, but the reader should keep in mind that this assumption can be weakened in the spirit of Remark 2.17. Fix a smooth submanifold

$$Z \subset M^m.$$

Then for any $J \in \mathcal{J}_\tau(M, \omega)$ or $\{J_s\} \in \mathcal{J}_\tau(M, \{\omega_s\})$, we define

$$\mathcal{M}_{g,m}(A; J; Z) = \mathrm{ev}^{-1}(Z) \subset \mathcal{M}_{g,m}(A; J),$$

and

$$\mathcal{M}_{g,m}(A; \{J_s\}; Z) = \{(s, u) \mid s \in [0, 1],\ u \in \mathcal{M}_{g,m}(A; J_s; Z)\},$$

with

$$\mathcal{M}_{g,m}(J; Z) = \bigcup_{A \in H_2(M)} \mathcal{M}_{g,m}(A; J; Z),$$

$$\mathcal{M}_{g,m}(\{J_s\}; Z) = \bigcup_{A \in H_2(M)} \mathcal{M}_{g,m}(A; \{J_s\}; Z).$$

In other words, the elements of $\mathcal{M}_{g,m}(J; Z)$ can be parametrized by J-holomorphic curves $u : \Sigma \to M$ with marked points $\zeta_1, \ldots, \zeta_m \subset \Sigma$ satisfying the constraint

$$(u(\zeta_1), \ldots, u(\zeta_m)) \in Z. \tag{2.6}$$

The most common special case we will need is the following: given points $p_1, \ldots, p_m \in M$, we can choose Z to be the 1-point set

$$Z = \{(p_1, \ldots, p_m)\},$$

and in this case use the notation

$$\mathcal{M}_{g,m}(A; J; p_1, \ldots, p_m) := \mathcal{M}_{g,m}(A; J; Z) = \mathrm{ev}^{-1}(p_1, \ldots, p_m),$$

so that elements u of $\mathcal{M}_{g,m}(A; J; p_1, \ldots, p_m)$ with marked points $\zeta_1, \ldots, \zeta_m$ satisfy

$$u(\zeta_i) = p_i \quad \text{for} \quad i = 1, \ldots, m.$$

If $u \in \mathcal{M}_{g,m}(A; J; Z)$ is Fredholm regular and has trivial automorphism group, then Theorem 2.11 says that the unconstrained moduli space $\mathcal{M}_{g,m}(A; J)$ is a smooth manifold of dimension equal to vir-dim $\mathcal{M}_{g,m}(A; J)$ near u. We will say that u is **Fredholm regular for the constrained problem** and write

$$u \in \mathcal{M}_{g,m}^{\mathrm{reg}}(A; J; Z)$$

if, in addition to the above conditions, u is a transverse intersection of the evaluation map $\mathrm{ev} : \mathcal{M}_{g,m}(A; J) \to M^m$ with the submanifold Z. Note that in the case $Z = \{(p_1, \ldots, p_m)\}$, this simply means $(p_1, \ldots, p_m)$ is a regular value of ev. The open subset $\mathcal{M}_{g,m}^{\mathrm{reg}}(A; J; Z) \subset \mathcal{M}_{g,m}(A; J; Z)$ is then also a smooth manifold, with dimension less than that of $\mathcal{M}_{g,m}^{\mathrm{reg}}(A; J)$ by $\mathrm{codim}\, Z$, so in particular $2nm$ if $Z = \{(p_1, \ldots, p_m)\}$. We thus define the virtual dimension of the constrained moduli space by

$$\mathrm{vir\text{-}dim}\, \mathcal{M}_{g,m}(A; J; Z) = \mathrm{vir\text{-}dim}\, \mathcal{M}_{g,m}(A; J) - \mathrm{codim}\, Z,$$

which we will sometimes also call the **constrained index** of any curve in $u \in \mathcal{M}_{g,m}(A; J; Z)$. In the case $Z = \{(p_1, \ldots, p_m)\}$, we have

$$\begin{aligned}
\mathrm{vir\text{-}dim}\, \mathcal{M}_{g,m}(A; J; p_1, \ldots, p_m) &= \mathrm{vir\text{-}dim}\, \mathcal{M}_{g,m}(A; J) - 2nm \\
&= (n-3)(2-2g) + 2c_1(A) - 2m(n-1) \\
&= \mathrm{ind}(u) - 2m(n-1).
\end{aligned} \tag{2.7}$$

The neighborhood of a constrained Fredholm regular curve $u \in \mathcal{M}_{g,m}^{\mathrm{reg}}(A; J; Z)$ can also be deformed smoothly under any small deformation of the data J and Z, in analogy with Theorem 2.15.

As usual, we shall denote by

$$\mathcal{M}_{g,m}^{*}(A; J; Z) \subset \mathcal{M}_{g,m}(A; J; Z)$$

$$\mathcal{M}_{g,m}^{*}(A; \{J_s\}; Z) \subset \mathcal{M}_{g,m}(A; \{J_s\}; Z)$$

the open subsets consisting of somewhere injective curves. Observe that if $J \in \mathcal{J}_{\tau}^{\mathrm{reg}}(M, \omega)$, then standard transversality results from differential topology imply that generic perturbations of the submanifold $Z \subset M^m$ make ev transverse to Z; in the case $Z = \{(p_1, \ldots, p_m)\}$, this is an immediate consequence of Sard's theorem. After such a perturbation, every curve in $\mathcal{M}_{g,m}^{*}(A; J; Z)$ is also Fredholm regular for the constrained problem. Alternatively, the arguments behind the transversality results in Sect. 2.1.3 above can be modified to prove the following statement, in which Z can be fixed in advance but J must be perturbed:

Theorem 2.21. *Given the submanifold* $Z \subset M^m$, *integers* $g, m \geqslant 0$ *and* $A \in H_2(M)$, *there exists a comeager subset*

$$\mathcal{J}_{\tau}^{\mathrm{reg}}(M, \omega; Z) \subset \mathcal{J}_{\tau}(M, \omega)$$

such that for all $J \in \mathcal{J}_\tau^{\mathrm{reg}}(M, \omega; Z)$, every somewhere injective curve in $\mathcal{M}_{g,m}(A; J; Z)$ is Fredholm regular for the constrained problem. In particular, $\mathcal{M}_{g,m}^*(A; J; Z)$ is a smooth oriented manifold with dimension equal to vir-dim $\mathcal{M}_{g,m}(A; J; Z)$.   □

Theorem 2.22. *Suppose M is a closed manifold with a smooth 1-parameter family of symplectic structures* $\{\omega_s\}_{s\in[0,1]}$, $J \in \mathcal{J}_\tau^{\mathrm{reg}}(M, \omega_0; Z)$, $J' \in \mathcal{J}_\tau^{\mathrm{reg}}(M, \omega_1; Z)$, $g, m \geqslant 0$ *are integers and* $A \in H_2(M)$. *Then there exists a comeager subset*

$$\mathcal{J}_\tau^{\mathrm{reg}}(M, \{\omega_s\}; Z; J, J') \subset \mathcal{J}_\tau(M, \{\omega_s\}; J, J')$$

such that for all $\{J_s\} \in \mathcal{J}_\tau^{\mathrm{reg}}(M, \{\omega_s\}; Z; J, J')$, *the space* $\mathcal{M}_{g,m}^*(A; \{J_s\}; Z)$ *is a smooth oriented manifold with boundary*

$$\partial \mathcal{M}_{g,m}^*(A; \{J_s\}; Z) = -\left(\{0\} \times \mathcal{M}_{g,m}^*(A; J_0; Z) \right) \cup \left(\{1\} \times \mathcal{M}_{g,m}^*(A; J_1; Z) \right),$$

and all critical values of the projection

$$\mathcal{M}_{g,m}^*(A; \{J_s\}; Z) \to [0, 1] : (s, u) \mapsto s$$

lie in the interior $(0, 1)$.   □

Once again the cases with negative dimension mean that the moduli space is empty. Let us state two special cases of this that will be useful in applications: first, set $Z = \{(p_1, \ldots, p_m)\}$.

Corollary 2.23. *Fix* $p_1, \ldots, p_m \in M$. *Then for generic* $J \in \mathcal{J}_\tau(M, \omega)$, *the space* $\mathcal{M}_{g,m}^*(A; J; p_1, \ldots, p_m)$ *is empty unless*

$$(n - 3)(2 - 2g) + 2c_1(A) \geqslant 2m(n - 1).$$

Similarly, given a smooth family $\{\omega_s\}_{s\in[0,1]}$ *of symplectic structures, for generic choices of* $J \in \mathcal{J}_\tau(M, \omega_0)$, $J' \in \mathcal{J}_\tau(M, \omega_1)$ *and* $\{J_s\} \in \mathcal{J}_\tau(M, \{\omega_s\}; J, J')$, *the parametric moduli space* $\mathcal{M}_{g,m}^*(A; \{J_s\}; p_1, \ldots, p_m)$ *is empty unless*[5]

$$(n - 3)(2 - 2g) + 2c_1(A) \geqslant 2m(n - 1) - 1.$$

□

Remark 2.24. The reader should be cautioned that in Corollary 2.23 and similar results such as Corollary 2.25 below, the definition of the word "generic" *depends on the choice of the points* $p_1, \ldots, p_m$. In most situations, there is no single J that can make the spaces $\mathcal{M}_{g,m}^*(A; J; p_1, \ldots, p_m)$ simultaneously smooth for all

---

[5]A version of Remark 2.20 also applies to this result since the left hand side of the inequality is always even, and this is useful in some applications—but again, this numerical coincidence does not usually occur in more general settings (cf. Remark 8.32).

possible choices of $p_1, \ldots, p_m$. Such a J would need to belong to an *uncountable* intersection of comeager subsets, which may in general be empty.

One can derive various closely related corollaries that have nothing directly to do with marked points. For instance, notice that if the points $p_1, \ldots, p_m$ are all distinct, then the image of the natural map

$$\mathcal{M}_{g,m}(A; J; p_1, \ldots, p_m) \to \mathcal{M}_g(A; J)$$

defined by forgetting the marked points consists of every curve $u \in \mathcal{M}_g(A; J)$ that passes through all of the points $p_1, \ldots, p_m$. For generic J, restricting this map to the somewhere injective curves gives a smooth map $\mathcal{M}^*_{g,m}(A; J; p_1, \ldots, p_m) \to \mathcal{M}^*_g(A; J)$, where

$$\dim \mathcal{M}^*_{g,m}(A; J; p_1, \ldots, p_m) = \dim \mathcal{M}^*_g(A; J) - 2m(n-1).$$

If $n \geqslant 2$, this means the domain is a manifold of dimension strictly smaller than the target, so that by Sard's theorem, the image misses almost every point in $\mathcal{M}^*_g(A; J)$, implying:

Corollary 2.25. *Given a closed symplectic manifold (M, ω) of dimension $2n \geqslant 4$ and a nonempty finite set of pairwise distinct points $p_1, \ldots, p_m$, there exists a comeager subset $\mathcal{J}^{\mathrm{reg}} \subset \mathcal{J}_\tau(M, \omega)$ such that for any $J \in \mathcal{J}^{\mathrm{reg}}$, $g \geqslant 0$ and $A \in H_2(M)$, the set of curves in $\mathcal{M}^*_g(A; J)$ whose images do not contain $\{p_1, \ldots, p_m\}$ is open and dense. Moreover, every curve u satisfying $\mathrm{ind}(u) < 2m(n-1)$ belongs to this subset.* □

The Sard's theorem trick can be used to deduce a result about **triple self-intersections** that will be useful in Chap. 7, i.e. we would like to exclude curves $u : \Sigma \to M$ that have three distinct points $z_1, z_2, z_3 \in \Sigma$ satisfying

$$u(z_1) = u(z_2) = u(z_3).$$

Suppose we are given $m \geqslant 0$ and a collection of pairwise disjoint submanifolds $Z_1, \ldots, Z_m \subset M$, set

$$Z := Z_1 \times \ldots \times Z_m \subset M^m,$$

and consider first the scenario in which a triple self-intersection occurs at three points that are distinct from the marked points. We can deal with this by defining the submanifolds

$$\Delta = \{(p, p, p) \mid p \in M\} \subset M \times M \times M$$

and

$$Z_\Delta = Z \times \Delta \subset M^{m+3}.$$

The latter has codim $Z_\Delta = \text{codim } Z + 4n$, thus

$$\dim \mathcal{M}^*_{g,m+3}(A; J; Z_\Delta) = \text{vir-dim } \mathcal{M}_{g,m+3}(A; J) - \text{codim } Z - 4n$$
$$= \text{vir-dim } \mathcal{M}_{g,m}(A; J) + 6 - \text{codim } Z - 4n$$
$$= \dim \mathcal{M}^*_{g,m}(A; J; Z) + 6 - 4n.$$

Applying Sard's theorem to the natural map $\mathcal{M}^*_{g,m+3}(A; J; Z_\Delta) \to \mathcal{M}^*_{g,m}(A; J; Z)$ that forgets the last three marked points, we deduce that if $n \geqslant 2$ so that $6 - 4n < 0$, an open and dense subset of the curves in $\mathcal{M}^*_{g,m}(A; J; Z)$ will have no triple self-intersections distinct from the marked points, and this open and dense subset will include everything if vir-dim $\mathcal{M}_{g,m}(A; J; Z) < 2(2n - 3)$. To allow for triple self-intersections involving marked points, we observe that since the submanifolds $Z_1, \ldots, Z_m$ are assumed disjoint, we only need to worry about the case $u(z_1) = u(z_2) = u(z_3)$ where z_1 is a marked point while z_2 and z_3 are not. We thus consider for each $k = 1, \ldots, m$ the submanifold

$$\Delta_k = \{(p_1, \ldots, p_{m+2}) \in M^{m+2} \mid p_k = p_{m+1} = p_{m+2}\} \subset M^{m+2},$$

which has codimension $4n$ and intersects $Z \times M \times M$ transversely, giving another submanifold

$$Z_k := (Z \times M \times M) \cap \Delta_k \subset M^{m+2}$$

of codimension codim $Z + 4n$. Now

$$\dim \mathcal{M}^*_{g,m+2}(A; J; Z_k) = \text{vir-dim } \mathcal{M}_{g,m+2}(A; J) - \text{codim } Z - 4n$$
$$= \text{vir-dim } \mathcal{M}_{g,m}(A; J) + 4 - \text{codim } Z - 4n$$
$$= \dim \mathcal{M}^*_{g,m}(A; J; Z) + 4 - 4n,$$

so Sard's theorem gives the same conclusion as long as $4 - 4n < 0$, and in fact the self-intersections in question are avoided altogether under a slightly weaker condition than before, namely vir-dim $\mathcal{M}_{g,m}(A; J; Z) < 4(n - 1)$. Letting k vary over $1, \ldots, m$ and putting these two cases together, we conclude:

Corollary 2.26. *Suppose (M, ω) is a closed symplectic manifold of dimension $2n \geqslant 4$, $m \geqslant 0$ is an integer,*

$$Z_1, \ldots, Z_m \subset M$$

is a collection of pairwise disjoint submanifolds, and $Z = Z_1 \times \ldots \times Z_m \subset M^m$. Then for generic $J \in \mathcal{J}_\tau(M, \omega)$ and any $g \geqslant 0$ and $A \in H_2(M)$, the set of *curves in $\mathcal{M}^*_{g,m}(A; J; Z)$ that have no triple self-intersections is open and dense. Moreover:*

- *If* vir-dim $\mathcal{M}_{g,m}(A; J; Z) < 4n - 6$, *then no curve in* $\mathcal{M}^*_{g,m}(A; J; Z)$ *has any triple self-intersections.*
- *If* vir-dim $\mathcal{M}_{g,m}(A; J; Z) < 4n - 4$, *then no curve in* $\mathcal{M}^*_{g,m}(A; J; Z)$ *has any triple self-intersections at points that include a marked point.*

□

Exercise 2.27. Derive the analogous corollary for the diagonal $\Delta \subset M \times M$ and use it to deduce that if dim $M \geq 6$, the injective curves generically form an open and dense subset of $\mathcal{M}^*_g(A; J)$. Notice however that you cannot prove this for dim $M = 4$; in fact it is false, as there are strict topological controls in dimension four that prevent intersections from just disappearing, see Sect. 2.2.2.

By a slight abuse of conventions, we can also apply the above trick to J-holomorphic curves whose domains have two connected components, producing statements about intersections between two distinct curves, such as the following.

Theorem 2.28. *Under the same assumptions as in Corollary 2.26, fix integers $k, g, h \geq 0$ with $k \leq m$, homology classes $A, B \in H_2(M)$, and let*

$$Z' = Z_1 \times \ldots \times Z_k \subset M^k, \qquad Z'' = Z_{k+1} \times \ldots \times Z_m \subset M^{m-k}.$$

Then for generic $J \in \mathcal{J}_\tau(M, \omega)$, there exists an open and dense subset of

$$\{(u, v) \in \mathcal{M}^*_{g,k}(A; J; Z') \times \mathcal{M}^*_{h,m-k}(B; J; Z'') \mid u \neq v\}$$

consisting of pairs (u, v) such that intersection points between u and v never coincide with the self-intersection points of each. Moreover, all pairs (u, v) of distinct somewhere injective curves have this property if the sum of their constrained indices is less than $4n - 6$.

□

2.1.5 Constraints on Derivatives

One can also impose constraints on derivatives of holomorphic curves at marked points, and some results of this type will be needed in Chap. 7. The contents of this section will not be needed for the proofs of the main theorems stated in Chap. 1, so you may prefer to skip it on first reading. We will continue under the assumption that M is closed, though this assumption can also be weakened if perturbations of J are restricted to a precompact open subset (see Remark 2.17).

To understand the moduli spaces of interest in this section, we can (at least locally) enhance the evaluation map ev : $\mathcal{M}_{g,m}(A; J) \to M^m$ with information that keeps track of the derivatives of curves at each marked point up to some prescribed order. This information is best expressed in the language of *jets*. Following [Zeh15], we define the space $\mathrm{Jet}^\ell_J(M)$ of **holomorphic ℓ-jets** in (M, J) to consist of equivalence classes of J-holomorphic maps from a neighborhood of the origin in $\mathbb{C}$ into M, where two such maps are considered equivalent if their values and derivatives up to order ℓ are the same at the origin. It is straightforward to check that the latter condition does not depend on any choices of local coordinates in M. This is immediate in the case $\ell = 1$, for which there is an obvious bijection with the vector bundle

$$\mathrm{Jet}^1_J(M) = \left\{ (p, \Phi) \mid p \in M \text{ and } \Phi : (\mathbb{C}, i) \to (T_pM, J) \text{ is complex linear} \right\}, \tag{2.8}$$

where p and Φ represent the value and first derivative respectively of a local J-holomorphic curve. The fact that *every* such pair (p, Φ) corresponds to a holomorphic 1-jet follows from a somewhat nontrivial result on the local existence of J-holomorphic curves with prescribed derivatives at a point, see [Wenc, §2.12] or [Zeh15, §2]. One can show in the same manner that the space of holomorphic ℓ-jets near a given point is in bijective correspondence with the set of holomorphic Taylor polynomials of degree ℓ, thus $\mathrm{Jet}^\ell_J(M)$ is a smooth manifold with

$$\dim \mathrm{Jet}^\ell_J(M) = 2n(\ell + 1).$$

We can view $\mathrm{Jet}^\ell_J(M)$ as a submanifold of the space $\mathrm{Jet}^\ell(M)$ of *smooth* ℓ-jets of maps $\mathbb{C} \to M$; the latter is locally in bijective correspondence with the set of *all* (not necessarily holomorphic) Taylor polynomials, thus it has dimension $2n(1 + 2 + 3 + \ldots + (\ell + 1)) = n(\ell + 1)(\ell + 2)$. For our applications, we will only need the case $\ell = 1$.

Recall from Sect. 2.1.3 that a neighborhood of any element in $\mathcal{M}^*_{g,m}(A; J)$ represented by a curve $u_0 : (\Sigma, j_0) \to (M, J)$ with marked points $\zeta_1, \ldots, \zeta_m \in \Sigma$ can be identified with the zero-set of a smooth Fredholm section $\bar\partial_J : \mathcal{B} \to \mathcal{E}$. Here $\mathcal{E}$ is a Banach space bundle over a Banach manifold $\mathcal{B}$ consisting of pairs (j, u), where j is a complex structure on Σ close to j_0 and $u : \Sigma \to M$ is a map close to u_0 in some Sobolev regularity class. One can define $\mathcal{B}$ so as to assume that the complex structures j all match j_0 near the marked points, and then fix holomorphic coordinate charts identifying a neighborhood of each marked point with a neighborhood of the origin in $\mathbb{C}$. If the Sobolev completion is chosen so that u is always of class C^ℓ or better for some $\ell \geq 1$, then for any integers $0 \leq \ell_j \leq \ell$ with $j = 1, \ldots, m$, these choices determine a jet evaluation map

$$\mathrm{ev} : \bar\partial_J^{-1}(0) \to \mathrm{Jet}^{\ell_1}_J(M) \times \ldots \times \mathrm{Jet}^{\ell_m}_J(M),$$

whose ith component at $(j, u) \in \bar\partial_J^{-1}(0)$ for $i = 1, \ldots, m$ is the holomorphic ℓ_i-jet represented by u in the chosen holomorphic coordinates at ζ_i. One can now show that for any smooth submanifold $Z' \subset \mathrm{Jet}^{\ell_1}_J(M) \times \ldots \times \mathrm{Jet}^{\ell_m}_J(M)$, generic[6] choices of J make ev transverse to Z', see e.g. [Wene, Appendix A]. This discussion depends on several choices and is thus difficult to express in a global way for the whole moduli space $\mathcal{M}^*_{g,m}(A; J)$, but certain geometrically meaningful submanifolds Z' can be defined in ways that are insensitive to these choices and thus give rise to global results. Rather than attempting to state a general theorem along these lines, let us single out two special cases that will be useful.

First, use the identification (2.8) to define

$$Z_{\text{crit}} = \left\{ (p, 0) \in \text{Jet}^1_J(M) \mid p \in M \right\} \subset \text{Jet}^1_J(M).$$

The resulting constrained moduli space

$$\mathcal{M}_{g,\text{crit}}(A; J) \subset \mathcal{M}_{g,1}(A; J)$$

is the space of equivalence classes of curves $u : (\Sigma, j) \to (M, J)$ with a marked point $\zeta \in \Sigma$ such that $du(\zeta) = 0$. We have

$$\text{codim } Z_{\text{crit}} = \dim \text{Jet}^1_J(M) - \dim Z_{\text{crit}} = 4n - 2n = 2n,$$

hence

$$\text{vir-dim} \, \mathcal{M}_{g,\text{crit}}(A; J) = \text{vir-dim} \, \mathcal{M}_{g,1}(A; J) - 2n = \text{vir-dim} \, \mathcal{M}_g(A; J) - 2(n-1).$$

More generally, one can combine the non-immersed point constraint with the pointwise constraints defined by a submanifold $Z \subset M^m$ as in the previous subsection. Here we have two cases to distinguish, depending whether the non-immersed point occurs at one of the m marked points constrained by Z or not. For the latter case, we define the submanifold

$$Z \times Z_{\text{crit}} \subset M^m \times \text{Jet}^1_J(M)$$

with codimension codim $Z + 2n$, giving rise to a moduli space

$$\mathcal{M}_{g,m,\text{crit}}(A; J; Z) \subset \mathcal{M}_{g,m+1}(A; J; Z \times M)$$

of curves $u : \Sigma \to M$ that have marked points $\zeta_1, \ldots, \zeta_m, \zeta' \in \Sigma$, with the first m marked points satisfying the constraint (2.6) and the last one satisfying $du(\zeta') = 0$. This space has virtual dimension

$$\text{vir-dim} \, \mathcal{M}_{g,m,\text{crit}}(A; J; Z) = \text{vir-dim} \, \mathcal{M}_{g,m+1}(A; J) - \text{codim } Z - 2n$$

$$= \text{vir-dim} \, \mathcal{M}_{g,m}(A; J; Z) - 2(n-1).$$

---

[6] As with Remark 2.24, it is important to understand that in this statement the definition of the word "generic" depends on the submanifold Z'.

Alternatively, one can stick with only m marked points subject to the constraint (2.6) but assume additionally that the kth marked point is non-immersed. For $k = 1$, this means considering the transverse intersection of $Z_{\mathrm{crit}} \times M^{m-1} \subset \mathrm{Jet}^1_J(M) \times M^{m-1}$ with

$$(\pi \times \mathrm{Id})^{-1}(Z) \subset \mathrm{Jet}^1_J(M) \times M^{m-1},$$

where $\pi : \mathrm{Jet}^1_J(M) \to M$ denotes the bundle projection. Performing the analogous construction for any $k = 1, \ldots, m$ gives a submanifold of codimension codim $Z +$ $2n$ in $M^{k-1} \times \mathrm{Jet}^1_J(M) \times M^{m-k}$, so the corresponding moduli space

$$\mathcal{M}_{g,m,\mathrm{crit}_k}(A; J; Z) \subset \mathcal{M}_{g,m}(A; J; Z)$$

satisfies

$$\mathrm{vir\text{-}dim}\,\mathcal{M}_{g,m,\mathrm{crit}_k}(A; J; Z) = \mathrm{vir\text{-}dim}\,\mathcal{M}_{g,m}(A; J) - \mathrm{codim}\,Z - 2n$$

$$= \mathrm{vir\text{-}dim}\,\mathcal{M}_{g,m}(A; J; Z) - 2n.$$

As usual, we shall denote the sets of somewhere injective curves in these moduli spaces by

$$\mathcal{M}^*_{g,m,\mathrm{crit}}(A; J; Z) \subset \mathcal{M}_{g,m,\mathrm{crit}}(A; J; Z), \qquad \mathcal{M}^*_{g,m,\mathrm{crit}_k}(A; J; Z) \subset \mathcal{M}_{g,m,\mathrm{crit}_k}(A; J; Z).$$

Theorem 2.29. *Given* $Z \subset M^m$, *for generic* $J \in \mathcal{J}_\tau(M, \omega)$, *the spaces* $\mathcal{M}^*_{g,m,\mathrm{crit}}(A; J; Z)$ *and* $\mathcal{M}^*_{g,m,\mathrm{crit}_k}(A; J; Z)$ *for* $k = 1, \ldots, m$ *are all smooth manifolds with dimension matching their virtual dimensions.* $\square$

Applying Sard's theorem to the natural maps $\mathcal{M}^*_{g,m,\mathrm{crit}}(A; J; Z) \to$ $\mathcal{M}^*_{g,m}(A; J; Z)$ and $\mathcal{M}^*_{g,m,\mathrm{crit}_k}(A; J; Z) \to \mathcal{M}^*_{g,m}(A; J; Z)$, where the former forgets the extra marked point, we deduce:

Corollary 2.30. *Suppose* (M, ω) *is a closed symplectic manifold of dimension* $2n \geq 4$, $m \geq 0$ *is an integer and* $Z \subset M^m$ *is a submanifold. Then for generic* $J \in \mathcal{J}_\tau(M, \omega)$ *and any* $g \geq 0$ *and* $A \in H_2(M)$, *the set of curves in* $\mathcal{M}^*_{g,m}(A; J; Z)$ *that are immersed is open and dense. Moreover:*

- *If* $\mathrm{vir\text{-}dim}\,\mathcal{M}_{g,m}(A; J; Z) < 2n - 2$, *then all curves in* $\mathcal{M}^*_{g,m}(A; J; Z)$ *are immersed.*
- *If* $\mathrm{vir\text{-}dim}\,\mathcal{M}_{g,m}(A; J; Z) < 2n$, *then no curve in* $\mathcal{M}^*_{g,m}(A; J; Z)$ *has a non-immersed point at any of its marked points.*

$\square$

Next, we consider the problem of ensuring that self-intersections of a simple curve in dimension four are transverse. Define the submanifold

$$Z_{\mathrm{tan}} = \left\{ ((p, \Phi), (p, \Psi)) \mid p \in M \text{ and } \Psi = c\Phi \neq 0 \text{ for some } c \in \mathbb{C} \right\}$$

$$\subset \mathrm{Jet}^1_J(M) \times \mathrm{Jet}^1_J(M),$$

which has codimension

$$\text{codim } Z_{\text{tan}} = 4n - 2. \tag{2.9}$$

The resulting constrained moduli space

$$\mathcal{M}_{g,\text{tan}}(A; J) \subset \mathcal{M}_{g,2}(A; J)$$

consists of equivalence classes of curves $u : (\Sigma, j) \to (M, J)$ with two marked points $\zeta_1, \zeta_2 \in \Sigma$, both of them immersed points for u, such that $u(\zeta_1) = u(\zeta_2)$ and $\text{im } du(\zeta_1) = \text{im } du(\zeta_2)$. We say in this case that u has a **tangential** self-intersection. In light of (2.9), we set

$$\text{vir-dim } \mathcal{M}_{g,\text{tan}}(A; J) = \text{vir-dim } \mathcal{M}_{g,2}(A; J) - (4n - 2)$$
$$= \text{vir-dim } \mathcal{M}_g(A; J) - (4n - 6).$$

One can just as easily combine this with the pointwise constraints defined via a submanifold $Z \subset M^m$ as in the previous subsection, though some extra assumption is necessary in order to ensure the necessary transversality. One reasonable option is to restrict our attention (as in Corollary 2.26) to submanifolds of the form

$$Z = Z_1 \times \ldots \times Z_m \subset M^m$$

for a collection of pairwise disjoint submanifolds $Z_1, \ldots, Z_m \subset M$. We can then define a moduli space

$$\mathcal{M}_{g,m,\text{tan}}(A; J; Z) \subset \mathcal{M}_{g,m+2}(A; J; Z \times M \times M)$$

with

$$\text{vir-dim } \mathcal{M}_{g,m,\text{tan}}(A; J; Z) = \text{vir-dim } \mathcal{M}_{g,m}(A; J; Z) - (4n - 6)$$

consisting of curves $u : \Sigma \to M$ with marked points $\zeta_1, \ldots, \zeta_m, \zeta', \zeta'' \in \Sigma$ satisfying (2.6) such that $u(\zeta') = u(\zeta'')$ is a tangential intersection. For each $k = 1, \ldots, m$, there is also a moduli space

$$\mathcal{M}_{g,m,\text{tan}_k}(A; J; Z) \subset \mathcal{M}_{g,m+1}(A; J; Z \times M)$$

with

$$\text{vir-dim } \mathcal{M}_{g,m,\text{tan}_k}(A; J; Z) = \text{vir-dim } \mathcal{M}_{g,m}(A; J; Z) - (4n - 4),$$

in which the tangential intersection involves the kth marked point and the extra marked point. There is no need to consider intersections involving two of the m

marked points since the constraint submanifolds $Z_1, \ldots, Z_m$ are all disjoint. We shall again use superscript asterisks to indicate the sets of somewhere injective curves.

Theorem 2.31. *Fix $m \geqslant 0$ and pairwise disjoint submanifolds $Z_1, \ldots, Z_m \subset M$, defining $Z = Z_1 \times \ldots \times Z_m \subset M^m$. Then for generic $J \in \mathcal{J}_\tau(M, \omega)$, the spaces $\mathcal{M}^*_{g,m,\tan}(A; J; Z)$ and $\mathcal{M}^*_{g,m,\tan_k}(A; J; Z)$ for each $k = 1, \ldots, m$ are all smooth manifolds with dimension matching their virtual dimensions.* $\square$

As with Theorem 2.28, one can also apply this technique to curves with domains having two connected components, producing results about pairs of J-holomorphic curves that intersect each other tangentially. Putting all this together and applying the usual Sard's theorem trick, we have:

Corollary 2.32. *Suppose (M, ω) is a closed symplectic manifold of dimension $2n \geqslant 4$, $g, h, m, k \geqslant 0$ are integers with $k \leqslant m$, $A, B \in H_2(M)$, and $Z_1, \ldots, Z_m \subset M$ are pairwise disjoint submanifolds defining $Z = Z_1 \times \ldots \times Z_m \subset M^m$ and*

$$Z' = Z_1 \times \ldots \times Z_k \subset M^k, \qquad Z'' = Z_{k+1} \times \ldots \times Z_m \subset M^{m-k}.$$

Then for generic $J \in \mathcal{J}_\tau(M, \omega)$:

- *The set of curves in $\mathcal{M}^*_{g,m}(A; J; Z)$ that have no tangential self-intersections is open and dense.*
- *If $\operatorname{vir-dim} \mathcal{M}_{g,m}(A; J; Z) < 4n - 6$, then no curve in $\mathcal{M}^*_{g,m}(A; J; Z)$ has tangential self-intersections.*
- *If $\operatorname{vir-dim} \mathcal{M}_{g,m}(A; J; Z) < 4n - 4$, then no curve in $\mathcal{M}^*_{g,m}(A; J; Z)$ has tangential self-intersections at points that include a marked point.*

Moreover, the set of pairs $(u, v) \in \mathcal{M}^*_{g,k}(A; J; Z') \times \mathcal{M}^*_{h,m-k}(B; J; Z'')$ for which u and v have no tangential intersections with each other is open and dense in the subset $\{(u, v) \mid u \neq v\}$, and all such pairs have this property if the sum of their constrained indices is less than $4n - 6$. $\square$

Remark 2.33. Corollary 2.32 is not very interesting when $\dim M > 4$, since one can use the simpler methods of Sect. 2.1.4 in that case to show that generic simple J-holomorphic curves are injective (cf. Exercise 2.27) and generic pairs of distinct simple curves are disjoint. In dimension four this is generally false, and the homological intersection number $A \cdot B \in \mathbb{Z}$ guarantees intersections whenever it is nonzero, but the corollary allows us to conclude that generically such intersections will be transverse.

2.1.6 Gromov Compactness and Singularities

The space $\mathcal{M}_{g,m}(A; J)$ is generally not compact: sequences of J-holomorphic curves can degenerate into non-smooth objects, but it is another deep consequence of elliptic regularity theory that one can characterize precisely what kinds of degenerations are possible. The result is that $\mathcal{M}_{g,m}(A; J)$ has a natural compactification, called the **Gromov compactification** $\overline{\mathcal{M}}_{g,m}(A; J)$, which contains $\mathcal{M}_{g,m}(A; J)$ as an open subset.

We define the moduli space of unparametrized **nodal J-holomorphic curves** with **arithmetic genus** g and m marked points representing the homology class $A \in H_2(M)$ as a space of equivalence classes

$$\overline{\mathcal{M}}_{g,m}(A; J) = \{(S, j, u, (\zeta_1, \ldots, \zeta_m), \Delta)\} \Big/ \sim,$$

where the various symbols have the following meaning. We assume (S, j) is a closed (but not necessarily connected) Riemann surface, and $u : (S, j) \to (M, J)$ is a J-holomorphic curve such that if $S_1, \ldots, S_p$ denote the connected components of S,

$$[u] := \sum_{i=1}^{p} u_*[S_i] = A \in H_2(M).$$

The marked points $(\zeta_1, \ldots, \zeta_m)$ are again an ordered set of distinct points in S. The **nodes** are encoded by Δ, which is a finite unordered set of unordered pairs

$$\Delta = \{\{\hat{z}_1, \check{z}_1\}, \ldots, \{\hat{z}_r, \check{z}_r\}\}$$

of points in S, such that all the points $\hat{z}_1, \check{z}_1, \ldots, \hat{z}_r, \check{z}_r, \zeta_1, \ldots, \zeta_m$ are distinct and

$$u(\hat{z}_i) = u(\check{z}_i) \text{ for } i = 1, \ldots, r. \tag{2.10}$$

We shall usually refer to the individual points $\hat{z}_i, \check{z}_i$ as **nodal points**, while referring to each pair $\{\hat{z}_i, \check{z}_i\} \in \Delta$ as a **node**. The arithmetic genus condition means the following: first, one can define a compact surface $\overline{S}$ with boundary by replacing each of the nodal points $\hat{z}_i, \check{z}_i$ in Δ with circles $\hat{C}_i$ and $\check{C}_i$. Next, for each $i = 1, \ldots, r$, glue together the circles $\hat{C}_i$ and $\check{C}_i$ in order to form a closed oriented surface $\hat{S}$; see Fig. 2.2. We require $\hat{S}$ to be a *connected* surface with *genus* g. Observe that in light of (2.10), $u : S \to M$ naturally determines a continuous map $\hat{u} : \hat{S} \to M$.

Finally, we define $(S, j, u, (\zeta_1, \ldots, \zeta_m), \Delta) \sim (S', j', u', (\zeta'_1, \ldots, \zeta'_m), \Delta')$ if the double points Δ' can be written as

$$\Delta' = \{\{\hat{z}'_1, \check{z}'_1\}, \ldots, \{\hat{z}'_r, \check{z}'_r\}\}$$

so that there exists a biholomorphic map $\varphi : (S, j) \to (S', j')$ with $u = u' \circ \varphi$, $\varphi(\zeta_i) = \zeta'_i$ for $i = 1, \ldots, m$ and $\varphi(\hat{z}_i) = \hat{z}'_i$, $\varphi(\check{z}_i) = \check{z}'_i$ for $i = 1, \ldots, r$. Taking the union for all homology classes, we denote

$$\overline{\mathcal{M}}_{g,m}(J) = \bigcup_{A \in H_2(M)} \overline{\mathcal{M}}_{g,m}(A; J),$$

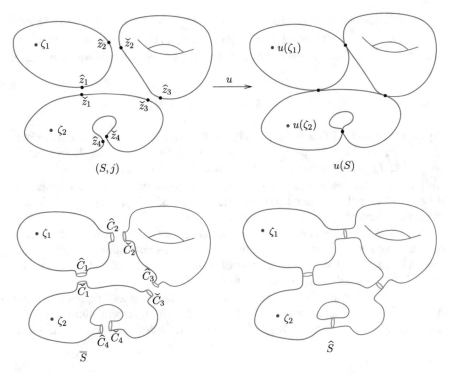

Fig. 2.2 Four ways of viewing a nodal holomorphic curve with arithmetic genus 3 and two marked points. At the upper left, we see the disconnected Riemann surface (S, j) with marked points (ζ_1, ζ_2) and nodal pairs $\{\hat{z}_i, \check{z}_i\}$ for $i = 1, 2, 3, 4$. To the right of this is a possible picture of the image of the nodal curve, with nodal pairs always mapped to identical points. The bottom right shows the surface $\overline{S}$ with boundary, obtained from S by replacing the points $\hat{z}_i, \check{z}_i$ with circles $\hat{C}_i, \check{C}_i$. Gluing these pairs of circles together gives the closed connected surface $\hat{S}$ at the bottom right, whose genus is by definition the arithmetic genus of the nodal curve

and the case with no marked points will sometimes be abbreviated

$$\overline{\mathcal{M}}_g(A; J) := \overline{\mathcal{M}}_{g,0}(A; J), \qquad \overline{\mathcal{M}}_g(J) := \overline{\mathcal{M}}_{g,0}(J).$$

We shall sketch below a notion of convergence for a sequence of smooth holomorphic curves approaching a nodal curve. There's a slight subtlety if one wants to use this notion to define a Hausdorff topology on $\overline{\mathcal{M}}_{g,m}(A; J)$, as the most obvious definition produces limits that are not necessarily unique. This problem is fixed by the following additional definition, due to Kontsevich.

Definition 2.34. A nodal curve $(S, j, u, (\zeta_1, \ldots, \zeta_m), \Delta)$ is said to be **stable** if, after removing all the marked points $\zeta_1, \ldots, \zeta_m$ and nodal points Δ from S to produce a punctured surface $\dot{S}$, every connected component of $\dot{S}$ on which u is constant has negative Euler characteristic.

With this notion in place, the definition of $\overline{\mathcal{M}}_{g,m}(A; J)$ is supplemented by the requirement that all elements $[(S, j, u, (\zeta_1, \ldots, \zeta_m), \Delta)] \in \overline{\mathcal{M}}_{g,m}(A; J)$ should be stable, and $\overline{\mathcal{M}}_{g,m}(A; J)$ then turns out to be a metrizable space. It admits a natural inclusion

$$\mathcal{M}_{g,m}(A; J) \subset \overline{\mathcal{M}}_{g,m}(A; J)$$

by regarding each $[(\Sigma, j, u, (\zeta_1, \ldots, \zeta_m))] \in \mathcal{M}_{g,m}(A; J)$ as a nodal curve with $\Delta = \emptyset$; we will call these the **smooth** (or sometimes *non-nodal*) curves in $\overline{\mathcal{M}}_{g,m}(A; J)$. More generally, one can take any $[(S, j, u, (\zeta_1, \ldots, \zeta_m), \Delta)] \in \overline{\mathcal{M}}_{g,m}(A; J)$ and restrict u, j and the marked point set to each connected component of S, producing the so-called **smooth components** of $[(S, j, u, (\zeta_1, \ldots, \zeta_m), \Delta)]$, which are elements of spaces $\mathcal{M}_{h,k}(B; J)$.

Spherical components of S on which u is constant play somewhat of a special role in Gromov compactness: they are referred to as **ghost bubbles**. The stability condition requires that the total number of nodal points and marked points on each ghost bubble should always be at least three.

We can now state the version of Gromov's compactness theorem that we will need. For any symplectic manifold (M, ω) with an ω-tame almost complex structure J, we define the **energy** of a closed J-holomorphic curve $u : \Sigma \to M$ by

$$E_\omega(u) = \int_\Sigma u^* \omega.$$

This is nonnegative in general, and it vanishes if and only if u is constant (cf. Proposition 2.8). Moreover, it depends only on $[u] \in H_2(M)$ and $[\omega] \in H_{dR}^2(M)$, so in particular, the energy of all curves in $\mathcal{M}_{g,m}(A; J)$ for a fixed $A \in H_2(M)$ is uniformly bounded.

Theorem 2.35. *Suppose M is a closed manifold with a sequence of symplectic structures ω_k converging in C^∞ to a symplectic structure ω, $J_k \in \mathcal{J}_\tau(\omega_k)$ is a sequence of tame almost complex structures converging in C^∞ to $J \in \mathcal{J}_\tau(\omega)$, and $u_k \in \mathcal{M}_{g,m}(J_k)$ is a sequence of nonconstant holomorphic curves satisfying a uniform energy bound*

$$E_{\omega_k}(u_k) \leqslant C$$

for some constant $C > 0$. Then u_k has a subsequence that converges to a stable nodal curve in $\overline{\mathcal{M}}_{g,m}(J)$. □

The precise definition of convergence to a nodal curve is somewhat complicated to state, but for our purposes it suffices to have the following description. Recall from the above discussion that any given nodal curve $[(S, j, u, (\zeta_1, \ldots, \zeta_m), \Delta)] \in \overline{\mathcal{M}}_{g,m}(J)$ determines a closed oriented surface $\hat{S}$ which is connected and has genus g, and $u : S \to M$ determines a continuous map

$$\hat{u} : \hat{S} \to M$$

which is constant on each of the special circles where $\overline{S}$ is glued to form $\hat{S}$; we shall denote the union of all these circles by

$$C \subset \hat{S}.$$

Since the marked points $\zeta_1, \ldots, \zeta_m$ are disjoint from Δ, they can also be said to lie on $\hat{S} \backslash C$ in a natural way. Now if $[(\Sigma_k, j_k, u_k, (\zeta_1^k, \ldots, \zeta_m^k))] \in \mathcal{M}_{g,m}(J_k)$ is a sequence coverging to

$$[(S, j, u, (\zeta_1, \ldots, \zeta_m), \Delta)] \in \overline{\mathcal{M}}_{g,m}(J),$$

the crucial fact is that there exists a sequence of diffeomorphisms

$$\varphi_k : \hat{S} \to \Sigma_k$$

such that $\varphi_k(\zeta_i) = \zeta_i^k$ for $i = 1, \ldots, m$, while $u_k \circ \varphi_k$ converges to $\hat{u}$ in both $C^0(\hat{S}, M)$ and in $C^\infty_{\text{loc}}(\hat{S} \backslash C, M)$.

Observe that by this notion of convergence, the homology class $[u] \in H_2(M)$ of the limit $u \in \overline{\mathcal{M}}_{g,m}(J)$ must match $[u_k]$ for all sufficiently large k, and the evaluation map (2.3) admits a natural continuous extension

$$\text{ev} : \overline{\mathcal{M}}_{g,m}(J) \to M^m.$$

A related result which plays a major role in proving Gromov's compactness theorem is the theorem on *removal of singularities*. We state it here in a local form that will also be applicable to punctured curves in Chap. 8.

Theorem 2.36. *Suppose (M, ω) is a symplectic manifold with $J \in \mathcal{J}_\tau(M, \omega)$, and $u : \mathbb{D}^2 \backslash \{0\} \to M$ is a punctured J-holomorphic disk that satisfies $\int_{\mathbb{D}^2 \backslash \{0\}} u^* \omega < \infty$ and has image contained in a compact subset. Then u extends smoothly to a J-holomorphic disk $u : \mathbb{D}^2 \to M$.* □

2.1.7 Gluing

It is sometimes also possible to describe the local topological structure of the compactification $\overline{\mathcal{M}}_{g,m}(A; J)$, analogously to our local description of $\mathcal{M}_{g,m}(A; J)$ via the implicit function theorem in Sect. 2.1.3. This subject is known as "gluing," because describing the smooth curves in the neighborhood of a nodal curve requires some procedure for reversing the degeneration in Gromov's compactness theorem, i.e. for gluing the smooth components of the nodal curve back together. We now sketch the simplest case of this, since it will be needed in Chap. 7.

Assume $[(S, j_0, u_0, (\zeta_1, \ldots, \zeta_m), \Delta)] \in \overline{\mathcal{M}}_{g,m}(A; J)$ is a nodal curve that has exactly one node and two smooth components u_0^+ and u_0^-, which live in spaces $\mathcal{M}_{g^+,m^+}(A^+; J)$ and $\mathcal{M}_{g^-,m^-}(A^-; J)$ respectively with $g^+ + g^- = g$, $m^+ + m^- = m$ and $A^+ + A^- = A$. Denote the domains of $u^\pm$ (i.e. the connected components of S) by $\Sigma^\pm$, so by assumption, Δ consists of a single pair $\{z^+, z^-\}$ with $z^\pm \in \Sigma^\pm$, and z^+ and z^- are both distinct from any of the marked points $\zeta_1, \ldots, \zeta_m$. Any nodal curve $u \in \overline{\mathcal{M}}_{g,m}(A; J)$ in a neighborhood of u_0 can be written as a nearby map on the same domain S with a nearby complex structure and the same marked points and nodal points, and if we view its smooth components $u^\pm$ as elements of $\mathcal{M}_{g^\pm,m^\pm+1}(A^\pm; J)$ with $z^\pm$ as the extra marked point, it satisfies the incidence relation

$$\mathrm{ev}_{m^++1}(u^+) = \mathrm{ev}_{m^-+1}(u^-). \tag{2.11}$$

Definition 2.37. Given a pair $[(\Sigma^\pm, j^\pm, u^\pm, (\zeta_1^\pm, \ldots, \zeta_{m^\pm}^\pm, z^\pm))] \in \mathcal{M}_{g^\pm,m^\pm+1}$ $(A^\pm; J)$ satisfying the incidence relation (2.11), let

$$u^+ \# u^- \in \overline{\mathcal{M}}_{g,m}(A; J)$$

denote the nodal curve consisting of the disjoint union of the maps $u^+ :$ $(\Sigma^+, j^+) \rightarrow (M, J)$ and $u^- : (\Sigma^-, j^-) \rightarrow (M, J)$, with marked points $\zeta_1^+, \ldots, \zeta_{m^+}^+, \zeta_1^-, \ldots, \zeta_{m^-}^-$ and node $\Delta = \{\{z^+, z^-\}\}$.

Since a neighborhood of u_0 in $\overline{\mathcal{M}}_{g,m}(A; J)$ may contain both smooth curves and nodal curves, an obvious prerequisite for having a nice description of this neighborhood is that the set of nodal curves in it should on its own have a nice structure. To this end, we impose the following conditions on u_0:

(1) u_0^+ and u_0^- are both somewhere injective and are not identical curves (up to parametrization);
(2) u_0^+ and u_0^- are both Fredholm regular;
(3) The intersection at (u_0^+, u_0^-) between $\mathrm{ev}_{m^++1} : \mathcal{M}_{g^+,m^+-1}(A^+; J) \rightarrow M$ and $\mathrm{ev}_{m^-+1} : \mathcal{M}_{g^-,m^-+1}(A^-; J) \rightarrow M$ is transverse.

Note that the third condition is equivalent to requiring the map

$$(\mathrm{ev}_{m^++1}, \mathrm{ev}_{m^-+1}) : \mathcal{M}_{g^+,m^++1}(A^+; J) \times \mathcal{M}_{g^-,m^-+1}(A^-; J) \rightarrow M \times M$$

to be transverse to the diagonal. If the $u_0^\pm$ are distinct curves and both are simple, then the results stated in Sects. 2.1.3 and 2.1.4 imply that the required transversality conditions can all be assumed for generic $J \in \mathcal{J}_\tau(M, \omega)$, and the set

$$\{(u^+, u^-) \in \mathcal{M}_{g^+,m^++1}(A^+; J) \times \mathcal{M}_{g^-,m^-+1}(A^-; J) \mid \mathrm{ev}_{m^++1}(u^+)$$
$$= \mathrm{ev}_{m^-+1}(u^-)\} \tag{2.12}$$

is a smooth manifold near (u_0^+, u_0^-) with dimension

$$\text{vir-dim}\,\mathcal{M}_{g^+,m^+}(A^+; J) + \text{vir-dim}\,\mathcal{M}_{g^-,m^-}(A^-; J) + 4 - 2n$$

$$= \text{vir-dim}\,\mathcal{M}_{g,m}(A; J) - 2. \tag{2.13}$$

Theorem 2.38. *Under the transversality conditions listed above, there exists a neighborhood $\mathcal{U}$ of (u_0^+, u_0^-) in the space (2.12) and a smooth embedding*

$$\Psi : \mathcal{U} \times [0, \infty) \times S^1 \hookrightarrow \mathcal{M}_{g,m}(A; J)$$

whose image contains every smooth curve in some neighborhood of the nodal curve $u_0 \in \overline{\mathcal{M}}_{g,m}(A; J)$. Moreover, Ψ admits a continuous extension

$$\Psi : \mathcal{U} \times [0, \infty] \times S^1 \to \overline{\mathcal{M}}_{g,m}(A; J)$$

such that for each $\theta \in S^1$ and $(u^+, u^-) \in \mathcal{U}$,

$$\Psi((u^+, u^-), \infty, \theta) = u^+ \# u^-,$$

and the maps

$$\text{ev} \circ \Psi(\cdot, R, \cdot) : \mathcal{U} \times S^1 \to M^m$$

are C^1-convergent to $\text{ev} \circ \Psi(\cdot, \infty, \cdot)$ as $R \to \infty$. □

We call the map Ψ in this theorem a **gluing map**. Notice that by (2.13), the domain and target space of Ψ have the same dimension. The two extra parameters $(R, \theta) \in [0, \infty) \times S^1$ in the domain can be understood as follows. The first step in defining Ψ is to define a so-called **pre-gluing map** whose image is not in $\mathcal{M}_{g,m}(A; J)$ but is close to it; in other words, we associate to each nodal curve with smooth components $u^+ : (\Sigma^+, j^+) \to (M, J)$ and $u^- : (\Sigma^-, j^-) \to (M, J)$ various *approximately* J-holomorphic curves that are close to degenerating to $u^+ \# u^-$. One way to do this starts with fixing holomorphic cylindrical coordinates on $\Sigma^\pm$ near each of the nodal points $z^\pm$ so that punctured neighborhoods of z^+ and z^- are identified with $[0, \infty) \times S^1$ and $(-\infty, 0] \times S^1$ respectively; here S^1 is defined as $\mathbb{R}/\mathbb{Z}$ and the standard complex structure on $\mathbb{R} \times S^1$ is defined such that the diffeomorphism

$$\mathbb{R} \times S^1 \to \mathbb{C}\backslash\{0\} : (s, t) \mapsto e^{2\pi(s+it)}$$

is biholomorphic. Writing $(s, t) \in \mathbb{R} \times S^1$ for the cylindrical coordinates, the maps $u^+(s, t)$ and $u^-(s, t)$ are approximately equal to the same constant $p := u^+(z^+) = u^-(z^-)$ when $|s|$ is large, so for any sufficiently large constant $R > 0$, one obtains approximately J-holomorphic maps $\hat{u}^\pm$ by multiplying $u^\pm$ with a cutoff function in coordinates near p so that $\widehat{u}^{\pm}(s,t) = u(s,t)$ for $|s| \leqslant R - 1$ but $\widehat{u}^{\pm}(s,t) = p$ for $|s| \geqslant R$. After this modification, one can glue (Σ^+, j^+) and (Σ^-, j^-) together to form a Riemann surface (Σ_R, j_R) of genus g by truncating the ends of Σ^+ and Σ^- to $[0, 2R] \times S^1$ and $[-2R, 0] \times S^1$ respectively, then identifying them via the shift map $(s,t) \mapsto (s - 2R, t)$ so that $\widehat{u}^+$ and $\widehat{u}^-$ glue together to form a smooth and approximately J-holomorphic map

$$u^+ \#^R u^- : \Sigma_R \to M.$$

This construction is designed so that $u^+ \#^R u^-$ degenerates to the nodal curve $u^+ \# u^-$ as $R \to \infty$. As described above, however, the construction depends somewhat arbitrarily on the choice of holomorphic coordinate charts near z^+ and z^-, and in particular, one obtains an equally valid construction by rotating one of these charts. This freedom introduces an extra parameter $\theta \in S^1$, which can be incorporated into the above picture by generalizing our identification of $[0, 2R] \times S^1$ with $[-2R, 0] \times S^1$ to allow rotated shift maps $(s,t) \mapsto (s - 2R, t + \theta)$, resulting in a more general family of glued Riemann surfaces $(\Sigma_{(R,\theta)}, j_{(R,\theta)})$ and approximately J-holomorphic maps

$$u^+ \#^{(R,\theta)} u^- : \Sigma_{(R,\theta)} \to M.$$

This description accounts for both of the extra parameters in the domain of the gluing map Ψ. The proof of Theorem 2.38 then proceeds via a delicate application of the implicit function theorem (or equivalently the contraction mapping principle) to show that when $R > 0$ is large enough, one can perturb each $u^+ \#^{(R,\theta)} u^-$ to a unique element of $\mathcal{M}_{g,m}(A; J)$ by moving a small distance in the space of maps $\Sigma_{(R,\theta)} \to M$ in a direction orthogonal to $\mathcal{M}_{g,m}(A; J)$. The analytical details are carried out (in a slightly different setting) in [MS12, Chap. 10].

The specific case of Theorem 2.38 that will be needed later is the following.

Corollary 2.39. *Suppose (M, ω) is a symplectic 4-manifold, $J \in \mathcal{J}_\tau(M, \omega)$, $m \geqslant 0$ is an integer, $\{p_1, \ldots, p_m\} \subset M$ is a finite set partitioned into two subsets $\{p_1^{\pm}, \ldots, p_{m\pm}^{\pm}\}$, and $u \in \overline{\mathcal{M}}_{g,m}(A; J; p_1, \ldots, p_m)$ is a nodal J-holomorphic curve having exactly one node, which connects two inequivalent smooth components $u^{\pm} \in \mathcal{M}_{g\pm,m\pm}(A^{\pm}; J; p_1^{\pm}, \ldots, p_{m\pm}^{\pm})$, both of them simple and Fredholm regular for the constrained problem, with constrained index 0, and intersecting each other transversely at the node. Then every other curve in some neighborhood of u in $\overline{\mathcal{M}}_{g,m}(A; J; p_1, \ldots, p_m)$ is smooth, and this neighborhood has the structure of a 2-dimensional topological manifold.*

Proof. The constrained regularity and index assumptions mean that both of the evaluation maps $\mathrm{ev} : \mathcal{M}_{g\pm,m\pm}(A^{\pm}; J) \to M^{m\pm}$ are diffeomorphisms between neighborhoods of $u^{\pm}$ and $(p_1^{\pm}, \ldots, p_{m\pm}^{\pm})$. Since $\dim M = 4$, having u^+ and u^- intersect transversely at the node means moreover that the transverse evaluation map condition needed for Theorem 2.38 is satisfied, and since isolated transverse intersections cannot be perturbed away, a neighborhood $\mathcal{U}$ of (u^+, u^-) in the space (2.12) is in this case simply a neighborhood in $\mathcal{M}_{g^+, m^+}(A^+; J) \times \mathcal{M}_{g^-, m^-}(A^-; J)$. Choose the neighborhood $\mathcal{U}$ so that it contains no other pair of curves satisfying the same marked point constraints. Then if $\Psi : \mathcal{U} \times [0, \infty) \times S^1 \to \mathcal{M}_{g,m}(A; J)$ is the gluing map, the C^1-convergence of the evaluation map as $R \to \infty$ implies that for any (R, θ) with R sufficiently large, there will also be exactly one element $\psi(R, \theta) \in \mathcal{U}$ such that

$$\Psi(\psi(R, \theta), R, \theta) \in \text{ev}^{-1}(p_1, \ldots, p_m),$$

and the map $\mathcal{U} \to M^m : (v^+, v^-) \mapsto \text{ev}(\Psi((v^+, v^-), R, \theta))$ is also a local diffeomorphism near $\psi(R, \theta)$. Thus for $R_0 > 0$ sufficiently large, the embedding

$$[R_0, \infty) \times S^1 \hookrightarrow \mathcal{M}_{g,m}(A; J; p_1, \ldots, p_m) : (R, \theta) \mapsto \Psi(\psi(R, \theta), R, \theta)$$

parametrizes the set of all smooth curves in $\overline{\mathcal{M}}_{g,m}(A; J; p_1, \ldots, p_m)$ near u. Identifying $[R_0, \infty) \times S^1$ with a punctured disk via $(R, \theta) \mapsto e^{-2\pi(R+i\theta)}$, we obtain a homeomorphism between the 2-dimensional disk and a neighborhood of u in $\overline{\mathcal{M}}_{g,m}(A; J; p_1, \ldots, p_m)$. $\square$

Remark 2.40. We are intentionally avoiding any claims about a smooth structure for $\overline{\mathcal{M}}_{g,m}(A; J; p_1, \ldots, p_m)$ in Corollary 2.39, as the question of smoothness as $R \to \infty$ is fairly subtle. This detail will not make much difference for the application we have in mind, since the moduli space is only 2-dimensional.

2.1.8 Orientations

The theorems of Sects. 2.1.3 and 2.1.4 on smooth manifold structures for moduli spaces all included the word "oriented," which we snuck in although it does not follow from the implicit function theorem in infinite dimensions. In fact, there is no sensible notion of orientation on infinite-dimensional Banach manifolds, thus some additional ideas are required in order to see why certain finite-dimensional submanifolds carry natural orientations. Let us briefly sketch how this works for a moduli space of constrained somewhere injective curves

$$\mathcal{M}_{g,m}^*(A; J; Z) = \text{ev}^{-1}(Z) \subset \mathcal{M}_{g,m}^*(A; J)$$

as in Sect. 2.1.4, where $Z \subset M^m$ is a smooth oriented submanifold and J is chosen generically with respect to Z. (Similar ideas apply also for parametric moduli spaces as in Theorem 2.15, and for spaces satisfying jet constraints as in Sect. 2.1.5.)

Since M is naturally oriented and an orientation is also assumed on Z, it suffices to orient $\mathcal{M}_{g,m}^*(A; J)$, as the co-orientation of Z then determines an orientation of the submanifold $\mathcal{M}^*_{g,m}(A; J; Z)$. Recall the functional analytic picture sketched in Sect. 2.1.3: locally near a curve $u_0 : (\Sigma, j_0) \to (M, J)$ with marked points $\zeta_1, \ldots, \zeta_m \in \Sigma$, the space $\mathcal{M}^*_{g,m}(A; J)$ can be identified with $\bar{\partial}_J^{-1}(0)/G$, where the nonlinear Cauchy-Riemann operator

$$\bar{\partial}_J : \mathcal{T} \times \mathcal{B} \to \mathcal{E} : (j, u) \mapsto Tu + J \circ Tu \circ j$$

is a smooth Fredholm section of an infinite-dimensional Banach space bundle $\mathcal{E} \to \mathcal{T} \times \mathcal{B}$, and G is the automorphism group of the domain $(\Sigma, j_0, (\zeta_1, \ldots, \zeta_m))$, hence a finite-dimensional and naturally complex manifold. The base of the bundle $\mathcal{E} \to \mathcal{T} \times \mathcal{B}$ consists of a smooth finite-dimensional manifold $\mathcal{T}$ of complex structures near j_0 that parametrize the Teichmüller space of Σ with its marked points, plus an infinite-dimensional Banach manifold $\mathcal{B}$ containing maps $u : \Sigma \to M$ in some Sobolev regularity class. For notational simplicity, let us pretend for the moment that the group G is trivial and the identification is global, so that elements of $\mathcal{M}^*_{g,m}(A; J)$ can be written as pairs $(j, u) \in \bar{\partial}_J^{-1}(0)$. We say that $(j, u) \in \mathcal{M}^*_{g,m}(A; J)$ is *regular* if the linearization $D\bar{\partial}_J(j, u) : T_{(j,u)}(\mathcal{T} \times \mathcal{B}) \to \mathcal{E}_{(j,u)}$ is surjective, in which case the implicit function theorem implies that a neighborhood of (j, u) in $\mathcal{M}^*_{g,m}(A; J)$ is a smooth finite-dimensional manifold and its tangent space at (j, u) has a natural identification

$$T_{(j,u)}\mathcal{M}^*_{g,m}(A; J) = \ker D\bar{\partial}_J(j, u). \tag{2.14}$$

Differentiating the evaluation map ev $: \bar{\partial}_J^{-1}(0) \to M^m : (j, u) \mapsto (u(\zeta_1), \ldots, u(\zeta_m))$ gives the linearized evaluation map

$$d(\mathrm{ev})(j, u) : \ker D\bar{\partial}_J(j, u) \to T_{\mathrm{ev}(j,u)}M^m : (y, \eta) \mapsto (\eta(\zeta_1), \ldots, \eta(\zeta_m)),$$

and (j, u) is also regular for the constrained problem if the image of this linear map is transverse to $T_{\mathrm{ev}(j,u)}Z$.

If J is generic so that every $(j, u) \in \bar{\partial}_J^{-1}(0)$ is regular, then orienting $\mathcal{M}^*_{g,m}(A; J)$ is a matter of choosing an orientation for the kernel of each of the Fredholm operators

$$\mathbf{L}_{(j,u)} := D\bar{\partial}_J(j, u) : T_j\mathcal{T} \oplus T_u\mathcal{B} \to \mathcal{E}_{(j,u)},$$

varying continuously with $(j, u) \in \bar{\partial}_J^{-1}(0)$. This operator takes the form

$$\mathbf{L}_{(j,u)}(y, \eta) = J \circ Tu \circ y + \mathbf{D}_u\eta,$$

where $\mathbf{D}_u$ is the extension to a suitable Sobolev space setting of a real-linear *Cauchy-Riemann type* operator

$$\mathbf{D}_u : \Gamma(u^*TM) \to \Omega^{0,1}(\Sigma, u^*TM) : \eta \mapsto \nabla\eta + J \circ \nabla\eta \circ j + (\nabla_\eta J) \circ Tu \circ j.$$

Here a symmetric connection ∇ is chosen in order to write down the operator, but one can check that $\mathbf{D}_u$ does not depend on this choice. We now observe two things about $\mathbf{L}_{(j,u)}$: first, its domain and target are both naturally complex vector spaces, though in general $\mathbf{L}_{(j,u)}$ is only a *real*-linear map. On the other hand, the first term $y \mapsto J \circ Tu \circ y$ in $\mathbf{L}_{(j,u)}$ is complex linear, as a consequence of the assumption that $\bar{\partial}_J(j, u) = 0$. If $\mathbf{D}_u$ happens also to be complex linear, then $\ker \mathbf{L}_{(j,u)}$ is a complex vector space and thus inherits a natural orientation from its complex structure. This observation actually solves the whole problem in certain settings, e.g. if the almost complex structure J on M is integrable, then $\mathbf{D}_u$ is always complex linear and the moduli space $\mathcal{M}^*_{g,m}(A; J)$ inherits a natural orientation as a consequence.

When $\mathbf{D}_u$ is not complex linear, it nonetheless has a well-defined *complex-linear part*

$$\mathbf{D}^{\mathbb{C}}_u \eta := \frac{1}{2} \left(\mathbf{D}\eta - J\mathbf{D}_u(J\eta) \right),$$

which is also a Cauchy-Riemann type operator and thus Fredholm, and in fact the difference between $\mathbf{D}_u$ and $\mathbf{D}^{\mathbb{C}}_u$ is a zeroth-order term, so as a bounded linear operator between the relevant Sobolev spaces of sections, it is compact.[7] This implies that there is a natural homotopy through Fredholm operators from $\mathbf{D}_u$ to $\mathbf{D}^{\mathbb{C}}_u$. Combining this with the first term in $\mathbf{L}_{(j,u)}$ yields in turn a canonical homotopy through Fredholm operators from $\mathbf{L}_{(j,u)}$ to its complex-linear part $\mathbf{L}^{\mathbb{C}}_{(j,u)}$. If we could prove that all the operators in this homotopy are surjective, then we would be done: the natural prescription for orienting $\mathcal{M}^*_{g,m}(A; J)$ would then be to assign to each $\ker \mathbf{L}_{(j,u)}$ whichever orientation extends along the homotopy to reproduce the natural complex orientation of $\ker \mathbf{L}^{\mathbb{C}}_{(j,u)}$. The trouble with this idea is that, usually, we have no way of ensuring that the Fredholm operators along the homotopy from $\mathbf{L}_{(j,u)}$ to $\mathbf{L}^{\mathbb{C}}_{(j,u)}$ will remain surjective; in general their kernels will have jumping dimensions and the notion of "continuously" extending an orientation of the kernel ceases to be well defined. An elegant solution to this problem is provided by the notion of the **determinant line bundle**. Given a real-linear Fredholm operator $\mathbf{T}$ between Banach spaces, its **determinant line** is defined to be the real 1-dimensional vector space

$$\det(\mathbf{T}) = (\Lambda^{\max} \ker \mathbf{T}) \otimes (\Lambda^{\max} \operatorname{coker} \mathbf{T})^*,$$

with the convention that $\det(\mathbf{T})$ is canonically defined as $\mathbb{R}$ if the kernel and cokernel are both trivial. Choosing an orientation for $\det(\mathbf{T})$ is equivalent to choosing an orientation for $\ker \mathbf{T} \oplus \operatorname{coker} \mathbf{T}$, or simply $\ker \mathbf{T}$ whenever $\mathbf{T}$ is surjective. Note that if the domain and target of $\mathbf{T}$ have complex structures and $\mathbf{T}$ is also

---

[7] This is the major detail that will differ when we replace closed Riemann surfaces with punctured surfaces in Sect. 8.3.6: zeroth-order terms over noncompact surfaces are not compact operators, thus the orientation story in the punctured case is more complicated.

complex linear, then ker $\mathbf{T}$ and coker $\mathbf{T}$ both inherit complex structures, hence $\det(\mathbf{T})$ is naturally oriented; in the case where $\mathbf{T}$ is a complex-linear isomorphism, the natural orientation of $\det(\mathbf{T})$ is taken to be the standard orientation of $\mathbb{R}$. The usefulness of determinant lines comes from the following standard result, proofs of which can be found e.g. in [MS12, Appendix A.2], [Wend, Chap. 11] or [Zin16].

Theorem 2.41. *Fix real Banach spaces X and Y and let* $\mathrm{Fred}_{\mathbb{R}}(X, Y)$ *denote the space of Fredholm operators* $X \to Y$, *with its usual topology as an open subset of the space of all bounded linear operators* $X \to Y$. *Then there exists a topological vector bundle*

$$\det(X, Y) \xrightarrow{\pi} \mathrm{Fred}_{\mathbb{R}}(X, Y)$$

of real rank 1 such that $\pi^{-1}(\mathbf{T}) = \det(\mathbf{T})$ *for each* $\mathbf{T} \in \mathrm{Fred}_{\mathbb{R}}(X, Y)$.   □

With this result in hand, the prescription for orienting $\mathcal{M}^*_{g,m}(A; J)$ is to assign to each of the real 1-dimensional vector spaces $\det(\mathbf{L}_{(j,u)})$ whichever orientation extends continuously along the homotopy from $\mathbf{L}_{(j,u)}$ to reproduce the natural orientation of $\det(\mathbf{L}^{\mathbb{C}}_{(j,u)})$ arising from the fact that $\mathbf{L}^{\mathbb{C}}_{(j,u)}$ is complex linear. In light of Theorem 2.41, the question of whether each operator in the homotopy is surjective no longer plays any role.

Remark 2.42. We made two simplifying assumptions in the above discussion. One was that the correspondence between pairs $(j, u) \in \bar{\partial}_J^{-1}(0)$ and elements of $\mathcal{M}^*_{g,m}(A; J)$ is global, but it is easy to check that the prescription described above gives a well-defined orientation for $\mathcal{M}^*_{g,m}(A; J)$ without this assumption. The other was that the group G of automorphisms of the domain is trivial. Here the key fact is that every positive-dimensional Lie group that can appear in this context is naturally complex, hence so its Lie algebra, and its impact is therefore to replace the right hand side of (2.14) with a quotient by a complex subspace. The only modification required in the rest of the discussion is thus to include the canonical orientation of this complex subspace in the picture.

For most applications it suffices to know that $\mathcal{M}^*_{g,m}(A; J)$ and $\mathcal{M}^*_{g,m}(A; J; Z)$ have orientations without worrying about where those orientations come from, but we will encounter a situation in Chap. 7 where slightly more information is needed. To set up the statement, observe that whenever vir-dim $\mathcal{M}^*_{g,m}(A; J; Z) = 0$, orienting $\mathcal{M}^*_{g,m}(A; J; Z)$ means assigning a sign to each element $u \in \mathcal{M}^*_{g,m}(A; J; Z)$ that is Fredholm regular for the constrained problem. Given an orientation of $\mathcal{M}^*_{g,m}(A; J)$, this sign must match the sign of the isolated transverse intersection at u between ev $: \mathcal{M}^*_{g,m}(A; J) \to M^m$ and $Z \subset M^m$. While the hypotheses in the following result may appear improbable out of context, we will be able to verify them in certain low-dimensional settings using automatic transversality (see Proposition 2.47).

Lemma 2.43. *Suppose $Z \subset M^m$ is an almost complex submanifold, $\mathcal{M}^*_{g,m}(A; J; Z)$ has virtual dimension 0, and $u : (\Sigma, j) \to (M, J)$ with marked points $\zeta_1, \ldots, \zeta_m \in \Sigma$ represents an element of $\mathcal{M}^*_{g,m}(A; J; Z)$ that is Fredholm regular for the constrained problem. Let $\{\mathbf{L}^\tau_u\}_{\tau \in [0,1]}$ denote the canonical homotopy of Fredholm operators $T_j\mathcal{T} \oplus T_u\mathcal{B} \to \mathcal{E}_{(j,u)}$ from $\mathbf{L}^1_u := D\bar{\partial}_J(j, u)$ to its complex-linear part $\mathbf{L}^0_u$. Assume additionally that the following hold for every $\tau \in [0, 1]$:*

(1) $\mathbf{L}^\tau_u$ is surjective;
(2) The linear evaluation map

$$\ker \mathbf{L}^\tau_u \to T_{\mathrm{ev}(u)} M^m : (y, \eta) \mapsto (\eta(\zeta_1), \ldots, \eta(\zeta_m))$$

is transverse to $T_{\mathrm{ev}(u)} Z$.

Then the sign of u determined by the natural orientation on $\mathcal{M}^*_{g,m}(A; J; Z)$ is positive.

Proof. The surjectivity of $\mathbf{L}^\tau_u$ for every $\tau \in [0, 1]$ means that the spaces $\ker \mathbf{L}^\tau_u$ vary continuously with τ and they have orientations determined by the orientations of $\det(\mathbf{L}^\tau_u)$, which vary continuously and become the natural complex orientation at $\tau = 0$. In light of the transversality condition for the linear evaluation maps, it follows that the sign of the intersection at u between ev and Z matches the sign of the intersection between $T_{\mathrm{ev}(u)} Z$ and the linear evaluation map at $\tau = 0$. The latter is positive since the linear evaluation map at $\tau = 0$ is a complex-linear map and Z is almost complex. $\square$

2.2 Dimension Four

The theory of holomorphic curves has some special features in dimension four, mainly as consequences of the fact that Riemann surfaces are 2-dimensional and $4 = 2 + 2$. Throughout this section, we shall assume (M, J) is an almost complex manifold of real dimension four.

2.2.1 Automatic Transversality

The most important difference between the following result and Theorem 2.12 above is that there is no need to perturb J. This is one of the few situations where it is possible to verify explicitly that transversality is achieved.

Theorem 2.44. *Suppose (M, J) is an almost complex 4-manifold and $u \in \mathcal{M}_g(J)$ is an immersed J-holomorphic curve satisfying $\mathrm{ind}(u) > 2g - 2$. Then u is Fredholm regular.* $\square$

Corollary 2.45. *In any almost complex 4-manifold, every immersed pseudoholomorphic sphere with nonnegative index is Fredholm regular.*   □

Theorem 2.44 was first stated by Gromov [Gro85] and full details were later written down by Hofer-Lizan-Sikorav [HLS97]. There are also generalizations for non-immersed curves [IS99] and for punctured curves with finite energy in symplectic cobordisms (see Sect. 8.3.7 or [Wen10a]), but all of them depend crucially on the assumption that dim $M = 4$.

We will need the higher genus case of the above result only once in this book, namely for Theorem 7.36, which strictly speaking is a digression, unconnected to our main applications. But we will frequently need the following extension of the genus zero case to curves with pointwise constraints, as defined in Sect. 2.1.4.

Theorem 2.46. *Suppose* (M, J) *is an almost complex 4-manifold,* $p_1, \ldots, p_m \in M$ *are arbitrary points for* $m \geq 0$, *and* $u \in \mathcal{M}_{0,m}(J; p_1, \ldots, p_m)$ *is an immersed* J-*holomorphic sphere satisfying* $\mathrm{ind}(u) \geq 2m$. *Then* u *is Fredholm regular for the constrained problem, i.e.* $u \in \mathcal{M}_{0,m}^{\mathrm{reg}}(J; p_1, \ldots, p_m)$.

Notice that by (2.7), the virtual dimension of $\mathcal{M}_{0,m}(J; p_1, \ldots, p_m)$ is $\mathrm{ind}(u) - 2m$, so the index condition just means vir-dim $\mathcal{M}_{0,m}(J; p_1, \ldots, p_m) \geq 0$. (This simple interpretation is unique to the genus zero case—the generalization of Theorem 2.46 for higher genus curves requires a more stringent index condition as in Theorem 2.44.)

Let us sketch a proof of Theorem 2.46. There are three essential assumptions:

    (1) dim $M = 4$;
    (2) u is immersed;
    (3) $g = 0$ and $\mathrm{ind}(u) \geq 2m$.

The first two imply that one can split the bundle $u^*TM \to \Sigma$ into a sum of two complex line bundles $T\Sigma \oplus N_u$, where $N_u \to \Sigma$ denotes the normal bundle. We thus have

$$c_1([u]) = \chi(S^2) + c_1(N_u), \tag{2.15}$$

where $c_1(N_u)$ is an abbreviation for the first Chern *number* $\langle c_1(N_u), [S^2] \rangle \in \mathbb{Z}$. One can then study nearby curves in the moduli space by identifying them (up to parametrization) with small sections of N_u. One can show that at the linearized level, such sections $\eta \in \Gamma(N_u)$ represent curves in $\mathcal{M}_{0,m}(J; p_1, \ldots, p_m)$ if and only if they vanish at the m marked points and satisfy the linearized Cauchy-Riemann equation,

$$\mathbf{D}_u^N \eta = 0,$$

where $\mathbf{D}_u^N$ is just the restriction of the usual linearized operator $D\bar{\partial}_J(j, u)$ to sections of the normal bundle. Moreover, Fredholm regularity can now be restated as the condition that $\mathbf{D}_u^N$, operating on a suitable Banach space of sections of N_u

vanishing at the marked points, must be surjective. The domain of $\mathbf{D}_u^N$ is thus a subspace of codimension $2m$ in the domain of the *unconstrained* linearized problem, which by the Riemann-Roch formula has index $\chi(S^2) + 2c_1(N_u)$, thus

$$\text{ind}(\mathbf{D}_u^N) = \chi(S^2) + 2c_1(N_u) - 2m.$$

If you inspect (2.5) with $n = 2$ and plug in (2.15), you'll find that this equals $\text{ind}(u) - 2m$. The third assumption therefore means

$$c_1(N_u) \geq m - 1, \quad \text{and} \quad \text{ind}(\mathbf{D}_u^N) \geq 0.$$

By the definition of the Fredholm index, we have $\dim \ker \mathbf{D}_u^N \geq \text{ind}(\mathbf{D}_u^N)$, with equality if and only if $\mathbf{D}_u^N$ is surjective. Thus the result follows if we can show that

$$\dim \ker \mathbf{D}_u^N \leq 2 + 2c_1(N_u) - 2m \quad \text{whenever} \quad c_1(N_u) \geq m - 1. \quad (2.16)$$

This will follow essentially from the fact that every nontrivial section of N_u satisfying a linear Cauchy-Riemann type equation can have only *isolated* and *positive* zeroes. The technical ingredient behind this fact is known as the *similarity principle* [Wenc, §2.7], though one can also prove it using Aronszajn's theorem, see [MS12]. Given this fact, it's easy to see why (2.16) holds when $c_1(N_u) = m - 1$: in this case $\text{ind}(\mathbf{D}_u^N) = 0$, and since any nontrivial section $\eta \in \ker \mathbf{D}_u^N$ is constrained to vanish at m points, there can be no such sections, hence $\mathbf{D}_u^N$ is injective and therefore also surjective. If $c_1(N_u) \geq m$, we set $k = c_1(N_u) - m + 1$ and argue as follows: choose any set of distinct points $z_1, \ldots, z_k \in \Sigma$ that are disjoint from the marked points, and define the linear "evaluation" map

$$\text{ev} : \ker \mathbf{D}_u^N \to (N_u)_{z_1} \oplus \ldots \oplus (N_u)_{z_k} : \eta \mapsto \big(\eta(z_1), \ldots, \eta(z_k)\big).$$

Any nontrivial $\eta \in \ker \mathbf{D}_u^N$ has only positive zeroes, and already vanishes at the m marked points, so it can have at most $k - 1$ additional zeroes since $c_1(N_u) = m + k - 1$. It follows that the right hand side can only be trivial if η is trivial, showing that ev is injective. Since the target of this map is a real vector space of dimension $2k = 2c_1(N_u) - 2m + 2$, the inequality (2.16) follows, completing the proof of Theorem 2.46.

The proof of Theorem 2.44 for higher genus curves also uses the similarity principle, but requires one additional idea: every Cauchy-Riemann type operator $\mathbf{D}$ on a complex line bundle E has a *formal adjoint* $\mathbf{D}^*$, which is equivalent to a Cauchy-Riemann type operator on some other bundle E' and satisfies $\ker \mathbf{D}^* \cong \text{coker} \, \mathbf{D}$. Thus if the goal is to prove that $\mathbf{D}$ is surjective, it is just as well to prove that $\mathbf{D}^*$ is injective, which will be true whenever $c_1(E') < 0$ since nontrivial sections in $\ker \mathbf{D}^*$ must always have a nonnegative count of zeroes. The sufficient condition stated in Theorem 2.44 turns out to be equivalent to the condition $c_1(E') < 0$ in the case where E is the normal bundle of an immersed J-holomorphic curve.

The methods in our proof of Theorem 2.46 above can also be applied to understand orientations in the space $\mathcal{M}_{0,m}(J; p_1, \ldots, p_m)$. Notice that in counting the positive zeroes of sections in $\ker \mathbf{D}_u^N$, we do not use any special facts about $\mathbf{D}_u^N$ except that it is a Cauchy-Riemann type operator (and thus satisfies the similarity principle). The only other important detail is the value of $c_1(N_u)$, which means that the same argument proves surjectivity for *all* Cauchy-Riemann type operators homotopic to $\mathbf{D}_u^N$; in particular, the conclusion holds for every operator in the canonical homotopy from $\mathbf{D}_u^N$ to its complex-linear part. One can transform this into a statement about $D\bar{\partial}_J(j, u)$ and linear evaluation maps using the setup in [Wen10a, §3.4], and this is enough to establish the hypotheses of Lemma 2.43, proving:

Proposition 2.47. *Under the assumptions of Theorem 2.46, suppose additionally that* $\operatorname{ind}(u) = 2m$, *so* vir-dim $\mathcal{M}_{0,m}([u]; J; p_1, \ldots, p_m) = 0$. *Then for the natural orientation of* $\mathcal{M}_{0,m}([u]; J; p_1, \ldots, p_m)$, *u has positive sign.*   □

Remark 2.48. We can now outline a proof of the claim from Remark 1.16 that the map

$$\pi : \operatorname{Diff}_+(S^2) \to \mathcal{J}(S^2) : \varphi \mapsto \varphi^* i$$

is a Serre fibration. The claim means that π satisfies the homotopy lifting property with respect to disks (see e.g. [Hat02, §4.2]), so for every integer $k \geq 0$, given continuous (in the C^∞-topology) families of diffeomorphisms $\{\varphi_\tau \in \operatorname{Diff}_+(S^2)\}_{\tau \in \{0\} \times \mathbb{D}^k}$ and complex structures $\{j_\tau \in \mathcal{J}(S^2)\}_{\tau \in [0,1] \times \mathbb{D}^k}$ satisfying $j_\tau = \varphi_\tau^* i$ for all $\tau \in \{0\} \times \mathbb{D}^k$, we need to extend the family of diffeomorphisms to all $\tau \in [0, 1] \times \mathbb{D}^k$ so that $j_\tau = \varphi_\tau^* i$ is always satisfied. This would mean that the maps

$$\varphi_\tau^{-1} : (S^2, i) \to (S^2, j_\tau)$$

are all pseudoholomorphic, and by Proposition 2.5, any pseudoholomorphic map $S^2 \to S^2$ with degree 1 is automatically a diffeomorphism, hence the real task here is to understand how the space of j-holomorphic spheres of degree 1 in S^2 behaves under deformations of the complex structure $j \in \mathcal{J}(S^2)$ on the target space. Any such map $u : (S^2, i) \to (S^2, j)$ can naturally be regarded as an element of the moduli space $\mathcal{M}_{0,3}([S^2]; j)$, where the purpose of the three marked points is to eliminate the reparametrization freedom; indeed, each element $u \in \mathcal{M}_{0,3}([S^2]; j)$ has a unique parametrization that places the marked points at $0, 1, \infty$, and we shall refer to this in the following as the *standard* parametrization. By (2.4),

$$\text{vir-dim}\,\mathcal{M}_{0,3}([S^2]; j) = -2\chi(S^2) + 2c_1(u^* T S^2) + 6 = 6$$

for a chosen map $u : S^2 \to S^2$ parametrizing an element of $\mathcal{M}_{0,3}([S^2], j)$, as $\deg(u) = 1$ implies $c_1(u^* T S^2) = c_1(T S^2) = \chi(S^2)$. Any $u \in \mathcal{M}_{0,3}([S^2]; j)$ is then also an element of the constrained moduli space $\mathcal{M}_{0,3}([S^2]; j; p_1, p_2, p_3)$, with the constraint points $p_1, p_2, p_3 \in S^2$ chosen to be the images under u of the three marked points, and we have

$$\text{vir-dim}\,\mathcal{M}_{0,3}([S^2]; j; p_1, p_2, p_3) = 0.$$

We claim that every element of this space, for any $j \in \mathcal{J}(S^2)$ and any three distinct constraint points p_1, p_2, p_3, is Fredholm regular for the constrained problem. This is most easily seen by fixing standard parametrizations, thus identifying $\mathcal{M}_{0,3}([S^2]; j)$ with the degree 1 component of the zero-set of a smooth section

$$\bar{\partial}_j : \mathcal{B} \to \mathcal{E} : u \mapsto Tu + j \circ Tu \circ i,$$

where $\mathcal{B}$ is a Banach manifold of maps $S^2 \to S^2$ in some Sobolev regularity class, e.g. $W^{k,p}$ for $k \geq 1$ and $kp > 2,$[8] and $\mathcal{E} \to \mathcal{B}$ is a Banach space bundle whose fiber $\mathcal{E}_u$ over each $u \in \mathcal{B}$ is a Sobolev space of sections (e.g. of class $W^{k-1,p}$) of the vector bundle $\overline{\text{Hom}}_{\mathbb{C}}((TS^2, i), (u^*TS^2, j))$ of complex-antilinear maps $(TS^2, i) \to (u^*TS^2, j)$. The linearization of $\bar{\partial}_j$ at some $u \in \bar{\partial}_j^{-1}(0)$ is a linear Cauchy-Riemann type operator

$$\mathbf{D}_u : \Gamma(u^*TS^2) \to \Gamma\left(\overline{\text{Hom}}_{\mathbb{C}}((TS^2, i), (u^*TS^2, j))\right),$$

extended to a bounded linear operator on the relevant Sobolev spaces of sections, and the Riemann-Roch formula gives its Fredholm index as

$$\text{ind}\,\mathbf{D}_u = \chi(S^2) + 2c_1(u^*TS^2) = 3\chi(S^2) = 6.$$

As in the proof of Theorem 2.46, we can now consider the linearized evaluation map

$$\text{ev} : \ker \mathbf{D}_u \to T_{p_1}S^2 \oplus T_{p_2}S^2 \oplus T_{p_3}S^2 : \eta \mapsto (\eta(0), \eta(1), \eta(\infty)), \qquad (2.17)$$

and observe that since all zeroes of nontrivial sections $\eta \in \ker \mathbf{D}_u$ are isolated and positive, with total signed count equal to $c_1(u^*TS^2) = \chi(S^2) = 2 < 3$, this map is injective. It follows that $\dim \ker \mathbf{D}_u \leq 6 = \text{ind}\,\mathbf{D}_u$, hence $\mathbf{D}_u$ has trivial cokernel and u is therefore Fredholm regular. Moreover, (2.17) is also the derivative at u of the nonlinear evaluation map

$$\text{ev} : \mathcal{M}_{0,3}([S^2]; j) \to S^2 \times S^2 \times S^2 : u \mapsto (u(0), u(1), u(\infty)),$$

and the above argument shows that it is an injection between equidimensional spaces and therefore an isomorphism, proving the claim about regularity for the constrained problem.

---

[8]The condition $kp > 2$ is related to the Sobolev embedding theorem, as it guarantees that all maps $S^2 \to S^2$ of class $W^{k,p}$ will be continuous, see e.g. [AF03].

It follows now from the implicit function theorem that the given family of maps $u_{(0,\tau)} : (S^2, i) \to (S^2, j_{(0,\tau)})$ defined for $\tau \in \mathbb{D}^k$ admits an extension to a family

$$\left\{ u_{(s,\tau)} : (S^2, i) \to (S^2, j_{(s,\tau)}) \right\}_{(s,\tau) \in [0,\epsilon) \times \mathbb{D}^k}$$

for $\epsilon > 0$ sufficiently small, and the extension is uniquely determined if we impose the constraint

$$u_{(s,\tau)}(\zeta) = u_{(0,\tau)}(\zeta) \quad \text{for} \quad \zeta = 0, 1, \infty.$$

To show that this extension actually exists for all $s \in [0, 1]$, we combine the above with a compactness argument: it is an easy consequence of Gromov's compactness theorem that for any C^∞-convergent sequence $j_k \to j \in \mathcal{J}(S^2)$, a sequence $u_k \in \mathcal{M}_{0,3}([S^2]; j_k)$ has a subsequence converging to an element of $\mathcal{M}_{0,3}([S^2]; j)$, as any nodal curve appearing in such a limit would need to have smooth components whose degrees are all nonnegative and add up to 1, and the stability condition precludes the existence of a nodal curve with arithmetic genus zero having ghost bubbles in addition to a single degree 1 component. One can therefore extend the family to $s = \epsilon$ and use the implicit function theorem again to go beyond this, showing that there is no upper bound for $s \in [0, 1]$ beyond which the family cannot be extended.

2.2.2 Positivity of Intersections and Adjunction

The single most useful feature of the case dim $M = 4$ is that intersections between two distinct holomorphic curves may be transverse, and even when they are not, they can be counted. We shall denote by

$$H_2(M) \times H_2(M) \to \mathbb{Z} : (A, B) \mapsto A \cdot B$$

the homological intersection pairing on $H_2(M)$, which is well defined only when M is 4-dimensional. A local statement of the result known as "positivity of intersections" asserts that if $u, v : \mathbb{D}^2 \to M$ are any two J-holomorphic disks in an almost complex 4-manifold (M, J) with $u(0) = v(0)$, then unless they have identical images near the intersection (which would mean they are locally both reparametrizations or branched covers of the same curve), the intersection must be isolated and count positively; in fact, its local intersection index is always at least 1, with equality if and only if the intersection is transverse. In particular, a non-transverse intersection of two holomorphic curves cannot be perturbed away, but will always give rise to even more transverse intersections under a generic perturbation. This implies the following global result:

Theorem 2.49. *Suppose (M, J) is an almost complex 4-manifold and $u : \Sigma \to M$, $v : \Sigma' \to M$ are both closed and connected J-holomorphic curves whose images are not identical. Then u and v have finitely many points of intersection, and*

$$[u] \cdot [v] \geqslant \left| \{ (z, z') \in \Sigma \times \Sigma' \mid u(z) = v(z') \} \right|,$$

with equality if and only if all the intersections are transverse. In particular, $[u] \cdot [v] = 0$ if and only if the images of u and v are disjoint, and $[u] \cdot [v] = 1$ if and only if there is exactly one intersection and it is transverse. □

If $u : \Sigma \to M$ is a closed J-holomorphic curve which is *simple*, then Proposition 2.6 implies that it also intersects *itself* in at most finitely many places, and one can count (with signs) the number of self-intersections. In particular if u is simple and immersed, one defines the integer

$$\delta(u) = \frac{1}{2} \sum_{u(z)=u(z')} i(z, z'), \qquad (2.18)$$

where the sum ranges over all ordered pairs (z, z') of distinct points in Σ at which $u(z) = u(z')$, and $i(z, z') \in \mathbb{Z}$ is defined to be the algebraic index of the isolated intersection between $u(\mathcal{U}_z)$ and $u(\mathcal{U}_{z'})$ for sufficiently small neighborhoods $z \in \mathcal{U}_z \subset \Sigma$ and $z' \in \mathcal{U}_{z'} \subset \Sigma$. Positivity of intersections implies that $i(z, z') > 0$ for all such pairs, so it follows that $\delta(u) \geqslant 0$, with equality if and only if u is embedded.

If u is simple but not immersed, one can define $\delta(u)$ to be $\delta(u')$, where u' is a C^∞-close immersed perturbation of u. This can be chosen in a sufficiently canonical way so that $\delta(u')$ does not depend on the choice and, moreover, is strictly positive whenever u has non-immersed points (see [MW95] or [Wenf]).

Remark 2.50. There is no sensible way to define $\delta(u)$ when u is a multiple cover, thus the important Theorem 2.51 below applies *only* to somewhere injective curves.

One can also relate $\delta(u)$ to the obviously homotopy invariant numbers $[u] \cdot [u]$ and $c_1([u])$, producing the so-called **adjunction formula**,[9] originally due to McDuff [McD94] and Micallef-White [MW95]. The idea in the immersed case with transverse self-intersections is easy to understand: $[u] \cdot [u]$ can be computed by counting the algebraic number of intersections between u and a small perturbation u_ϵ of it, defined via a generic section of its normal bundle, see Fig. 2.3. Each transverse self-intersection of u contributes 2 to this

---

[9]In some sources, notably in [MS12], the adjunction formula is stated as an adjunction *inequality*, which is equivalent to our Theorem 2.51 but looks a bit different. The reason for the discrepancy is that some authors prefer a different and simpler definition of $\delta(u)$, in which it just counts the number of geometric self-intersections, without worrying about the local intersection index or requiring an immersed perturbation. Our version of $\delta(u)$ is generally larger than this one, but equals it if and only if u is immersed and all its self-intersections are transverse.

Fig. 2.3 Computing the homological self-intersection number of an immersed holomorphic curve

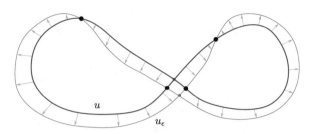

count, while the zeroes of a section of N_u contribute $c_1(N_u)$, which equals the homotopy invariant quantity $c_1([u]) - \chi(\Sigma)$. The definition of $\delta(u)$ in the non-immersed case is conceived to ensure that the resulting relation will remain true in general:

Theorem 2.51. *If* $u : \Sigma \to M$ *is a closed somewhere injective* J-*holomorphic curve in an almost complex* 4-*manifold* (M, J), *then*

$$[u] \cdot [u] = 2\delta(u) + c_1([u]) - \chi(\Sigma),$$

where $\delta(u) \in \mathbb{Z}$ *satisfies* $\delta(u) \geqslant 0$, *with equality if and only if* u *is embedded.* □

Observing that every term in this formula other than $\delta(u)$ manifestly depends only on the homology class $[u] \in H_2(M)$ and genus of Σ, we obtain:

Corollary 2.52. *If* (M, J) *is a* 4-*dimensional almost complex manifold and* $u \in \mathcal{M}_g(A; J)$ *is embedded, then every other somewhere injective curve in* $\mathcal{M}_g(A; J)$ *is also embedded.* □

2.2.3 An Implicit Function Theorem for Embedded Spheres with Constraints

We now explain a more specific result that will be useful in our main applications. Observe that for any embedded curve $u \in \mathcal{M}_g(J)$ and any set of pairwise distinct points $p_1, \ldots, p_m$ in the image of u, one can also regard u naturally as an element of the constrained moduli space $\mathcal{M}_{g,m}(J; p_1, \ldots, p_m)$ we defined in Sect. 2.1.4.

Proposition 2.53. *Suppose* (M, J) *is an almost complex* 4-*manifold and* $u : S^2 \to M$ *is an embedded* J-*holomorphic sphere with* $[u] \cdot [u] = m \geqslant 0$. *Then for any choice of pairwise distinct points* $p_1, \ldots, p_m \in u(S^2)$, *the curve*

$$u \in \mathcal{M}_{0,m}(J; p_1, \ldots, p_m)$$

is Fredholm regular for the constrained problem, and a neighborhood $\mathcal{U} \subset \mathcal{M}_{0,m}(J; p_1, \ldots, p_m)$ *of* u *admits the structure of a smooth* 2-*dimensional manifold such that the following conditions are satisfied:*

> (1) *Each curve $v \in \mathcal{U}$ is embedded, and any two curves $v, w \in \mathcal{U}$ intersect each other only at the points $p_1, \ldots, p_m$, with all intersections transverse;*
>
> (2) *The images $v(S^2) \backslash \{p_1, \ldots, p_m\}$ for $v \in \mathcal{U}$ form the leaves of a smooth foliation of some open neighborhood of $u(S^2) \backslash \{p_1, \ldots, p_m\}$ in $M \backslash \{p_1, \ldots, p_m\}$.*

Sketch of the Proof. Writing $A := [u]$, the space $\mathcal{M}_{0,m}(A; J; p_1, \ldots, p_m)$ has virtual dimension given by (2.7), where we can deduce $c_1(A)$ from the adjunction formula (Theorem 2.51). Indeed, since u is embedded and satisfies $[u] \cdot [u] = m$, adjunction gives

$$m = c_1(A) - \chi(S^2),$$

hence $c_1(A) = m + 2$. Plugging this and $n = 2$ into (2.7) gives

$$\text{vir-dim}\, \mathcal{M}_{0,m}(A; J; p_1, \ldots, p_m) = -2 + 2c_1(A) - 2m = 2.$$

Theorem 2.46 then implies that u is Fredholm regular for the constrained problem. It will be helpful to review why this is true in the case at hand: since $u : S^2 \to M$ is embedded, we can define its normal bundle $N_u \to S^2$ and identify nearby curves $v \in \mathcal{M}_{0,m}(A; J; p_1, \ldots, p_m)$ with small sections η of N_u satisfying the linear Cauchy-Riemann type equation

$$\mathbf{D}_u^N \eta = 0.$$

In light of the constraints $v(\zeta_i) = p_i$, we can also set up the identification so that these sections vanish at all of the marked points $\zeta_1, \ldots, \zeta_m$. The restriction of $\mathbf{D}_u^N$ to the space of sections satisfying this constraint is then a Fredholm operator of index 2. We claim now that nontrivial sections in the kernel of this operator vanish *only* at $\zeta_1, \ldots, \zeta_m$, and their zeroes at these points are simple. Since solutions of $\mathbf{D}_u^N \eta = 0$ have only isolated positive zeroes, the claim follows from the observation that, in light of the computation $c_1(A) = m + 2$ above,

$$c_1(N_u) = c_1(A) - \chi(S^2) = m.$$

With this understood, it follows that the vector space of sections $\eta \in \Gamma(N_u)$ satisfying $\mathbf{D}_u^N \eta = 0$ and vanishing at the marked points is at most 2-dimensional: indeed, choosing any point $z \in S^2 \backslash \{\zeta_1, \ldots, \zeta_m\}$, the linear evaluation map taking a section η to $\eta(z) \in (N_u)_z$ is now an injective map into a real 2-dimensional vector space. Since this space is the kernel of an operator with Fredholm index 2, it follows that the operator is surjective, i.e. transversality is achieved.

The remaining statements in Proposition 2.53 follow essentially from the observation that, by the above discussion, not only is $\mathcal{M}_{0,m}(A; J; p_1, \ldots, p_m)$ smooth near u but its tangent space $T_u\mathcal{M}_{0,m}(A; J; p_1, \ldots, p_m)$ is naturally isomorphic to the space of sections $\eta \in \Gamma(N_u)$ satisfying $\mathbf{D}_u^N \eta = 0$ and vanishing at the marked points—and as we just showed, these sections have zeroes only at $\zeta_1, \ldots, \zeta_m$, all of them simple. This may be thought of as the "linearization" of the statement that the curves in $\mathcal{M}_{0,m}(A; J; p_1, \ldots, p_m)$ near u foliate an open subset of $M \backslash \{p_1, \ldots, p_m\}$.

One can also use Theorem 2.49 to analyze the intersections of any two distinct curves $v, w \in \mathcal{M}_{0,m}(A; J; p_1, \ldots, p_m)$ near u: since

$$[v] \cdot [w] = A \cdot A = m$$

and the two curves are already *forced* to intersect at the points $p_1, \ldots, p_m$, there can be no additional intersections, and the m forced intersections are all transverse. $\square$

Chapter 3
Blowups and Lefschetz Fibrations

In this chapter we discuss a few standard topics from symplectic topology that are mostly independent of holomorphic curves. We begin with the blowup operation, which was sketched already in Example 1.6 and appears in the statements of several of the main theorems in Sect. 1.2. It can be defined in both the complex and the symplectic category, and for our purposes it will be important to understand both definitions and their relation to each other. We then discuss Lefschetz pencils and Lefschetz fibrations, their topological properties, and how they determine deformation classes of symplectic structures.

3.1   The Complex Blowup

Suppose M is a complex n-dimensional manifold with $n \geq 2$, and $z \in M$ is a point. As a set, we define the **complex blowup** of M at z to be

$$\widetilde{M} = (M \backslash \{z\}) \cup \mathbb{P}(T_z M),$$

where $\mathbb{P}(T_z M) \cong \mathbb{CP}^{n-1}$ is the space of complex lines in $T_z M$. For example, blowing up $\mathbb{C}^n$ at the origin produces a set that can be naturally identified with the **tautological line bundle**

$$\widetilde{\mathbb{C}}^n = \left\{ (\ell, v) \in \mathbb{CP}^{n-1} \times \mathbb{C}^n \mid v \in \ell \right\},$$

where elements of $\mathbb{CP}^{n-1}$ are regarded as complex lines $\ell \subset \mathbb{C}^n$. Using the holomorphic coordinate charts on $\mathbb{CP}^{n-1}$ discussed in Example 1.4, one can easily construct local trivializations that give the projection

$$\pi : \widetilde{\mathbb{C}}^n \to \mathbb{CP}^{n-1} : (\ell, v) \mapsto \ell$$

© Springer International Publishing AG, part of Springer Nature 2018
C. Wendl, *Holomorphic Curves in Low Dimensions*, Lecture Notes
in Mathematics 2216, https://doi.org/10.1007/978-3-319-91371-1_3

the structure of a holomorphic line bundle, such that the so-called **blowdown map**

$$\beta : \widetilde{\mathbb{C}}^n \to \mathbb{C}^n : (\ell, v) \mapsto v$$

is a holomorphic map of equidimensional complex manifolds. Notice that away from the zero-section $\mathbb{CP}^{n-1} := \mathbb{CP}^{n-1} \times \{0\} \subset \widetilde{\mathbb{C}}^n$, β restricts to a biholomorphic diffeomorphism

$$\widetilde{\mathbb{C}}^n \backslash \mathbb{CP}^{n-1} \xrightarrow{\beta} \mathbb{C}^n \backslash \{0\}, \tag{3.1}$$

but it collapses the entirety of the zero-section to the origin in $\mathbb{C}^n$. This discussion carries over to the blowup of an arbitrary complex manifold M at $z \in M$ by choosing holomorphic coordinates near z: the blowup $\widetilde{M}$ thus becomes a complex n-dimensional manifold, and the natural blowdown map

$$\beta : \widetilde{M} \to M$$

collapsing $\mathbb{P}(T_z M)$ to z while identifying $\widetilde{M} \backslash \mathbb{P}(T_z M)$ with $M \backslash \{z\}$ is holomorphic. The projectivization $\mathbb{P}(T_z M) \subset \widetilde{M}$ forms a complex hypersurface biholomorphic to $\mathbb{CP}^{n-1}$ and is called an **exceptional divisor**. When $n = 2$, it is a copy of $\mathbb{CP}^1 \cong S^2$ and is thus also called an **exceptional sphere**. Exercise 3.2 below implies that its self-intersection number in this case is -1. One can use the following exercise to show that the smooth and complex structures on $\widetilde{M}$ do not depend on the choice of holomorphic coordinates near p; moreover, it is clear that if M is connected, then moving p does not change the diffeomorphism type of $\widetilde{M}$.

Exercise 3.1. Suppose $\mathcal{U}, \mathcal{U}' \subset \mathbb{C}^n$ are two neighborhoods of the origin, $\widetilde{\mathcal{U}}, \widetilde{\mathcal{U}}' \subset \widetilde{\mathbb{C}}^n$ denote the corresponding neighborhoods of the zero-section $\mathbb{CP}^{n-1} \subset \widetilde{\mathbb{C}}^n$ under the identification (3.1), and $f : \mathcal{U} \to \mathcal{U}'$ is a biholomorphic diffeomorphism with $f(0) = 0$. Show that the map $\widetilde{f} : \widetilde{\mathcal{U}} \backslash \mathbb{CP}^{n-1} \to \widetilde{\mathcal{U}}' \backslash \mathbb{CP}^{n-1}$ defined by restricting f to $\mathcal{U} \backslash \{0\}$ and then using the identification (3.1) has a unique continuous extension $\widetilde{f} : \widetilde{\mathcal{U}} \to \widetilde{\mathcal{U}}'$ which preserves $\mathbb{CP}^{n-1}$ and is biholomorphic. Conversely, show that every biholomorphic diffeomorphism $\widetilde{f} : \widetilde{\mathcal{U}} \to \widetilde{\mathcal{U}}'$ that preserves $\mathbb{CP}^{n-1}$ arises in this way from some biholomorphic diffeomorphism $f : \mathcal{U} \to \mathcal{U}'$ that fixes 0.

Exercise 3.2. Show (e.g. by counting the zeroes of a section) that the tautological line bundle over $\mathbb{CP}^1$ has first Chern number -1.

Exercise 3.3. Show that if $\widetilde{M}$ is the blowup of a complex surface M at a point, then $\widetilde{M}$ admits an orientation-preserving diffeomorphism to $M \# \overline{\mathbb{CP}}^2$, where the bar over $\mathbb{CP}^2$ indicates a reversal of its usual orientation. *Hint: if $B_\epsilon^4 \subset \mathbb{C}^2$ denotes the ϵ-ball and $\widetilde{B}_\epsilon^4 \subset \widetilde{\mathbb{C}}^2$ is its blowup at the origin, you need to show that the closure of $\widetilde{B}_\epsilon^4$ is orientation-reversing diffeomorphic to the complement of a ball in $\mathbb{CP}^2$. The latter is equivalently a tubular neighborhood of the sphere at infinity $\mathbb{CP}^1 \subset \mathbb{CP}^2$. Compare the normal bundle of the latter with that of the zero-section $\mathbb{CP}^1 \subset \widetilde{\mathbb{C}}^2$: what are their Euler classes?*

Remark 3.4. One can generalize the exercise above to show that if $\dim_{\mathbb{C}} M = n \geqslant$ 2, the blowup of M is diffeomorphic to $M \# \overline{\mathbb{CP}}^n$.

The inverse of this operation, the **complex blowdown**, can be performed on any complex manifold $\tilde{M}$ containing a complex hypersurface $E \subset \tilde{M}$ with a neighborhood biholomorphically identified with a neighborhood of $\mathbb{CP}^{n-1}$ in $\tilde{\mathbb{C}}^n$. The blowdown is then the complex manifold M obtained by replacing this neighborhood with the corresponding neighborhood of the origin in $\mathbb{C}^n$, the effect of which is to collapse E to a point. One can again use Exercise 3.1 to show that this construction does not depend on any choices beyond the exceptional divisor $E \subset \tilde{M}$.

Remark 3.5. As a set, the blowup of M at $z \in M$ depends on the complex structure only at z, while the definition of its smooth structure requires an integrable complex structure on a neighborhood of z. We can therefore define $\tilde{M}$ as a *smooth* (but not necessarily complex) manifold given only the germ of a complex structure near z, and $\tilde{M}$ then inherits the germ of a complex structure near the exceptional divisor. In our applications, we will have occasion to define the blowup of an *almost* complex manifold (M, J) at a point $z \in M$ under the assumption that J is integrable near z. This gives rise to an almost complex manifold $(\tilde{M}, \tilde{J})$ and a pseudoholomorphic blowdown map

$$\beta : (\tilde{M}, \tilde{J}) \to (M, J). \tag{3.2}$$

Similarly, one can blow down an almost complex manifold $(\tilde{M}, \tilde{J})$ along any exceptional divisor $E \subset \tilde{M}$ such that $\tilde{J}$ is integrable and can be identified with the standard complex structure of $\tilde{\mathbb{C}}^n$ on some neighborhood of E.

Notice that if $\tilde{u} : (\Sigma, j) \to (\tilde{M}, \tilde{J})$ is a $\tilde{J}$-holomorphic curve, then $u := \beta \circ \tilde{u} : (\Sigma, j) \to (M, J)$ is J-holomorphic, and conversely, removal of singularities (Theorem 2.36) defines a unique lift of any nonconstant J-holomorphic curve $u : (\Sigma, j) \to (M, J)$ to a $\tilde{J}$-holomorphic curve $\tilde{u} : (\Sigma, j) \to (\tilde{M}, \tilde{J})$ such that $u = \beta \circ \tilde{u}$.

Exercise 3.6. Given pseudoholomorphic curves $\tilde{u}$ in $(\tilde{M}, \tilde{J})$ and $u = \beta \circ \tilde{u}$ in its almost complex blowdown (M, J) along an exceptional divisor $E \subset \tilde{M}$ in the sense of Remark 3.5, show that if $\tilde{u}$ is immersed and transverse to E, then u is also immersed. If additionally $\dim_{\mathbb{R}} M = 4$, show that the normal bundles N_u and $N_{\tilde{u}}$ are related by

$$c_1(N_u) = c_1(N_{\tilde{u}}) + [\tilde{u}] \cdot [E],$$

and that u is embedded and passes through $\beta(E)$ if and only if $\tilde{u}$ is embedded with $[\tilde{u}] \cdot [E] = 1$, in which case

$$[u] \cdot [u] = [\tilde{u}] \cdot [\tilde{u}] + 1.$$

3.2   The Symplectic Blowup

The starting point of the symplectic blowup construction is the definition of a
suitable symplectic form on the total space of the tautological line bundle $\pi : \widetilde{\mathbb{C}}^n \to \mathbb{CP}^{n-1}$. Using the fact that π and the complex blowdown map $\beta : \widetilde{\mathbb{C}}^n \to \mathbb{C}^n$ are
both holomorphic, it is not hard to show that for any constant $R > 0$,

$$\omega_R := \beta^* \omega_{\text{st}} + R^2 \pi^* \omega_{\text{FS}}$$

is a Kähler form on $\widetilde{\mathbb{C}}^n$, meaning it is symplectic and compatible with the natural
complex structure. This depends on the fact that ω_{st} and ω_{FS} are also Kähler forms
on $\mathbb{C}^n$ and $\mathbb{CP}^{n-1}$ respectively. The zero-section is now a symplectic submanifold
of $(\widetilde{\mathbb{C}}^n, \omega_R)$ endowed with a canonical symplectomorphism to $(\mathbb{CP}^{n-1}, R^2 \omega_{\text{FS}})$. In
the following, for each $r > 0$ we shall denote by $B_r^{2n} \subset \mathbb{C}^n$ the open ball of radius
r about the origin, and denote the corresponding neighborhood of $\mathbb{CP}^{n-1}$ in $\widetilde{\mathbb{C}}^n$ by

$$\widetilde{B}_r^{2n} := \beta^{-1}(B_r^{2n}) \subset \widetilde{\mathbb{C}}^n.$$

It turns out that neighborhoods of $\mathbb{CP}^{n-1}$ in $(\widetilde{\mathbb{C}}^n, \omega_R)$ with the zero-section
removed are naturally symplectomorphic to annular regions in $(\mathbb{C}^n, \omega_{\text{st}})$. To see
this, let us use the natural identification $\widetilde{\mathbb{C}}^n \backslash \mathbb{CP}^{n-1} = \mathbb{C}^n \backslash \{0\}$ and write down ω_R
in "cylindrical coordinates" on the latter, i.e. using the diffeomorphism

$$\Upsilon : \mathbb{R} \times S^{2n-1} \to \mathbb{C}^n \backslash \{0\} : (t, z) \mapsto e^{t/2} z. \tag{3.3}$$

This map is chosen to have the convenient property that $V_{\text{st}} := \partial_t \Upsilon$ is the so-called
standard Liouville vector field on $\mathbb{C}^n \backslash \{0\}$, which in coordinates $(z_1, \ldots, z_n) = (p_1 + iq_1, \ldots, p_n + iq_n)$ takes the form

$$V_{\text{st}} = \frac{1}{2} \sum_{j=1}^{n} \left(p_j \frac{\partial}{\partial p_j} + q_j \frac{\partial}{\partial q_j} \right).$$

Being a Liouville vector field means that it satisfies $\mathcal{L}_{V_{\text{st}}} \omega_{\text{st}} = \omega_{\text{st}}$, so by Cartan's
formula, the corresponding **Liouville form**

$$\lambda_{\text{st}} := \omega_{\text{st}}(V_{\text{st}}, \cdot) = \frac{1}{2} \sum_{j=1}^{n} (p_j \, dq_j - q_j \, dp_j)$$

satisfies $d\lambda_{\text{st}} = \omega_{\text{st}}$ and

$$\mathcal{L}_{V_{\text{st}}} \lambda_{\text{st}} = d\left(\iota_{V_{\text{st}}} \lambda_{\text{st}}\right) + \iota_{V_{\text{st}}} d\lambda_{\text{st}} = d\left(\omega_{\text{st}}(V_{\text{st}}, V_{\text{st}})\right) + \omega_{\text{st}}(V_{\text{st}}, \cdot) = \lambda_{\text{st}}.$$

The time t flow of V_{st} thus dilates λ_{st} by a factor of e^t, and we therefore have $\Upsilon^*\lambda_{st} = e^t\alpha_{st}$, where α_{st} is defined on S^{2n-1} by restriction to the unit sphere,

$$\alpha_{st} = \lambda_{st}|_{TS^{2n-1}},$$

and consequently

$$\Upsilon^*\omega_{st} = d(e^t\alpha_{st}).$$

Using the characterization of the Fubini-Study form in (1.2), we also have $\Upsilon^*(\pi^*\omega_{FS}) = d\alpha_{st}$, hence

$$\Upsilon^*\omega_R = d(e^t\alpha_{st}) + R^2\, d\alpha_{st} = d\left((e^t + R^2)\alpha_{st}\right). \tag{3.4}$$

Notice that $(e^t + R^2)\alpha_{st} = F^*(e^t\alpha_{st})$ if we define the embedding $F : \mathbb{R} \times S^{2n-1} \hookrightarrow \mathbb{R} \times S^{2n-1}$ by $F(t, z) := (\log(e^t + R^2), z)$, thus $\Upsilon^*\omega_R = F^*d(e^t\alpha_{st})$. This leads us to write down the symplectomorphism

$$\Phi_R := \Upsilon \circ F \circ \Upsilon^{-1} : \left(\widetilde{B}_r^{2n}\backslash\mathbb{CP}^{n-1}, \omega_R\right) \to \left(B^{2n}_{\sqrt{R^2+r^2}}\backslash\overline{B}_R^{2n}, \omega_{st}\right) \tag{3.5}$$

$$z \mapsto \sqrt{|z|^2 + R^2}\,\frac{z}{|z|},$$

where on the left hand side we are implicity identifying $\widetilde{B}_r^{2n}\backslash\mathbb{CP}^{n-1}$ with $B_r^{2n}\backslash\{0\} \subset \mathbb{C}^n$ via the blowdown map.

Definition 3.7. Given a symplectic manifold (M, ω) of dimension $2n \geqslant 4$ and a symplectic embedding $\psi : (B^{2n}_{R+\epsilon}, \omega_{st}) \hookrightarrow (M, \omega)$ for some constants $R, \epsilon > 0$, the **symplectic blowup of** (M, ω) **with weight** R **along** ψ is defined by deleting the image of $\overline{B}_R^{2n}$ from M and gluing in a sufficiently small neighborhood of $\mathbb{CP}^{n-1}$ in $(\widetilde{\mathbb{C}}^n, \omega_R)$ via the symplectic map Φ_R in (3.5), that is,

$$(\widetilde{M}, \widetilde{\omega}) := \left(M\backslash\psi(\overline{B}_R^{2n}), \omega\right) \cup_{\psi\circ\Phi_R} \left(\widetilde{B}_\delta^{2n}, \omega_R\right),$$

with $\delta > 0$ chosen so that $\sqrt{R^2 + \delta^2} \leqslant R+\epsilon$. The resulting symplectic submanifold $(\mathbb{CP}^{n-1}, R^2\omega_{FS}) \subset (\widetilde{B}_\delta^{2n}, \omega_R)$ in $(\widetilde{M}, \widetilde{\omega})$ is called the **exceptional divisor**.

It is clear that the choice of $\delta > 0$ in this definition does not matter, so long as it's small enough to fit $\Phi_R(\widetilde{B}_\delta^{2n})$ inside $B^{2n}_{\sqrt{R^2+\epsilon^2}}\backslash\overline{B}_R^{2n}$. The choice of $\epsilon > 0$ also does not especially matter, in the sense that if $\psi : (B^{2n}_{R+\epsilon}, \omega_{st}) \hookrightarrow (M, \omega)$ is given, then we are free to shrink ϵ and restrict ψ to a smaller neighborhood of $\overline{B}_R^{2n}$; any two blowups produced in this way by alternative choices of $\epsilon > 0$ will be naturally symplectomorphic. The diffeomorphism type of $\widetilde{M}$ is also independent of the weight

$R > 0$, though the symplectic structure $\widetilde{\omega}$ depends on this parameter in essential ways, e.g. it determines the cohomology class $[\widetilde{\omega}] \in H^2_{\mathrm{dR}}(\widetilde{M})$, in particular its evaluation on the exceptional divisor. The dependence of $(\widetilde{M}, \widetilde{\omega})$ on the embedding ψ is a slightly subtle issue, mostly because basic questions about the topology of the space of symplectic embeddings $(B^{2n}_{R+\epsilon}, \omega_{\mathrm{st}}) \hookrightarrow (M, \omega)$ are typically difficult to answer—nonetheless it is straightforward to show that smooth deformations of such embeddings yield symplectic deformation equivalences for the resulting blowups. The second part of the following statement is based on this, in combination with the Moser stability theorem.

Theorem 3.8. *Suppose* $\psi_\tau : (B^{2n}_{R_\tau + \epsilon}, \omega_{\mathrm{st}}) \hookrightarrow (M, \omega)$ *for* $\tau \in [0, 1]$ *is a smooth 1-parameter family of symplectic embeddings of standard Darboux balls of varying radii* $R_\tau + \epsilon$, *and* $(\widetilde{M}_\tau, \widetilde{\omega}_\tau)$ *denotes the symplectic blowup of* (M, ω) *with weight* $R_\tau > 0$ *along* ψ_τ. *Then for each* $\tau \in [0, 1]$ *there exists a diffeomorphism* $\varphi_\tau : \widetilde{M}_0 \to \widetilde{M}_\tau$ *such that the family of symplectic forms* $\varphi_\tau^* \widetilde{\omega}_\tau$ *on* $\widetilde{M}_0$ *depends smoothly on* τ; *in particular, all of the* $(\widetilde{M}_\tau, \widetilde{\omega}_\tau)$ *are symplectically deformation equivalent. Moreover, if the weights* $R_\tau > 0$ *are constant in* τ, *then one can arrange for each* φ_τ *to be a symplectomorphism* $(\widetilde{M}_0, \widetilde{\omega}_0) \to (\widetilde{M}_\tau, \widetilde{\omega}_\tau)$. $\qquad\square$

Observe that for any given $R > 0$, the existence of a symplectic embedding $(B^{2n}_{R+\epsilon}, \omega_{\mathrm{st}}) \hookrightarrow (M, \omega)$ for some $\epsilon > 0$ is a nontrivial condition, though of course it is always satisfied for R sufficiently small. It is not hard to show that $\epsilon > 0$ plays no role in this condition, i.e. any symplectic embedding of the closed ball $(\overline{B}^{2n}_R, \omega_{\mathrm{st}}) \hookrightarrow (M, \omega)$ can be extended symplectically over a slightly larger ball $B^{2n}_{R+\epsilon}$. Similarly, while it is difficult in general to tell whether two embeddings $(\overline{B}^{2n}_R, \omega_{\mathrm{st}}) \hookrightarrow (M, \omega)$ are symplectically isotopic, any symplectic isotopy of such embeddings can be extended over slightly larger balls for the sake of applying Theorem 3.8. It turns out that allowing the weight R to vary simplifies matters considerably: if M is connected, then any two symplectic embeddings $\psi_i : (\overline{B}^{2n}_{R_i}, \omega_{\mathrm{st}}) \hookrightarrow (M, \omega)$ for $i = 0, 1$ can be related by a smooth family $\psi_\tau : (\overline{B}^{2n}_{R_\tau}, \omega_{\mathrm{st}}) \hookrightarrow (M, \omega)$ for $\tau \in [0, 1]$ with varying (possibly very small) weights $R_\tau > 0$. This surprisingly simple fact is based on the observation that, since every symplectic embedding $\psi : (\overline{B}^{2n}_R, \omega_{\mathrm{st}}) \hookrightarrow (\mathbb{R}^{2n}, \omega_{\mathrm{st}})$ fixing the origin is symplectically isotopic via $\psi_\tau(z) := \frac{1}{\tau} \psi(\tau x)$ to its linearization at 0, the space of symplectic embeddings $(B^{2n}_R, \omega_{\mathrm{st}}) \hookrightarrow (\mathbb{R}^{2n}, \omega_{\mathrm{st}})$ is connected, cf. [MS17, Exercise 7.1.27]. We conclude:

Corollary 3.9. *For any connected symplectic manifold* (M, ω), *the symplectic blowup* $(\widetilde{M}, \widetilde{\omega})$ *of* (M, ω) *is independent of all choices up to symplectic deformation equivalence.* $\qquad\square$

To discuss the blowdown of a $2n$-dimensional symplectic manifold $(\widetilde{M}, \widetilde{\omega})$, we call a symplectic submanifold E in $(\widetilde{M}, \widetilde{\omega})$ an **exceptional divisor** if it is symplectomorphic to $(\mathbb{CP}^{n-1}, R^2 \omega_{\mathrm{FS}})$ for some $R > 0$ and its symplectic normal bundle has first Chern class equal to minus the canonical generator of $H^2(\mathbb{CP}^{n-1})$. In dimension four, this reduces to the condition that E is a symplectically embedded sphere with

$$[E] \cdot [E] = -1,$$

and we call it an **exceptional sphere**. The symplectic neighborhood theorem [MS17, §3.4] then identifies a neighborhood $\widetilde{\mathcal{U}} \subset \widetilde{M}$ of E symplectically with $(\widetilde{B}_\delta^{2n}, \omega_R)$ for sufficiently small $\delta > 0$, where the constant $R > 0$ is uniquely determined by the symplectic volume of E. Using the map Φ_R in (3.5), $(\widetilde{\mathcal{U}} \backslash E, \widetilde{\omega})$ is thus symplectomorphic to the annular region $(B_{\sqrt{R^2 + \delta^2}}^{2n} \backslash \overline{B}_R^{2n}, \omega_{\mathrm{st}})$.

Definition 3.10. Given a symplectic manifold $(\widetilde{M}, \widetilde{\omega})$ of dimension $2n \geqslant 4$ and a symplectic embedding $\psi : (\mathbb{CP}^{n-1}, R^2\omega_{\mathrm{FS}}) \hookrightarrow (\widetilde{M}, \widetilde{\omega})$ whose image is an exceptional divisor E, the **symplectic blowdown of** $(\widetilde{M}, \widetilde{\omega})$ **along** E is defined by deleting E and gluing in a standard symplectic ball $(\overline{B}_R^{2n}, \omega_{\mathrm{st}})$. More precisely, we choose an extension of $\psi : \mathbb{CP}^{n-1} \to E$ to a symplectic embedding $\Psi : (\widetilde{B}_\delta^{2n}, \omega_R) \hookrightarrow (\widetilde{M}, \widetilde{\omega})$ and define

$$(M, \omega) := \left(\widetilde{M} \backslash E, \widetilde{\omega} \right) \cup_{\Psi \circ \Phi_R^{-1}} \left(B_{R+\epsilon}^{2n}, \omega_{\mathrm{st}} \right),$$

where Φ_R^{-1} is restricted to $B_{R+\epsilon}^{2n} \backslash \overline{B}_R^{2n}$ and $\epsilon > 0$ is chosen so that $R + \epsilon \leqslant \sqrt{R^2 + \delta^2}$.

Theorem 3.11. *Up to symplectomorphism, the blowdown* (M, ω) *in Definition 3.10 depends only on the symplectic isotopy class of the embedding* $\psi : (\mathbb{CP}^{n-1}, R^2\omega_{\mathrm{FS}}) \hookrightarrow (\widetilde{M}, \widetilde{\omega})$ *that parametrizes the exceptional divisor* E.

Proof. If $\psi_\tau : (\mathbb{CP}^n, R^2\omega_{\mathrm{FS}}) \hookrightarrow (\widetilde{M}, \widetilde{\omega})$ is a smooth 1-parameter family of symplectic embeddings for $\tau \in [0, 1]$ whose images are exceptional divisors, then we can choose a smooth family of extensions $\Psi_\tau : (\widetilde{B}_\delta^{2n}, \omega_R) \hookrightarrow (\widetilde{M}, \widetilde{\omega})$ for $\delta > 0$ sufficiently small. This follows by a parametric version of the symplectic neighborhood theorem; the proof is the same as in the usual version, using the Moser deformation trick (cf. [MS17, Theorem 3.4.10]), one only needs to keep track of the extra parameter. With this understood, the embeddings Ψ_τ can be used to construct a 1-parameter family of blowdowns (M_τ, ω_τ) via Definition 3.10 above, and one can construct a family of diffeomorphisms $\varphi_\tau : M_0 \to M_\tau$ such that the symplectic forms $\varphi_\tau^* \omega_\tau$ on M_0 depend smoothly on τ. Moreover, these symplectic forms will be cohomologous, so the Moser stability theorem can be applied to change φ_τ into a family of symplectomorphisms $(M_0, \omega_0) \to (M_\tau, \omega_\tau)$.

It therefore remains only to check that the symplectomorphism type of (M, ω) in Definition 3.10 does not depend on any of the auxiliary choices, namely on the extension $\Psi : (\widetilde{B}_\delta^{2n}, \omega_R) \hookrightarrow (\widetilde{M}, \widetilde{\omega})$ of $\psi : (\mathbb{CP}^n, R^2\omega_{\mathrm{FS}}) \hookrightarrow (\widetilde{M}, \widetilde{\omega})$, and on $\epsilon > 0$. Given Ψ, it is easy to see that δ and ϵ can each be shrunk without changing the construction, i.e. given two blowdowns produced by alternative choices of constants with ϵ small enough so that $\Phi_R^{-1}(B_{R+\epsilon}^{2n} \backslash \overline{B}_R^{2n})$ fits inside $\widetilde{B}_\delta^{2n}$, there exists a natural symplectomorphism between them.

To see why there is also no dependence on the extension of ψ, suppose $\Psi_i :$ $(\widetilde{B}^{2n}_\delta, \omega_R) \hookrightarrow (\widetilde{M}, \widetilde{\omega})$ for $i = 0, 1$ are two choices of such extensions, and let (M_i, ω_i) denote the resulting blowdowns. Differentiating each Ψ_i at the zero-section gives a pair of symplectic bundle isomorphisms from $\widetilde{\mathbb{C}}^n$ to the normal bundle of E which we shall denote by $D\Psi_i$. Since the linear symplectic group on $\mathbb{R}^2$ is retractible to $U(1) \cong S^1$, the space of such bundle isomorphisms is retractible to the space of smooth maps $\mathbb{CP}^{n-1} \to S^1$, which is connected since $\pi_1(\mathbb{CP}^{n-1}) = 0$ implies that maps $\mathbb{CP}^{n-1} \to S^1$ always admit lifts $\mathbb{CP}^{n-1} \to \mathbb{R}$ to the universal cover of S^1. We can therefore pick a smooth homotopy from $D\Psi_0$ to $D\Psi_1$ and, using the symplectic neighborhood theorem (after possibly shrinking $\delta > 0$), accompany this with a smooth family of symplectic embeddings $\widetilde{\Psi}_\tau : (\widetilde{B}^{2n}_\delta, \omega_R) \hookrightarrow (\widetilde{M}, \widetilde{\omega})$ for $\tau \in [0, 1]$ such that $\widetilde{\Psi}_0 = \Psi_0$ and the derivative of $\widetilde{\Psi}_1$ at the zero-section matches $D\Psi_1$. In this case the argument of the previous paragraph shows that the blowdowns constructed via each $\widetilde{\Psi}_\tau$ are all symplectomorphic.

The above discussion allows us now to assume without loss of generality that $D\Psi_0 = D\Psi_1$, in which case, after shrinking $\delta > 0$ further, Ψ_0 and Ψ_1 may be assumed to be arbitrarily C^1-close. For $\epsilon \in (0, \delta]$ and $i = 0, 1$, denote

$$\mathcal{U}^\epsilon_i := \Psi_i(\widetilde{B}^{2n}_\epsilon) \subset \widetilde{M}.$$

Choosing a vector field whose flow gives $\Psi_1 \circ \Psi_0^{-1}$ on its domain of definition and then multiplying it by a suitable cutoff function, we can now find a diffeomorphism $\widetilde{\varphi} : \widetilde{M} \to \widetilde{M}$ that is globally C^1-close to the identity, has compact support in $\mathcal{U}^\delta_0$, and matches $\Psi_1 \circ \Psi_0^{-1}$ on $\mathcal{U}^{\delta'}_0$ for some $\delta' \in (0, \delta)$. This in turn gives rise to a diffeomorphism

$$\varphi : M_0 \to M_1$$

which is the identity (hence symplectic) on the glued in ball $\overline{B}^{2n}_R$ and matches $\widetilde{\varphi}$ on $M_0 \backslash \overline{B}^{2n}_R = \widetilde{M}_0 \backslash E$. The symplectic form $\varphi^* \omega_1$ on M_0 is then C^0-close to ω_0 and matches it precisely outside the region $\mathcal{U}^\delta_0 \backslash \overline{\mathcal{U}}^{\delta'}_0$. The latter has the homotopy type of S^{2n-1}, thus $\omega_0 - \varphi^* \omega_1$ is exact and we can linearly interpolate between ω_0 and $\varphi^* \omega_1$ by cohomologous symplectic forms, producing a symplectomorphism $(M_0, \omega_0) \to (M_1, \omega_1)$ via the Moser stability theorem.   □

Definition 3.12. As in Chap. 1, we will say more generally that (M_1, ω_1) is a **blowup** or **blowdown** of (M_0, ω_0) whenever the former can be produced from the latter by a finite sequence of symplectic blowup or blowdown operations respectively.

One disadvantage of the symplectic blowdown in comparison to its complex counterpart is that there is no natural notion of a "blowdown map" $\beta : (\widetilde{M}, \widetilde{\omega}) \to (M, \omega)$, cf. (3.2). When dealing with pseudoholomorphic curves, in particular, this makes it more convenient to apply the almost complex blowup and blowdown operations described in Remark 3.5. The next two results give us a means of interpreting these operations symplectically.

Theorem 3.13. *Suppose* (M, ω) *is a symplectic manifold of dimension* $2n \geqslant 4$, $\psi : (B^{2n}_{R+\epsilon}, \omega_{\mathrm{st}}) \hookrightarrow (M, \omega)$ *is a symplectic embedding of a standard symplectic ball for some* $R, \epsilon > 0$, *and* J *denotes the integrable complex structure* $\psi_* i$ *on* $\mathcal{U} := \psi(B^{2n}_{R+\epsilon}) \subset M$. *Let*

$$\widetilde{M} := (M \backslash \{z\}) \cup \mathbb{P}(T_z M)$$

denote the complex blowup of M *at* $z := \psi(0)$ *with respect to the complex structure* J, *which inherits a blowdown map* $\beta : \widetilde{M} \to M$ *and an integrable complex structure* $\widetilde{J}$ *on the region* $\widetilde{\mathcal{U}} := \beta^{-1}(\mathcal{U}) \cong \widetilde{B}^{2n}_{R+\epsilon}$. *Then* $\widetilde{M}$ *admits a symplectic form* $\widetilde{\omega}$ *that matches* ω *on* $\widetilde{M} \backslash \widetilde{\mathcal{U}} = M \backslash \mathcal{U}$ *and is compatible with* $\widetilde{J}$ *on* $\widetilde{\mathcal{U}}$, *such that* $(\widetilde{M}, \widetilde{\omega})$ *is symplectomorphic to the symplectic blowup of* (M, ω) *along* ψ *with weight* R.

Theorem 3.14. *Suppose* $(\widetilde{M}, \widetilde{\omega})$ *is a symplectic manifold of dimension* $2n \geqslant 4$, $E \subset \widetilde{M}$ *is an exceptional divisor,* $\Psi : (\widetilde{B}^{2n}_{\delta}, \omega_R) \hookrightarrow (\widetilde{M}, \widetilde{\omega})$ *is a symplectic embedding for some* $\delta > 0$ *and* $R > 0$ *with* $\Psi(\mathbb{CP}^{n-1}) = E$, *and* $\widetilde{J}$ *denotes the integrable complex structure* $\Psi_* i$ *on* $\widetilde{\mathcal{U}} := \Psi(\widetilde{B}^{2n}_{\delta}) \subset \widetilde{M}$. *Let*

$$M := \left(\widetilde{M} \backslash E \right) \cup \{z\}$$

denote the complex blowdown of $\widetilde{M}$ *along* E *with respect to the complex structure* $\widetilde{J}$, *which inherits a blowdown map* $\beta : \widetilde{M} \to M$ *and an integrable complex structure* J *on the region* $\mathcal{U} := \beta(\widetilde{\mathcal{U}}) \cong B^{2n}_{\delta}$. *Then* M *admits a symplectic form* ω *that matches* $\widetilde{\omega}$ *on* $M \backslash \mathcal{U} = \widetilde{M} \backslash \widetilde{\mathcal{U}}$ *and is compatible with* J *on* $\mathcal{U}$, *such that* (M, ω) *is symplectomorphic to the symplectic blowdown of* $(\widetilde{M}, \widetilde{\omega})$ *along* E.

We will prove both of these theorems by working in the "cylindrical coordinates" defined via the diffeomorphism $\Upsilon : \mathbb{R} \times S^{2n-1} \to \mathbb{C}^n \backslash \{0\} = \widetilde{\mathbb{C}}^n \backslash \mathbb{CP}^{n-1}$ in (3.3). The next two exercises provide us with a fairly general family of symplectic structures that are compatible with the standard complex structure on $\mathbb{R} \times S^{2n-1}$. We will see them again when we discuss the energy of punctured pseudoholomorphic curves in Sect. 8.3.1.

Exercise 3.15. Show that the complex structure $J := \Upsilon^* i$ on $\mathbb{R} \times S^{2n-1}$ has the following properties:

- It is invariant under the natural $\mathbb{R}$-action by translations of the first factor;
- It preserves the subbundle $\xi_{\mathrm{st}} := \ker \alpha_{\mathrm{st}} \subset TS^{2n-1}$;
- The pairing $(X, Y) \mapsto d\alpha_{\mathrm{st}}(X, JY)$ for $X, Y \in \xi_{\mathrm{st}}$ defines a bundle metric on ξ_{st};
- $J \partial_t$ is always tangent to the levels $\{t\} \times S^{2n-1}$ and defines a vector field on S^{2n-1} such that $d\alpha_{\mathrm{st}}(J \partial_t, \cdot) \equiv 0$ and $\alpha_{\mathrm{st}}(J \partial_t)$ is a positive constant.

Exercise 3.16. Using only the properties of $J = \Upsilon^* i$ on $\mathbb{R} \times S^{2n-1}$ established in Exercise 3.15, show that for any smooth function $f : \mathbb{R} \to \mathbb{R}$ with $f' > 0$ everywhere,

$$\omega_f := d\left(e^{f(t)} \alpha_{\mathrm{st}}\right) \tag{3.6}$$

defines a Kähler form on $(\mathbb{R} \times S^{2n-1}, J)$.

Proofs of Theorems 3.13 and 3.14. As in the statement of Theorem 3.13, denote $\mathcal{U} = \psi(B_{R+\epsilon}^{2n}) \subset M$, $z = \psi(0)$, $J = \psi_* i$, $\widetilde{\mathcal{U}} = \beta^{-1}(\mathcal{U}) \subset \widetilde{M}$, and $\widetilde{J} = \beta^* J$. Let us use $\Upsilon : \mathbb{R} \times S^{2n-1} \to \mathbb{C}^n \backslash \{0\} = \widetilde{\mathbb{C}}^n \backslash \mathbb{C}\mathbb{P}^{n-1}$ to identify both $\mathcal{U} \backslash \{z\} \subset M$ and $\widetilde{\mathcal{U}} \backslash \mathbb{P}(T_z M) \subset \widetilde{M}$ with the cylinder $(-\infty, T) \times S^{2n-1}$, where $T \in \mathbb{R}$ is defined to satisfy $e^{T/2} = R + \epsilon$. Under this identification, we have

$$\omega_{\mathrm{st}} = d(e^t \alpha_{\mathrm{st}}) \quad \text{and} \quad \omega_R = d\left((e^t + R^2)\alpha_{\mathrm{st}}\right). \tag{3.7}$$

Since $e^T > R^2$, we can choose a smooth function $f : (-\infty, T) \to \mathbb{R}$ with $f' > 0$ such that $f(t) = t$ near $t = T$ and $e^{f(t)} = e^t + R^2$ near $t = -\infty$, then set

$$\widetilde{\omega} = d(e^f \alpha_{\mathrm{st}}) \quad \text{on} \quad (-\infty, T) \times S^{2n-1} = \widetilde{\mathcal{U}} \backslash \mathbb{P}(T_z M) \subset \widetilde{M}.$$

By construction, $\widetilde{\omega}$ extends to a smooth symplectic form matching ω on $\widetilde{M} \backslash \widetilde{\mathcal{U}} = M \backslash \mathcal{U}$, and it also extends smoothly over $\mathbb{P}(T_z M)$ since it matches ω_R nearby. Exercise 3.16 implies that $\widetilde{\omega}$ is compatible with $\widetilde{J}$ on $\widetilde{\mathcal{U}}$. To see that $(\widetilde{M}, \widetilde{\omega})$ is symplectomorphic to a symplectic blowup of (M, ω), we observe that $(\widetilde{\mathcal{U}}, \widetilde{\omega})$ is symplectomorphic to $(\widetilde{B}_\delta^{2n}, \omega_R)$ where $\delta := \sqrt{(R + \epsilon)^2 - R^2}$; indeed, a natural symplectomorphism from the former to the latter is defined by identifying both with suitable subsets of $\mathbb{R} \times S^{2n-1}$ and writing a diffeomorphism of the form

$$G : (-\infty, T) \times S^{2n-1} \to (-\infty, T') \times S^{2n-1} : (t, z) \mapsto (g(t), z),$$

where $g : (-\infty, T) \to \mathbb{R}$ is chosen such that $e^g = e^f - R^2$, hence $G^*((e^t + R^2)\alpha_{\mathrm{st}}) = e^f \alpha_{\mathrm{st}}$. We can therefore view $(\widetilde{M}, \widetilde{\omega})$ as obtained from (M, ω) by replacing a copy of $(B_{R+\epsilon}^{2n}, \omega_{\mathrm{st}})$ with a copy of $(\widetilde{B}_\delta^{2n}, \omega_R)$, where these are identified with each other by the natural symplectomorphism near their boundaries, thus matching the definition of the symplectic blowup along ψ with weight R. This proves Theorem 3.13.

Theorem 3.14 is proved by a similar trick: given $\widetilde{\mathcal{U}} \subset \widetilde{M}$, $\widetilde{J}$, $\mathcal{U} \subset M$ and J as in the statement, we use Υ to identify both $\widetilde{\mathcal{U}} \backslash E$ and $\mathcal{U} \backslash \{z\}$ with $(-\infty, T) \times S^{2n-1}$, where $e^{T/2} = \delta$, and again ω_R and ω_{st} are given by the formulas in (3.7). Choose a smooth function $f : (-\infty, T) \to \mathbb{R}$ with $f' > 0$ such that $e^{f(t)} = e^t + R^2$ near $t = T$ and $f(t) = t$ near $t = -\infty$, then set

$$\omega = d(e^f \alpha_{\mathrm{st}}) \quad \text{on} \quad (-\infty, T) \times S^{2n-1} = \mathcal{U} \backslash \{z\} \subset M.$$

Analogous arguments as in the previous paragraph show that this extends smoothly to a symplectic form on M satisfying all of the desired properties. □

Remark 3.17. We will not need it, but with a little more care, one can tweak the proof of Theorem 3.14 to produce a blowdown containing a Darboux ball of radius greater than R on which the complex structure is also standard. This is achieved via a more precise reversal of the proof of Theorem 3.13.

3.3 Smooth Topology of Lefschetz Pencils and Fibrations

Lefschetz pencils and Lefschetz fibrations are each "fibration-like" objects that facilitate the description of $2n$-dimensional manifolds in terms of $(2n - 2)$-dimensional data. This is especially useful in the study of symplectic 4-manifolds since the classification of symplectic structures in dimension two is trivial. We already saw two examples in the introduction:

- Any smooth fibration $\pi : M \to \Sigma$ of an oriented 4-manifold over an oriented surface is also a Lefschetz fibration (with no singular fibers).
- The holomorphic spheres in $\mathbb{CP}^2$ described in Example 1.4 form the fibers of a Lefschetz pencil with one base point (and no singular fibers).

Lefschetz fibrations generalize the first example by allowing a finite set of fibers to be singular, where the degeneration from a smooth fiber to a singular one can be likened to the convergence of smooth J-holomorphic curves to a nodal curve in the Gromov compactification (see Sect. 2.1.6). Singular fibers will thus provide a convenient topological description of the degenerations of holomorphic curves that occur in our proofs of Gromov's and McDuff's results stated in Chap. 1. It is not strictly necessary to understand Lefschetz fibrations in order to prove most of those results, but we will find that they provide valuable intuition. Pencil singularities (see Definition 3.24 below) will also play a crucial role in our arguments for the case $[S] \cdot [S] > 0$.

Definition 3.18. Suppose M and Σ are closed, connected, oriented, smooth manifolds of dimensions 4 and 2 respectively. A **Lefschetz fibration** of M over Σ is a smooth map

$$\pi : M \to \Sigma$$

with finitely many critical points $M_{\text{crit}} := \text{Crit}(\pi) \subset M$ and critical values $\Sigma_{\text{crit}} := \pi(M_{\text{crit}}) \subset \Sigma$ such that near each point $p \in M_{\text{crit}}$, there exists a complex coordinate chart (z_1, z_2) and a corresponding complex coordinate z on a neighborhood of $\pi(p) \in \Sigma$ in which π locally takes the form

$$\pi(z_1, z_2) = z_1^2 + z_2^2. \tag{3.8}$$

Remark 3.19. In the above definition, and also in Definition 3.24 below, we always assume that the orientation induced by any choice of local complex coordinates on M or Σ matches the given orientation. The more general object obtained by dropping this orientation condition is called an *achiral* Lefschetz fibration.

Remark 3.20. One can think of critical points in a Lefschetz fibration as satisfying a complex Morse condition: the coordinates near M_{crit} and Σ_{crit} determine integrable complex structures on these neighborhoods such that π becomes a holomorphic map near M_{crit} and its critical points are nondegenerate. The proof of the standard Morse lemma (see [Mil63]) can be adapted to show that any nondegenerate critical point of a holomorphic map $\mathbb{C}^2 \to \mathbb{C}$ looks like (3.8) in some choice of holomorphic coordinates.

Exercise 3.21. Show that near any Lefschetz critical point, one can also choose complex coordinates such that π takes the form $\pi(z_1, z_2) = z_1 z_2$.

We shall denote the fibers of a Lefschetz fibration $\pi : M \to \Sigma$ by

$$M_z := \pi^{-1}(z) \subset M$$

for $z \in \Sigma$. The **regular fibers** M_z for $z \in \Sigma \backslash \Sigma_{\mathrm{crit}}$ are closed oriented surfaces, which we shall always assume to be connected (see Exercise 3.22 below). For $z \in \Sigma_{\mathrm{crit}}$, the **singular fiber** M_z is generally an immersed closed oriented surface with transverse self-intersections; it may in fact be a union of several **irreducible components** that have transverse double points and intersect each other transversely, and its *arithmetic genus* (defined the same as for nodal holomorphic curves, cf. Sect. 2.1.6) matches the genus of the regular fibers. Figure 3.1 shows some examples of what singular fibers can look like and how they relate to nearby regular fibers. It follows from the orientation condition (Remark 3.19 above) that the self-intersections of a Lefschetz singular fiber are always not only transverse but also positive.

Exercise 3.22. Assuming M and Σ are closed and connected, show that if $\pi : M \to \Sigma$ is a Lefschetz fibration with disconnected fibers, then one can write $\pi = \varphi \circ \pi'$ where $\varphi : \Sigma' \to \Sigma$ is a finite covering map of degree at least 2 and $\pi' : M \to \Sigma'$ is a Lefschetz fibration with connected fibers.

Example 3.23. To make the relationship between Lefschetz singular fibers and nodal holomorphic curves more explicit, let us write down a family of parametrizations of the fibers near a critical point. For this purpose it is convenient to use the model from Exercise 3.21, so assume $\pi : \mathbb{C}^2 \to \mathbb{C}$ is the map $\pi(z_1, z_2) = z_1 z_2$, and let $\overline{B}_\epsilon^4 \subset \mathbb{C}^2$ denote the closed ϵ-ball about the origin for some $\epsilon > 0$. The portion of the singular fiber $\pi^{-1}(0)$ in $\overline{B}_\epsilon^4$ is then the union of the transversely intersecting images of two embedded holomorphic disks

$$u_0^\pm : \mathbb{D}^2 \to \mathbb{C}^2, \qquad u_0^+(z) = (\epsilon z, 0) \quad \text{and} \quad u_0^-(z) = (0, \epsilon z).$$

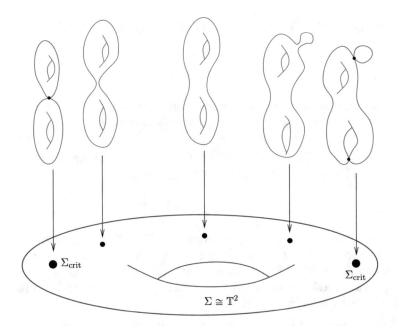

Fig. 3.1 A Lefschetz fibration over the torus, with fibers of genus 2. The singular fibers shown in this example each have two irreducible components

For any $w = re^{2\pi i\theta} \in \mathbb{C}\backslash\{0\}$ close enough to 0, $\pi^{-1}(w) \cap \overline{B}_\epsilon^4$ is the image of the embedded holomorphic annulus

$$u_w : [-R, R] \times S^1 \to \mathbb{C}^2 : (s, t) \mapsto \left(\sqrt{r}e^{2\pi(s+it)}, \sqrt{r}e^{2\pi i\theta}e^{-2\pi(s+it)}\right)$$

where

$$R := \frac{1}{4\pi} \cosh^{-1}\left(\frac{\epsilon^2}{2r}\right). \tag{3.9}$$

Fixing θ and letting $r \to 0$ in $w = re^{2\pi i\theta}$, the domains of these annuli expand toward infinite length and we have

$$\lim_{r\to 0} u_w(s + R, t) = u_0^+\left(e^{2\pi(s+it)}\right) \qquad \text{for } (s, t) \in (-\infty, 0] \times S^1,$$

$$\lim_{r\to 0} u_w(s - R, t) = u_0^-\left(e^{-2\pi(s+i(t-\theta))}\right) \qquad \text{for } (s, t) \in [0, \infty) \times S^1,$$

with C_{loc}^∞-convergence on the half-cylinder in each case. Moreover, for any sequence $w_k \in \mathbb{C}\backslash\{0\}$ with $r_k := |w_k| \to 0$ and R_k related to r_k via (3.9), together with a sequence $(s_k, t_k) \in [-R_k, R_k] \times S^1$ such that $s_k + R_k \to \infty$ and $R_k - s_k \to \infty$, we find

$$u_{w_k}(s_k, t_k) \to 0.$$

In particular, if we consider the compact topological annulus $\overline{Z}$ constructed by gluing $[0, \infty] \times S^1$ to $[-\infty, 0] \times S^1$ along the obvious homeomorphism $\{\infty\} \times S^1 \to \{-\infty\} \times S^1$, there is a continuous map $\bar{u}_\theta : \overline{Z} \to \mathbb{C}^2$ that matches $u_0^+(e^{2\pi(s+it)})$ on $[-\infty, 0] \times S^1$ and $u_0^-(e^{-2\pi(s+i(t-\theta))})$ on $[0, \infty] \times S^1$, and we can choose a smooth structure on $\overline{Z}$ and a smooth family of diffeomorphisms

$$\varphi_r : \overline{Z} \to [-R, R] \times S^1$$

such that for $w = re^{2\pi i\theta}$, $u_w \circ \varphi_r$ converges in $C^0(\overline{Z}, \mathbb{C}^2)$ to $\bar{u}_\theta$ as $r \to 0$. This is simply a local picture (near the node) of the notion of Gromov convergence that we described in Sect. 2.1.6: the annuli u_w are converging in the Gromov topology as $w \to 0$ to a "broken annulus" with smooth components u_0^+ and u_0^- that intersect each other transversely at a node.

A Lefschetz pencil is a further variation on the above which allows one more type of singularity. To motivate the definition, consider again the decomposition of $\mathbb{CP}^2$ in Example 1.4. The holomorphic spheres in that example are the fibers of the map

$$\pi : \mathbb{CP}^2\backslash\{[1:0:0]\} \to \mathbb{CP}^1 : [z_1 : z_2 : z_3] \mapsto [z_2 : z_3]. \qquad (3.10)$$

This defines a smooth fiber bundle structure on $\mathbb{CP}^2\backslash\{[1:0:0]\}$, but near the singular point $[1:0:0]$, where all the fibers intersect, we can choose a complex coordinate chart identifying $[1:z_1:z_2]$ with $(z_1, z_2) \in \mathbb{C}^2$ and write π in these coordinates as

$$\pi(z_1, z_2) = [z_1 : z_2].$$

This helps motivate the following notion.

Definition 3.24. Suppose M is a closed, connected, oriented, smooth manifold of dimension 4. A **Lefschetz pencil** on M is a Lefschetz fibration

$$\pi : M\backslash M_{\text{base}} \to \mathbb{CP}^1,$$

where $M_{\text{base}} \subset M$ is a finite subset, such that near each **base point** $p \in M_{\text{base}}$ there exists a complex coordinate chart (z_1, z_2) in which π locally takes the form

$$\pi(z_1, z_2) = [z_1 : z_2]. \qquad (3.11)$$

We shall regard the **fibers** of a Lefschetz pencil as the closures

$$M_z := \overline{\pi^{-1}(z)} \subset M$$

for $z \in \mathbb{CP}^1$. For $z \in \Sigma \backslash \Sigma_{\mathrm{crit}}$, these are embedded, closed, oriented surfaces that all contain M_{base} and intersect each other transversely (and positively) there. For $z \in \Sigma_{\mathrm{crit}}$, the singular fibers M_z have additional transverse (and positive) double points in $M \backslash M_{\mathrm{base}}$.

Exercise 3.25. Show that if M is closed and connected, then any Lefschetz pencil $\pi : M \backslash M_{\mathrm{base}} \to \mathbb{CP}^1$ with $M_{\mathrm{base}} \neq \varnothing$ has connected fibers.

Example 3.26. There is an obvious Lefschetz pencil on $S^2 \times S^2$, namely the trivial fibration over $S^2 = \mathbb{CP}^1$, but we can also define a more interesting Lefschetz pencil as follows. Regarding $S^2 = \mathbb{C} \cup \{\infty\}$ as the extended complex plane and identifying $\mathbb{CP}^1$ with S^2, we define

$$\pi : (S^2 \times S^2) \backslash \{(0,0),(\infty,\infty)\} \to S^2 : (z_1, z_2) \mapsto \frac{z_2}{z_1}. \tag{3.12}$$

Take a moment to convince yourself that this map can be written as (3.11) in suitable coordinates near the base points $(0,0)$ and (∞,∞). Moreover, the fibers $\pi^{-1}(z)$ for $z \in S^2 \backslash \{0,\infty\}$ are embedded spheres homologous to the diagonal, but there are two singular fibers

$$\pi^{-1}(0) = (S^2 \times \{0\}) \cup (\{\infty\} \times S^2)$$
$$\pi^{-1}(\infty) = (\{0\} \times S^2) \cup (S^2 \times \{\infty\}),$$

each with a single critical point at $(0,\infty)$ and $(\infty,0)$ respectively. A schematic picture is shown in Fig. 3.2.

Notation. For convenience, we shall wherever possible regard Lefschetz fibrations and Lefschetz pencils as special cases of the same type of object, which we'll denote by

$$\pi : M \backslash M_{\mathrm{base}} \to \Sigma.$$

Here it is to be understood that the finite subset $M_{\mathrm{base}} \subset M$ may in general be empty, but whenever it is nonempty, Σ must be $\mathbb{CP}^1$.

There is a lot that one could say about the topology of Lefschetz fibrations, most of which we do not have space for here (see e.g. [GS99] for much more on this topic). We will merely point out the features that figure into the main arguments on rational and ruled symplectic manifolds.

Fig. 3.2 A Lefschetz pencil on $S^2 \times S^2$ with two base points, two singular fibers, and genus 0 regular fibers isotopic to the diagonal

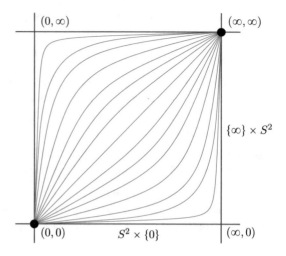

Proposition 3.27. *Suppose* $\pi : M \backslash M_{\text{base}} \to \mathbb{CP}^1$ *is a Lefschetz pencil with fibers diffeomorphic to* S^2, *having exactly one base point* $M_{\text{base}} = \{p\}$ *and no singular fibers. Then* M *admits a diffeomorphism to* $\mathbb{CP}^2$ *identifying* π *with* (3.10).

Proof. Let $\mathcal{U}_p \subset M$ denote an open neighborhood of the base point p that is identified with some standard ball $B_{\epsilon}^4 \subset \mathbb{C}^2$ in complex coordinates where π takes the form (3.11). Then the restriction of π to the sphere $\partial \overline{\mathcal{U}}_p \cong S^3$ is isomorphic to the Hopf fibration. This same fibration, with the orientation of its fibers reversed, is also the boundary of

$$\pi|_{M \backslash \mathcal{U}_p} : M \backslash \mathcal{U}_p \to \mathbb{CP}^1,$$

which is a disk bundle, and its first Chern number is therefore uniquely determined. Since complex line bundles over $\mathbb{CP}^1$ are classified by the first Chern number, it follows that the disk bundle $\pi|_{M \backslash \mathcal{U}_p}$ is isomorphic to the restriction of (3.10) to the complement of a neighborhood of its base point, and this isomorphism can then be extended via the local model (3.11) to a fiber-preserving diffeomorphism of M to $\mathbb{CP}^2$.   □

Observe that the complex coordinates defined on a neighborhood of any base point of a Lefschetz pencil $\pi : M \backslash M_{\text{base}} \to \mathbb{CP}^1$ determine an integrable complex structure on this neighborhood, so one can then define the complex blowup $\widetilde{M}$ of M at a base point $p \in M_{\text{base}}$ (cf. Remark 3.5). This means replacing p with the space of complex lines through p, forming an exceptional sphere $E \subset \widetilde{M}$. While any two distinct fibers of $\pi : M \backslash M_{\text{base}} \to \mathbb{CP}^1$ intersect each other transversely at p, it is not hard to show that they have lifts to $\widetilde{M}$ which pass through E at distinct points, resulting in a new Lefschetz pencil

$$\widetilde{\pi} : \widetilde{M} \backslash \widetilde{M}_{\text{base}} \to \mathbb{CP}^1,$$

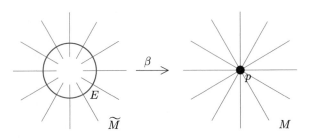

Fig. 3.3 The effect of blowing up a Lefschetz fibration at a base point p, shown here together with the blowdown map $\beta : \widetilde{M} \to M$

where $\widetilde{M}_{\text{base}} := M_{\text{base}} \setminus \{p\}$, see Fig. 3.3. To see this explicitly, observe that under the natural identification of $\mathbb{C}^2 \setminus \{0\}$ with the complement $\widetilde{\mathbb{C}}^2 \setminus \mathbb{CP}^1$ of the zero-section in the tautological line bundle, the model projection $\pi(z_1, z_2) = [z_1 : z_2]$ has a natural extension to a holomorphic map $\tilde{\pi} : \widetilde{\mathbb{C}}^2 \to \mathbb{CP}^1$ that is the identity at the zero-section. The exceptional sphere $E \subset \widetilde{M}$ is then a smooth section of $\tilde{\pi} : \widetilde{M} \setminus \widetilde{M}_{\text{base}} \to \mathbb{CP}^1$.

This process can also be reversed: suppose $\tilde{\pi} : \widetilde{M} \setminus \widetilde{M}_{\text{base}} \to \mathbb{CP}^1$ is a Lefschetz pencil with an **exceptional section** $E \subset \widetilde{M}$, i.e. a smooth section that satisfies $[E] \cdot [E] = -1$. The normal bundle $N_E \to E$ of $E \subset \widetilde{M}$ then has Euler number -1, so it is isomorphic as an oriented real vector bundle to the tautological line bundle $\widetilde{\mathbb{C}}^2 \to \mathbb{CP}^1$. Using the obvious identification of N_E with the vertical subbundle of the fibration along E, one can then identify a neighborhood $\mathcal{U}_E \subset \widetilde{M}$ of E with a neighborhood $\widetilde{B}^4_\delta \subset \widetilde{\mathbb{C}}^2$ of $\mathbb{CP}^1$ such that fibers of $\tilde{\pi}$ passing through $\mathcal{U}_E$ are identified with fibers of the tautological line bundle near its zero-section, hence $\tilde{\pi}$ in $\mathcal{U}_E$ becomes the bundle projection $\widetilde{B}^2_\delta \to \mathbb{CP}^1$. Replacing $\widetilde{B}^4_\delta$ with $B^4_\delta \subset \mathbb{C}^2$ to define the blowdown M thus produces a new Lefschetz pencil $\pi : M \setminus M_{\text{base}} \to \mathbb{CP}^1$, with $M_{\text{base}} := \widetilde{M}_{\text{base}} \cup \{p\}$ where $p \in M$ is the image of E under the blowdown map $\beta : \widetilde{M} \to M$.

Alternatively, one can consider the complex blowup at a regular point $p \in M \setminus (M_{\text{crit}} \cup M_{\text{base}})$ of a Lefschetz fibration/pencil $\pi : M \setminus M_{\text{base}} \to \Sigma$. To make sense of this, we can first choose complex local coordinates (compatible with the given orientations) near p and $\pi(p)$ such that p is the origin in $\mathbb{C}^2$ and π locally takes the form

$$\pi(z_1, z_2) = z_1.$$

Composing this with the standard blowdown map $\widetilde{\mathbb{C}}^2 \to \mathbb{C}^2 : ([z_1 : z_2], \lambda z_1, \lambda z_2) \mapsto (\lambda z_1, \lambda z_2)$ produces

$$\tilde{\pi} : \widetilde{\mathbb{C}}^2 \to \mathbb{C} : ([z_1 : z_2], (\lambda z_1, \lambda z_2)) \mapsto \lambda z_1. \tag{3.13}$$

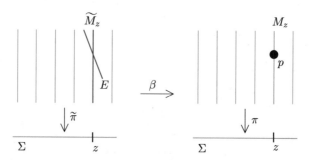

Fig. 3.4 The effect of blowing up a Lefschetz fibration $\pi : M \to \Sigma$ at a regular point $p \in M_z$, shown with the blowdown map $\beta : \widetilde{M} \to M$

Exercise 3.28. Show that the map (3.13) is holomorphic and is regular everywhere except at $[0 : 1] \in \mathbb{C}P^1 \subset \widetilde{\mathbb{C}}^2$, which is a Lefschetz critical point.

The exercise implies that the blown-up manifold $\widetilde{M}$ inherits a Lefschetz fibration/pencil $\widetilde{\pi} : \widetilde{M} \backslash \widetilde{M}_{\text{base}} \to \Sigma$ with $\widetilde{M}_{\text{base}} := M_{\text{base}}$ and with one extra critical point: the fiber $\widetilde{M}_z$ for $z := \pi(p)$ is now singular, and can be identified with the union of the original fiber M_z and an exceptional sphere E that intersects it transversely, see Fig. 3.4. Conversely, Proposition 3.30 below will enable us to blow down certain kinds of singular fibers so that they become regular.

Lemma 3.29. *Suppose* $\pi : M \backslash M_{\text{base}} \to \Sigma$ *is a Lefschetz fibration or pencil containing a singular fiber M_z with an irreducible component that is the image of an immersion $S \looparrowright M_z$ that intersects other irreducible components of M_z exactly k times, has d transversely immersed double points, and intersects M_{base} exactly b times. Then its normal bundle $N_S \to S$ satisfies $c_1(N_S) = -k - 2d + b$.*

Proof. This can be deduced from the observation that away from $M_{\text{crit}} \cup M_{\text{base}}$, any choice of oriented basis for $T_z\Sigma$ induces on M_z a canonical normal framing. One can then use the local models (3.8) and (3.11) to compute exactly how many zeroes appear if one extends this framing over the singularities to define a global section of N_S: each critical point (of which there are $k + 2d$ on S) contributes -1, while each base point contributes $+1$.   □

Proposition 3.30. *Suppose* $\pi : M \backslash M_{\text{base}} \to \Sigma$ *is a Lefschetz fibration or pencil containing a singular fiber M_z with an irreducible component $E \subset M_z$ that is disjoint from M_{base} and intersects other irreducible components of M_z exactly once. Then $[E] \cdot [E] = -1$.*

Proof. Assume E has $d \geq 0$ transversely immersed double points, so Lemma 3.29 gives $c_1(N_E) = -1 - 2d$. Then by the same argument as in the proof of the adjunction formula (see the paragraph preceding Theorem 2.51),

$$[E] \cdot [E] = 2d + c_1(N_E) = -1.$$

□

The operation of blowing up at a regular point can now be reversed as follows. Suppose $\tilde{\pi} : \tilde{M} \backslash \tilde{M}_{\text{base}} \to \Sigma$ has a singular fiber $\tilde{M}_z$ with an irreducible component $E \subset \tilde{M}_z$ that is an embedded sphere with only one critical point. A tubular neighborhood of E then admits a smooth orientation-preserving diffeomorphism to a neighborhood of the zero-section $\mathbb{CP}^1$ in the tautological line bundle $\tilde{\mathbb{C}}^2$, and we are free to choose this diffeomorphism so that the unique critical point in E is identified with $[0 : 1]$ and π matches (3.13) near this point for a suitably chosen local coordinate on Σ. Since all points in the rest of this neighborhood are regular and π sends the rest of E to the same point, we can then assume after a C^∞-small isotopy that (3.13) holds exactly on a sufficiently small neighborhood of the zero-section. Replacing this neighborhood with a ball then produces a blown-down manifold M with a Lefschetz fibration/pencil $\pi : M \backslash M_{\text{base}} \to \Sigma$ in which $M_{\text{base}} = \tilde{M}_{\text{base}}$ and M_z has one less critical point.

The blowup/blowdown operations described above depend on various choices, e.g. the integrable complex structure used for the blowup at a base point $M_{\text{base}} \subset M$ is guaranteed to exist, but it need not be unique, and the smooth structure of $\tilde{M}$ depends on this choice. This ambiguity should not be a cause for concern. We will show in the next section that compatible symplectic structures on Lefschetz pencils/fibrations can always be chosen such that the blowup and blowdown operations are equivalent to the symplectic operations in Sect. 3.2; as we have seen, these operations are well defined up to symplectic deformation equivalence (or symplectomorphism, in the case of the blowdown).

The following topological observation will play a crucial role in the proof of Theorems A and D.

Proposition 3.31. *There does not exist any closed oriented 4-manifold carrying a Lefschetz pencil with fibers diffeomorphic to S^2 and strictly more than one base point but no singular fibers.*

Proof. Arguing by contradiction, suppose $\pi : M \backslash M_{\text{base}} \to \mathbb{CP}^1$ is such a pencil, with $m \geqslant 2$ base points. Then the pencil $\tilde{\pi} : \tilde{M} \backslash \tilde{M}_{\text{base}} \to \mathbb{CP}^1$ that results from blowing up $m - 1$ of these base points is isomorphic to the standard pencil on $\mathbb{CP}^2$ by Proposition 3.27. But $\mathbb{CP}^2$ cannot be a blowup of M since it contains no exceptional sphere: indeed, $H_2(\mathbb{CP}^2)$ has a single generator $[\mathbb{CP}^1]$ satisfying $[\mathbb{CP}^1] \cdot [\mathbb{CP}^1] = 1$, hence there is no $A \in H_2(\mathbb{CP}^2)$ with $A \cdot A = -1$. $\qquad\square$

3.4 Symplectic Lefschetz Pencils and Fibrations

By a well known result of Thurston [Thu76] (see also [MS17, Theorem 6.1.4]), every 4-dimensional fibration over an oriented surface, with the property that the fibers are oriented and homologically non-torsion in the total space, admits a symplectic structure on its total space that makes the fibers into symplectic submanifolds. In this section we shall prove an extension of this result to Lefschetz fibrations

and pencils which is due to Gompf [GS99]. The prominent role played by Lefschetz
fibrations in symplectic topology since the 1990s is largely a consequence of this
theorem in combination with its (much harder) converse due to Donaldson [Don99],
which states that every closed symplectic manifold, after a small perturbation of
its symplectic form, admits a symplectic Lefschetz pencil. In dimension four, one
therefore obtains a topological characterization of the closed smooth manifolds that
admit symplectic structures: they are precisely those which admit Lefschetz pencils.

Definition 3.32. Given a Lefschetz fibration or pencil $\pi : M \backslash M_{\text{base}} \to \Sigma$, we
shall say that a symplectic structure ω on M is **compatible with** π if the following
conditions are satisfied:

(1) The smooth part of every fiber $M_z \backslash (M_{\text{base}} \cup M_{\text{crit}})$ for $z \in \Sigma$ is a
    symplectic submanifold;
(2) For any almost complex structure J defined near $M_{\text{base}} \cup M_{\text{crit}}$ that
    restricts to a smooth positively oriented complex structure on the smooth
    parts of all fibers $M_z \backslash (M_{\text{base}} \cup M_{\text{crit}})$, J is tamed by ω at $M_{\text{base}} \cup M_{\text{crit}}$.

Likewise, if (M, ω) is a symplectic manifold and $\pi : M \backslash M_{\text{base}} \to \Sigma$ is a Lefschetz
fibration or pencil, we call π a **symplectic Lefschetz fibration/pencil** if ω is
compatible with π.

Theorem 3.33 (Thurston [Thu76] and Gompf [GS99]). *Assume* π :
$M \backslash M_{\text{base}} \to \Sigma$ *is a Lefschetz fibration or pencil for which the fiber represents a non-
torsion class in* $H_2(M)$ *(cf. Proposition 3.34 below). Then M admits a symplectic
structure compatible with* π, *and any two such structures can be connected by a
smooth 1-parameter family of symplectic structures compatible with* π.

It is useful to note that the assumption on the homology class of the fiber is
always satisfied outside a very limited range of cases:

Proposition 3.34. *If $\pi : M \backslash M_{\text{base}} \to \Sigma$ is a Lefschetz fibration or pencil such that
the fiber represents a torsion class in* $H_2(M)$, *then $M_{\text{base}} = \emptyset$ and the fiber is a
torus.*

Proof. We use the fact that a torsion class necessarily has vanishing intersection
product with every other homology class: indeed, if $A, B \in H_2(M)$ with $kA = 0$
for some $k \in \mathbb{N}$, then

$$0 = kA \cdot B = k(A \cdot B),$$

implying $A \cdot B = 0$. Suppose $M_z \subset M$ is a smooth fiber. If there are $m > 0$ base
points in M_{base}, then by counting intersections of M_z with any other fiber $M_{z'} \subset M$
we have

$$[M_z] \cdot [M_{z'}] = m,$$

implying by the above observation that $[M_z] \in H_2(M)$ cannot be torsion. If $M_{\text{base}} = \varnothing$, then we instead argue as follows. Choosing complex coordinates (z_1, z_2) near each $p \in M_{\text{crit}}$ in which $\pi(z_1, z_2) = z_1^2 + z_2^2$, define a vector field σ near p by

$$\sigma(z_1, z_2) = (-\bar{z}_2, \bar{z}_1),$$

and notice that σ is tangent to every fiber in this neighborhood and vanishes only at p. Now extend σ to a global vector field on M that is everywhere tangent to the fibers, i.e. it is a smooth section of the vertical subbundle

$$VM \subset TM|_{M \backslash M_{\text{crit}}} \to M \backslash M_{\text{crit}},$$

whose fibers are $(VM)_p = T_p M_z$ for any regular point $p \in M_z$. After a generic perturbation that leaves σ unchanged near M_{crit}, we can assume it is transverse to the zero-section of VM, so its zero-set is the union of M_{crit} with a closed 2-dimensional submanifold

$$Z := \sigma^{-1}(0) \backslash M_{\text{crit}} \subset M.$$

By Sard's theorem, Z intersects almost every regular fiber M_z transversely, and the restriction of σ to such a fiber is then a smooth vector field on M_z with nondegenerate zeroes, whose signed count is $\chi(M_z)$. This proves

$$[M_z] \cdot [Z] = \chi(M_z),$$

which is nonzero unless $M_z \cong \mathbb{T}^2$, thus implying that $[M_z]$ is not torsion. □

Example 3.35. The following shows that Theorem 3.33 cannot always be applied for torus bundles. Let $\pi_0 : S^3 \to S^2$ denote the Hopf fibration, and define a torus bundle $\pi : S^1 \times S^3 \to S^2$ by

$$\pi(\theta, p) = \pi_0(p).$$

Observe that the fibers of $\pi_0 : S^3 \to S^2$ are nullhomologous since $H_1(S^3) = 0$, so it follows that the fibers of $\pi : S^1 \times S^3 \to S^2$ are also nullhomologous. And indeed, $S^1 \times S^3$ does not admit any symplectic structure since $H^2_{\text{dR}}(S^1 \times S^3) = 0$.

The rest of this section will be devoted mainly to the proof of Theorem 3.33. The main idea is to reduce the construction of symplectic forms in dimension 4 to a problem in dimension 2, where symplectic geometry is comparatively trivial. In particular, we will make essential use of the fact that on any given oriented surface, the space of symplectic forms is both nonempty and contractible—indeed, it is a convex subset of a vector space. As an intermediary between the two and four-dimensional settings, we introduce the following notion.

Definition 3.36. Assume $\pi : M\backslash M_{\text{base}} \to \Sigma$ is a Lefschetz fibration or pencil. A **fiberwise symplectic structure** on M with respect to π is a closed 2-form ω such that:

(1) ω restricts to a positive area form on the smooth part of every fiber $M_z\backslash(M_{\text{base}} \cup M_{\text{crit}})$;

(2) For any almost complex structure J defined near $M_{\text{base}} \cup M_{\text{crit}}$ that restricts to a smooth positively oriented complex structure on the smooth parts of all fibers $M_z\backslash(M_{\text{base}} \cup M_{\text{crit}})$, ω is nondegenerate and tames J at $M_{\text{base}} \cup M_{\text{crit}}$.

The only difference between this and Definition 3.32 is that we do not require ω to be nondegenerate except at $M_{\text{base}} \cup M_{\text{crit}}$; a 2-form ω is thus a symplectic structure compatible with π if and only if it is fiberwise symplectic and satisfies $\omega \wedge \omega > 0$. The advantage we gain in dropping nondegeneracy in the above definition is that for any fixed $\pi : M\backslash M_{\text{base}} \to \Sigma$, the space of fiberwise symplectic structures is convex; see Proposition 3.39 below. The following lemma shows that the apparent choice of an auxiliary almost complex structure in Definitions 3.32 and 3.36 is not actually a choice.

Lemma 3.37. *Given a Lefschetz pencil/fibration $\pi : M\backslash M_{\text{base}} \to \Sigma$, suppose J and J' are two almost complex structures defined on a neighborhood of $M_{\text{base}} \cup M_{\text{crit}}$ which both restrict to positively oriented complex structures on all the smooth fibers in this neighborhood. Then $J = J'$ at $M_{\text{base}} \cup M_{\text{crit}}$.*

Proof. Identifying a neighborhood of any point in $p \in M_{\text{base}} \cup M_{\text{crit}}$ with a neighborhood of the origin in $\mathbb{C}^2$ via (3.8) or (3.11), we see that in these coordinates, every complex 1-dimensional subspace of the standard $\mathbb{C}^2$ occurs as a tangent space to a fiber in every neighborhood of p. Since J and J' are continuous, it follows that every J-complex 1-dimensional subspace of T_pM is also J'-complex and both structures induce the same orientation, so the claim follows from Lemma 3.38 below.   □

Lemma 3.38. *Suppose J is a complex structure on the vector space $\mathbb{C}^n$ with $n > 1$ such that J preserves every i-complex line. Then $J = \pm i$.*

Proof. Let $(e_1, \ldots, e_n)$ denote the standard complex basis of $\mathbb{C}^n$. The assumption on J implies that there are numbers $a_j, b_j \in \mathbb{R}$ with $b_j \neq 0$ such that $Je_j = a_je_j + b_jie_j$ for $j = 1, \ldots, n$. Moreover, for $v := \sum_j e_j$, there are numbers $a, b \in \mathbb{R}$ with $b \neq 0$ such that $Jv = av + biv$. Since $(e_1, \ldots, e_n, ie_1, \ldots, ie_n)$ is a real basis of $\mathbb{C}^n$, combining these two formulas implies

$$a_1 = \ldots = a_n = a \quad \text{and} \quad b_1 = \ldots = b_n = b.$$

The condition $J^2 = -\mathbb{1}$ now implies $J(ie_j) = -\frac{1+a^2}{b}e_j - aie_j$ for $j = 1, \ldots, n$. Then for $w := e_1 + ie_2$, there are also constants $A, B \in \mathbb{R}$ with $B \neq 0$ such that

$Jw = Aw + Biw$, implying

$$J(e_1 + ie_2) = ae_1 + bie_1 - \frac{1+a^2}{b}e_2 - aie_2$$

$$= A(e_1 + ie_2) + B(-e_2 + ie_1)$$

and thus

$$a = A, \qquad b = B, \qquad -\frac{1+a^2}{b} = -B, \qquad -a = A.$$

We conclude $a = 0$ and $b = 1/b$, hence $b = \pm 1$, which proves $J = \pm i$. $\qquad\square$

Lemma 3.37 implies that when applying Definition 3.32 or 3.36, one can fix a suitable choice of almost complex structure near $M_{\text{base}} \cup M_{\text{crit}}$, and the resulting notions do not depend on this choice. Now observe that on any complex vector space (V, J), the space of antisymmetric real bilinear 2-forms Ω that satisfy

$$\Omega(v, Jv) > 0 \quad \text{for all } v \in V\backslash\{0\}$$

is convex, i.e. whenever Ω and Ω' are in this space, so are $s\Omega + (1-s)\Omega'$ for all $s \in [0, 1]$. Since the spaces of closed 2-forms and 2-forms that restrict positively to any given 2-dimensional subbundle are also convex, this implies:

Proposition 3.39. *On any Lefschetz fibration/pencil $\pi : M\backslash M_{\text{base}} \to \Sigma$, the space of fiberwise symplectic structures is convex.* $\qquad\square$

Lemma 3.40. *Assume $\pi : M\backslash M_{\text{base}} \to \Sigma$ is a Lefschetz fibration or pencil for which the fiber represents a non-torsion class in $H_2(M)$. Then M admits a fiberwise symplectic structure with respect to π.*

Before proving the lemma, let us go ahead and use it to prove Theorem 3.33. In order to turn fiberwise symplectic structures into actual symplectic forms, we use a trick originally due to Thurston [Thu76].

Suppose $\{\omega_s\}_{s\in[0,1]}$ is a smooth 1-parameter family of fiberwise symplectic structures. Fix an area form σ on Σ; in the case $M_{\text{base}} \neq \emptyset$, we shall assume specifically that σ is the *standard* symplectic form ω_{FS} on $\mathbb{CP}^1$. As we saw in Sect. 1.1, this can be defined by the property that if $\pi_S : S^3 \to S^3/S^1 = \mathbb{CP}^1$ denotes the Hopf fibration, then

$$\pi_S^* \omega_{\text{FS}} = d\alpha_{\text{st}},$$

where $\alpha_{\text{st}} := \lambda_{\text{st}}|_{TS^3}$ is the restriction to the unit sphere $S^3 \subset \mathbb{C}^2$ of the *standard Liouville form*

$$\lambda_{\text{st}} := \frac{1}{2}\sum_{j=1}^{2}(p_j\,dq_j - q_j\,dp_j)$$

on $\mathbb{C}^2$, using coordinates $z_j = p_j + iq_j$ for $j = 1, 2$.

Near each point in $M_{\mathrm{base}} \cup M_{\mathrm{crit}}$, fix a neighborhood with coordinates as in (3.8) or (3.11), and let J denote the integrable complex structure on a neighborhood of $M_{\mathrm{base}} \cup M_{\mathrm{crit}}$ determined by these coordinates. Note that for this choice, all smooth parts of fibers are J-complex curves wherever J is defined. For each $p \in M_{\mathrm{crit}}$, (3.8) also involves a complex coordinate chart near $\pi(p) \in \Sigma$, which determines a complex structure j_p near $\pi(p)$. For these choices, the map π is J-j_p-holomorphic near $p \in M_{\mathrm{crit}}$, and it is J-i-holomorphic near M_{base}, where i denotes the standard complex structure on $\mathbb{CP}^1$.

Since ω_s is fiberwise symplectic for all $s \in [0, 1]$ and tameness is an open condition, one can fix a small neighborhood $\mathcal{U} \subset M$ of $M_{\mathrm{base}} \cup M_{\mathrm{crit}}$ within the above coordinate neighborhoods on which ω_s is symplectic and tames J for all s. Denote the union of the connected components of this neighborhood containing M_{base} by $\mathcal{U}_{\mathrm{base}}$. Now choose a smooth function

$$\rho : M \rightarrow [0, 1]$$

such that $1 - \rho$ has compact support in $\mathcal{U}_{\mathrm{base}}$, while ρ vanishes near M_{base} and in $\mathcal{U}_{\mathrm{base}}$ it takes the form

$$\rho(z) = f(|z|), \qquad z := (z_1, z_2) \in \mathbb{C}^2$$

for some smooth function $f : [0, \infty) \rightarrow [0, 1]$ with $f' \geq 0$, using the coordinates of (3.11). Define the projection

$$\nu : \mathbb{C}^2 \backslash \{0\} \rightarrow S^3 : z \mapsto \frac{z}{|z|}.$$

This data allows us to define a closed 2-form on M by

$$\sigma_\rho := \begin{cases} d\left(\rho \cdot \nu^* \alpha_{\mathrm{st}}\right) & \text{on } \mathcal{U}_{\mathrm{base}}, \\ \pi^* \sigma & \text{on } M \backslash \mathcal{U}_{\mathrm{base}}. \end{cases} \tag{3.14}$$

Lemma 3.41. *There exists a constant $K_0 \geq 0$ such that for all constants $K \geq K_0$ and all $s \in [0, 1]$, the 2-forms*

$$\omega_s^K := \omega_s + K\sigma_\rho \tag{3.15}$$

are symplectic and compatible with π. Moreover, if ω_s is already symplectic for all $s \in [0, 1]$, then it suffices to take $K_0 = 0$.

Proof. We claim first that for all $K \geq 0$, ω_s^K is symplectic in the neighborhood $\mathcal{U}_{\mathrm{base}}$ and tames J. At M_{base} this is clear since $\rho = 0$ nearby and thus $\omega_s^K = \omega_s$. In $\mathcal{U}_{\mathrm{base}} \backslash M_{\mathrm{base}}$, we have

$$\omega_s^K = \omega_s + K\rho\, \nu^* d\alpha_{\mathrm{st}} + K\, d\rho \wedge \nu^* \alpha_{\mathrm{st}} = \omega_s + K\rho\, \pi^* \sigma + K\, d\rho \wedge \nu^* \alpha.$$

Applying this to a pair (X, JX) with $X \in T\mathcal{U}_{\text{base}}$ nonzero, the first term is positive, and the second is nonnegative since $\rho \geqslant 0$ and π is J-i-holomorphic:

$$\pi^*\sigma(X, JX) = \sigma(\pi_*X, \pi_*JX) = \sigma(\pi_*X, i\pi_*X) \geqslant 0.$$

The third term vanishes whenever $d\rho = 0$ and is also nonnegative when $d\rho \neq 0$: indeed, it is then positive on the complex lines spanned by vectors pointing radially outward, while its kernel contains all vectors that are orthogonal to these lines.

Similarly, we claim that ω_s^K is symplectic and tames J for all $K \geqslant 0$ on the components $\mathcal{U}_p \subset \mathcal{U}$ containing critical points $p \in M_{\text{crit}}$. Since π is J-j_p-holomorphic on such a neighborhood, we have for any nontrivial vector $X \in T\mathcal{U}_p$,

$$\omega_s^K(X, JX) = \omega_s(X, JX) + K\sigma(\pi_*X, \pi_*JX) = \omega_s(X, JX) + K\sigma(\pi_*X, j_p\pi_*X),$$

in which the first term is positive and the second is nonnegative.

Finally, consider ω_s^K on $M \backslash \mathcal{U}$. Since the fibers are all smooth in $M \backslash \mathcal{U}$, we can define the vertical subbundle $VM \subset TM|_{M \backslash \mathcal{U}}$, whose fibers are $V_pM = T_pM_z$ for $p \in M_z$. The fact that ω_s is fiberwise symplectic then implies $\omega_s^K|_{VM} = \omega_s|_{VM} > 0$, so VM is transverse to its ω_s-symplectic orthogonal complement, which is the subbundle $HM \subset TM|_{M \backslash \mathcal{U}}$ defined by

$$HM = \{X \in TM|_{M \backslash \mathcal{U}} \mid \omega_s(X, \cdot)|_{VM} = 0\}.$$

Observe that this is simultaneously the ω_s^K-symplectic orthogonal complement for every $K \in \mathbb{R}$, since for any $V \in VM$ and $H \in HM$ in the same tangent space,

$$\omega_s^K(V, H) = \omega_s(V, H) + K\pi^*\sigma(V, H) = 0.$$

Now ω_s^K is symplectic on $M \backslash \mathcal{U}$ if and only if $\omega_s^K|_{HM} > 0$, which is true for sufficiently large $K > 0$ since

$$\omega_s^K|_{HM} = K\left(\pi^*\sigma + \frac{1}{K}\omega_s\right)\bigg|_{HM},$$

and $\pi^*\sigma|_{HM} > 0$. It is also true for all $K \geqslant 0$ if $\omega_s|_{HM} > 0$, which is the case if and only if ω_s is symplectic. $\qquad\qquad\qquad\square$

Proof of Theorem 3.33. The existence of a symplectic form compatible with π follows immediately from Lemmas 3.40 and 3.41. Now suppose ω_0 and ω_1 are two such forms. By Proposition 3.39, we can then define a smooth family of fiberwise symplectic structures connecting these by

$$\omega_s = s\omega_1 + (1-s)\omega_0, \qquad s \in [0, 1].$$

Since nondegeneracy is an open condition, we can choose $\epsilon > 0$ sufficiently small so that ω_s is symplectic for all $s \in [0, \epsilon) \cup (1-\epsilon, 1]$. Now choose a smooth function $\beta : [0, 1] \to [0, 1]$ with compact support in $(0, 1)$ that is identically 1 on $[\epsilon, 1 - \epsilon]$. Defining ω_s^K as in (3.15), Lemma 3.41 then implies that for a sufficiently large constant $K > 0$,

$$\omega_s' := \omega_s^{K\beta(s)}$$

is a smooth family of symplectic forms compatible with π and satisfying $\omega_0' = \omega_0$, $\omega_1' = \omega_1$.   □

Remark 3.42. Given a section $S \subset M$ of the Lefschetz fibration/pencil $\pi : M \backslash M_{\text{base}} \to \Sigma$, we can always arrange the compatible symplectic structure on M in the above proof so that S becomes a symplectic submanifold. This is simply a matter of choosing the constant $K > 0$ in Lemma 3.41 sufficiently large.

It remains to prove Lemma 3.40 on the existence of a fiberwise symplectic structure. This is accomplished in two steps: first we choose a suitable cohomology class, and then we use a partition of unity to realize this class by a 2-form that is positive on all fibers. The first step is the only place where we will make a simplifying assumption: we shall assume $\pi : M \backslash M_{\text{base}} \to \Sigma$ has the property that no two critical points lie in a single fiber, i.e. $\pi|_{M_{\text{crit}}} : M_{\text{crit}} \to \Sigma_{\text{crit}}$ is a bijection. This assumption is not necessary, but doing without it would require some tedious linear algebra (see in particular [Gom05, Lemma 3.3]), so we will content ourselves with proving a slightly less general statement than would be possible, as it suffices in any case for the main applications.

Lemma 3.43. *Assume $\pi : M \backslash M_{\text{base}} \to \Sigma$ is a Lefschetz fibration or pencil for which the fiber represents a non-torsion class in $H_2(M)$. Then there exists a cohomology class $\beta \in H_{\text{dR}}^2(M)$ such that for all irreducible components $E \subset M$ of fibers,*

$$\int_E \beta > 0.$$

Proof (Assuming $\pi|_{M_{\text{crit}}} : M_{\text{crit}} \to \Sigma_{\text{crit}}$ Bijective). By Poincaré duality, it suffices to find a homology class $A \in H_2(M; \mathbb{R})$ such that $A \cdot [E] > 0$ for all irreducible components of fibers $E \subset M$, as then we can set β to be the Poincaré dual of A. Let $[F] \in H_2(M)$ denote the homology class of the fiber. We shall handle the cases with and without base points slightly differently.

If $M_{\text{base}} \neq \emptyset$, then $[F] \cdot [F] > 0$ is the number of base points, and all irreducible components $E \subset M$ of fibers satisfy $[F] \cdot [E] > 0$ except for those containing no base points, which satisfy $[F] \cdot [E] = 0$. Denote the collection of the latter components by Γ. By Proposition 3.30, each of them satisfies $[E] \cdot [E] = -1$ since it must be connected via a unique critical point to another irreducible component E'

which does contain a base point. Let

$$A = [F] - \frac{1}{2} \sum_{E \in \Gamma} [E] \in H_2(M; \mathbb{R}).$$

Then $A \cdot [F] = [F] \cdot [F] > 0$, establishing the desired result for all regular fibers as well as singular fibers with no irreducible components belonging to Γ. For any $E \in \Gamma$, we also have $A \cdot [E] = 1/2 > 0$ since none of these components intersect each other. Now if E' is the other irreducible component in the same fiber with some $E_0 \in \Gamma$, then $[F] \cdot [E'] \geq 1$ since E' necessarily contains base points, but the uniquess of the critical point in each fiber implies $[E'] \cdot [E_0] = 1$, so

$$A \cdot [E'] = [F] \cdot [E'] - \frac{1}{2} \sum_{E \in \Gamma} [E] \cdot [E'] \geq 1 - \frac{1}{2} > 0,$$

and we are done.

If $M_{\text{base}} = \varnothing$, then we instead appeal to the assumption that $[F] \in H_2(M)$ is not torsion, so by the nondegeneracy of the intersection product, there exists $A \in H_2(M)$ with $A \cdot [F] > 0$, and any class with this property will suffice for the regular fibers and singular fibers with only one irreducible component. By Proposition 3.30, any other singular fiber now consists of two irreducible components E_+ and E_- with $[E_+] \cdot [E_-] = 1$, $[E_\pm] \cdot [E_\pm] = -1$ and $[F] \cdot [E_\pm] = 0$, and we have

$$A \cdot [E_+] + A \cdot [E_-] > 0.$$

If either of the terms in this sum is nonpositive, we can choose the labeling so that without loss of generality $A \cdot [E_-] \leq 0$, and then it follows that

$$0 \leq -A \cdot [E_-] < A \cdot [E_+].$$

Choose $c \in \mathbb{R}$ with $-A \cdot [E_-] < c < A \cdot [E_+]$, and set

$$A' = A - c[E_-] \in H_2(M; \mathbb{R}).$$

Now $A' \cdot [F] = A \cdot [F] > 0$, $A' \cdot [E_+] = A \cdot [E_+] - c > 0$ and $A' \cdot [E_-] = A \cdot [E_-] + c > 0$. Repeating this procedure for every singular fiber eventually produces a homology class with the desired properties. $\qquad\square$

Proof of Lemma 3.40 (Assuming $\pi|_{M_{\text{crit}}} : M_{\text{crit}} \to \Sigma_{\text{crit}}$ Bijective). Fix a closed 2-form β representing the cohomology class provided by Lemma 3.43, and fix also a neighborhood $\mathcal{U}_{\text{base}} \subset M$ of M_{base} with complex coordinates in which π takes the form (3.11). Since all closed 2-forms are cohomologous in a ball, we may choose β without loss of generality (after shrinking $\mathcal{U}_{\text{base}}$ if necessary) to match ω_{st} in $\mathcal{U}_{\text{base}}$, where the latter denotes the standard symplectic form of $\mathbb{C}^2$, defined on $\mathcal{U}_{\text{base}}$ via the coordinates.

For every $z \in \Sigma \backslash \Sigma_{\mathrm{crit}}$, choose an area form ω_z on the smooth fiber M_z such that the restriction of ω_z to $M_z \cap \mathcal{U}_{\mathrm{base}}$ matches the restriction to this submanifold of ω_{st}, and $\int_{M_z} \omega_z = \int_{M_z} \beta$. Now for a sufficiently small neighborhood $\mathcal{U}_z \subset \Sigma$ of z, we can use a retraction of $\pi^{-1}(\mathcal{U}_z)$ to M_z to extend ω_z to a closed 2-form on $\pi^{-1}(\mathcal{U}_z)$ that has these same properties for every fiber $M_{z'}$ with $z' \in \mathcal{U}_z$.

We next do the same trick on neighborhoods of each singular fiber. For $z \in \Sigma_{\mathrm{crit}}$ and $p \in M_{\mathrm{crit}} \cap M_z$, fix a neighborhood $\mathcal{U}_p \subset M$ of p with complex coordinates in which π takes the form (3.8). Shrinking $\mathcal{U}_p$ if necessary, we can extend ω_{st} from $\mathcal{U}_p \cup \mathcal{U}_{\mathrm{base}}$ to a closed 2-form ω_z on $\pi^{-1}(\mathcal{U}_z)$ for some neighborhood $z \in \mathcal{U}_z \subset \Sigma$ which restricts to every irreducible component E of every fiber in $\pi^{-1}(\mathcal{U}_z)$ as an area form satisfying $\int_E \omega_z = \int_E \beta$.

Since Σ is compact, the open cover $\Sigma = \bigcup_{z \in \Sigma} \mathcal{U}_z$ has a finite subcover, which we shall denote by

$$\Sigma = \bigcup_{z \in I} \mathcal{U}_z.$$

Choose a partition of unity $\{\rho_z : \mathcal{U}_z \to [0, 1]\}_{z \in I}$ subordinate to this subcover. By construction, the 2-forms ω_z defined on $\pi^{-1}(\mathcal{U}_z)$ are cohomologous to the restrictions of β to these neighborhoods, so there exist smooth 1-forms α_z on $\pi^{-1}(\mathcal{U}_z)$ satisfying

$$\omega_z = \beta + d\alpha_z \quad \text{on } \pi^{-1}(\mathcal{U}_z).$$

Since ω_z and β both match ω_{st} in $\mathcal{U}_{\mathrm{base}}$, we may also assume without loss of generality that α_z vanishes in $\mathcal{U}_{\mathrm{base}}$. Now let

$$\omega = \beta + \sum_{z \in I} d\left((\rho_z \circ \pi)\alpha_z\right).$$

This matches ω_{st} on $\mathcal{U}_{\mathrm{base}}$, and at any $p \in M_{\mathrm{crit}}$, we have $d\pi = 0$ and thus

$$\omega = \sum_{z \in I} (\rho_z \circ \pi)(\beta + d\alpha_z) = \sum_{z \in I} (\rho_z \circ \pi)\omega_z = \omega_{\mathrm{st}}.$$

On any smooth piece of a fiber M_z, we have similarly $d(\rho_z \circ \pi)|_{TM_z} = 0$ and thus

$$\omega|_{TM_z} = \sum_{z \in I} (\rho_z \circ \pi)(\beta + d\alpha_z)|_{TM_z} = \sum_{z \in I} (\rho_z \circ \pi)\omega_z|_{TM_z} > 0,$$

so ω is a fiberwise symplectic structure. $\square$

Finally, let us revisit the blowup and blowdown operations that were defined for Lefschetz pencils and fibrations at the end of Sect. 3.3.

Theorem 3.44. *Assume $\pi : M \backslash M_{\text{base}} \to \Sigma$ is a Lefschetz pencil or fibration, $p \in M$ is either a regular point of π or a base point in M_{base}, and $\tilde{\pi} : \tilde{M} \backslash \tilde{M}_{\text{base}} \to \Sigma$ is the Lefschetz pencil or fibration obtained from π by blowing up M at p. Then there exist symplectic structures ω on M compatible with π and $\tilde{\omega}$ on $\tilde{M}$ compatible with $\tilde{\pi}$ such that $(\tilde{M}, \tilde{\omega})$ is symplectomorphic to the symplectic blowup of (M, ω) along some Darboux ball centered at p.*

Proof. Consider first the case where $p \in M_{\text{base}}$, so π is a Lefschetz pencil over $\Sigma := \mathbb{CP}^1$. The key observation is that in our proof of Theorem 3.33, we constructed a symplectic form ω compatible with $\pi : M \backslash M_{\text{base}} \to \mathbb{CP}^1$ which matches the standard symplectic form ω_{st} in some choice of holomorphic coordinates near p. Indeed, by Lemma 3.41 we can construct ω to be globally of the form $\omega_{\text{fib}} + K\sigma_\rho$ for some $K \gg 0$, where σ_ρ is a closed 2-form that vanishes near M_{base} (see (3.14)) and ω_{fib} is a fiberwise symplectic form as in the proof of Lemma 3.40, which was constructed in that proof to match ω_{st} near M_{base}. With these choices in place, we have a symplectic manifold (M, ω) with a symplectic Lefschetz pencil $\pi : M \backslash M_{\text{base}} \to \mathbb{CP}^1$ and a holomorphic symplectic embedding $\psi : (B^4_{R+\epsilon}, \omega_{\text{st}}) \hookrightarrow (M, \omega)$ centered at p, for any $R, \epsilon > 0$ sufficiently small. Now use Theorem 3.13 to define a symplectic form $\tilde{\omega}$ on the complex blowup $\tilde{M}$ such that $\tilde{\omega}$ is compatible with the complex structure near the exceptional divisor $E \subset \tilde{M}$ and matches ω elsewhere, and $(\tilde{M}, \tilde{\omega})$ is symplectomorphic to the symplectic blowup of (M, ω) along ψ with weight R. Compatibility of $\tilde{\omega}$ with the complex structure guarantees that all fibers of $\tilde{\pi} : \tilde{M} \backslash \tilde{M}_{\text{base}} \to \mathbb{CP}^1$ near E are also symplectic, hence $\tilde{\omega}$ is compatible with $\tilde{\pi}$.

In the case where $p \in M$ is a regular point of π, the same argument works after finding a symplectic embedding $\psi : (B^4_{R+\epsilon}, \omega_{\text{st}}) \hookrightarrow (M, \omega)$ with $\psi(0) = p$ such that, in complex coordinates (z_1, z_2) on $B^4_{R+\epsilon}$, $\pi \circ \psi(z_1, z_2) = z_1$. This again can be extracted from the proof of Theorem 3.33: indeed, writing $z = \pi(p)$, one can start by trivializing π over some neighborhood of z in $\Sigma \backslash \Sigma_{\text{crit}}$, and use this trivialization to select a fiberwise symplectic structure ω_{fib} that matches $dp_2 \wedge dq_2$ in some coordinates $(z_1, z_2) = (p_1 + iq_1, p_2 + iq_2)$ near p for which $\pi(z_1, z_2) = z_1$. We can then arrange without loss of generality for $\omega = \omega_{\text{fib}} + K\sigma_\rho$ to have the form $K\,dp_1 \wedge dq_1 + dp_2 \wedge dq_2$ in this same neighborhood, so a suitable rescaling of the z_1-coordinate produces the desired result. $\qquad\square$

One nice application of this result is to show that the minimal symplectic blowdown of a symplectic 4-manifold is not generally unique. We will see in Chap. 7 that this phenomenon can only occur in the world of rational and ruled symplectic 4-manifolds (cf. Corollary 7.9):

Proposition 3.45. *There exist two minimal symplectic 4-manifolds that are not homeomorphic but have blowups that are symplectomorphic.*

Proof. Recall from (3.10) and (3.12) respectively the Lefschetz pencils

$$\pi_1 : \mathbb{CP}^2 \backslash \{[1:0:0]\} \to \mathbb{CP}^1 \quad \text{and} \quad \pi_2 : (S^2 \times S^2) \backslash \{(0,0), (\infty, \infty)\} \to S^2 \cong \mathbb{CP}^1,$$

where both have fibers of genus zero, π_1 has no singular fibers, and π_2 has the two singular fibers $\pi_2^{-1}(0)$ and $\pi_2^{-1}(\infty)$, each of which has two irreducible components that are spheres with self-intersection number 0. Both Lefschetz pencils admit compatible symplectic structures by Theorem 3.33, which are necessarily minimal since neither $\mathbb{CP}^2$ nor $S^2 \times S^2$ contains any degree 2 homology class A with $A \cdot A = -1$. Now define

$$\tilde{\pi}_1 : \tilde{M}_1 \backslash \{p\} \to \mathbb{CP}^1$$

from π_1 by blowing up $\mathbb{CP}^2$ at two regular points of π_1 in distinct fibers; by Exercise 3.6, this produces two singular fibers that each have two irreducible components, both spheres, one with self-intersection 0 and the other an exceptional sphere. Define

$$\tilde{\pi}_2 : \tilde{M}_2 \backslash \{p\} \to \mathbb{CP}^1$$

in turn from π_2 by blowing up one of its base points, so $\tilde{\pi}_2$ also has two singular fibers, and they are of the same type as $\tilde{\pi}_1$, again due to Exercise 3.6; see Fig. 3.5. If we then regularize both of these singular fibers by blowing down the unique exceptional sphere in each of them, we obtain a Lefschetz pencil with one base point and no singular fibers, which by Proposition 3.27 is diffeomorphic to π_1. This

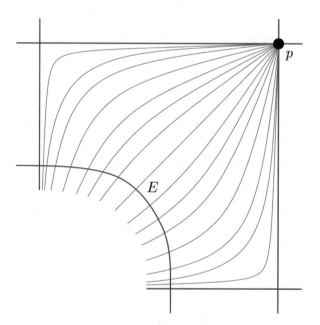

Fig. 3.5 The Lefschetz pencil $\tilde{\pi}_2 : \tilde{M}_2 \backslash \{p\} \to \mathbb{CP}^1$ in the proof of Proposition 3.45 is obtained from Fig. 3.2 by blowing up one of the base points, producing an exceptional section E

proves that $\tilde{\pi}_1$ and $\tilde{\pi}_2$ are diffeomorphic Lefschetz pencils on $\tilde{M}_1 \cong \mathbb{CP}^2 \# 2\overline{\mathbb{CP}}^2 \cong (S^2 \times S^2) \# \overline{\mathbb{CP}}^2 \cong \tilde{M}_2$. Now applying Theorem 3.44, we find compatible symplectic structures ω_1 on $\mathbb{CP}^2$, ω_2 on $S^2 \times S^2$, $\tilde{\omega}_1$ on $\tilde{M}_1$ and $\tilde{\omega}_2$ on $\tilde{M}_2$ such that $(\tilde{M}_1, \tilde{\omega}_1)$ is a symplectic blowup of $(\mathbb{CP}^2, \omega_1)$ and $(\tilde{M}_2, \tilde{\omega}_2)$ is a symplectic blowup of $(S^2 \times S^2, \omega_2)$. By the uniqueness statement in Theorem 3.33, $(\tilde{M}_1, \tilde{\omega}_1)$ and $(\tilde{M}_2, \tilde{\omega}_2)$ are also symplectically deformation equivalent. After choosing a diffeomorphism to identify them as smooth manifolds, the deformation from $\tilde{\omega}_1$ to $\tilde{\omega}_2$ gives rise via Theorem B to an isotopy of exceptional spheres and thus a symplectic deformation of the corresponding blowdowns, hence after a further deformation of either ω_1 or ω_2, we may assume that they have symplectic blowups which are symplectomorphic. □

Example 3.46. For an alternative proof of the above proposition, we claim that if Σ is any closed oriented surface, then there exist trivial and nontrival symplectic ruled surfaces $\Sigma \times S^2$ and $\Sigma \tilde{\times} S^2$ respectively which have symplectomorphic blowups. To see this, it suffices to think in terms of almost complex blowups and their effect on Lefschetz fibrations, since Theorem 3.44 can then be used to convert these into symplectic operations. With this understood, let $\tilde{M} \cong (\Sigma \times S^2) \# \overline{\mathbb{CP}}^2$ denote the blowup of $\Sigma \times S^2$ at a point p, which admits a symplectic Lefschetz fibration $\tilde{\pi} : \tilde{M} \to \Sigma$ with exactly one singular fiber. This fiber is a transverse union of two exceptional spheres E_0 and E_1, where we take E_1 to be the exceptional sphere produced by blowing up, and E_0 is the lift of the fiber in $\Sigma \times S^2$ through p via the blowdown map $\beta : \tilde{M} \to \Sigma \times S^2$. Now let M denote the result of blowing down $\tilde{M}$ along E_0, so it is also a symplectic ruled surface $\pi : M \to \Sigma$. Choose sections $S' \subset M$ of π and $S \subset \Sigma \times S^2$ of the trivial bundle such that both are disjoint from the blowup point: these then give rise to sections $S, S' \subset \tilde{M}$ of the Lefschetz fibration, with S intersecting E_0 but not E_1, and S' doing the opposite. One can check that in this situation, the self-intersection numbers $[S] \cdot [S]$ and $[S'] \cdot [S']$ differ by an odd number (see Exercise 7.66), so by Exercise 1.14, the fact that $\Sigma \times S^2$ is the trivial bundle implies that $\pi : M \to \Sigma$ is the nontrivial bundle $\Sigma \tilde{\times} S^2$. A special case of this phenomenon appeared in the proof of Proposition 3.45 above: since the blowup $\mathbb{CP}^2 \# \overline{\mathbb{CP}}^2$ is also the nontrivial rational ruled surface $S^2 \tilde{\times} S^2$, we have

$$(S^2 \times S^2) \# \overline{\mathbb{CP}}^2 \cong (\mathbb{CP}^2 \# \overline{\mathbb{CP}}^2) \# \overline{\mathbb{CP}}^2 \cong \mathbb{CP}^2 \# 2\overline{\mathbb{CP}}^2.$$

One can show that whenever Σ has positive genus, both $\Sigma \times S^2$ and $\Sigma \tilde{\times} S^2$ are minimal; see Exercise 7.7.

Exercise 3.47. Let $\Omega(\pi)$ denote the space of symplectic structures compatible with a given Lefschetz fibration or pencil $\pi : M \setminus M_{\text{base}} \to \Sigma$, with the natural C^∞-topology. Adapt the proof of Theorem 3.33 to show that $\pi_k(\Omega(\pi)) = 0$ for all integers $k \geqslant 0$. (We proved this above for $k = 0$.)

Remark 3.48. One can give the space $\Omega(\pi)$ in the previous exercise the structure of an infinite-dimensional metrizable Fréchet manifold, so a result of Palais [Pal66] implies that it has the homotopy type of a CW-complex. By Whitehead's theorem (see e.g. [Hat02]), the exercise therefore implies that $\Omega(\pi)$ is contractible.

Remark 3.49. The notions of Lefschetz pencils and Lefschetz fibrations can be generalized to dimensions greater than four, but things become much more difficult. A Lefschetz fibration or pencil on a $2n$-dimensional manifold (see e.g. [Gom04b]) has fibers of dimension $2n - 2$, thus it is no longer a trivial question whether the fibers even admit symplectic structures, which means there are no straightforward generalizations of Theorem 3.33 or Exercise 3.47. One can however generalize the notion of a *fiberwise* symplectic structure (Definition 3.36) and show that the obvious map from the space of compatible symplectic structures to the space of fiberwise symplectic structures is a homotopy equivalence—the caveat here is that the latter space may be quite hard to understand in general, e.g. it could be empty, or its homotopy type may be completely unknown. In another direction, one can consider fibration-like objects in which the fibers are 2-dimensional but the base has dimension $2n - 2$. Now the symplectic geometry of the fibers is trivial, but one must assume more about the base—one natural assumption is to take a standard symplectic manifold such as $\mathbb{CP}^{n-1}$ for the base, which leads to the notion of a *hyperpencil*. Gompf [Gom04a] has shown that a hyperpencil always gives rise to a distinguished deformation class of symplectic forms, but it is not known whether this result has a converse involving existence of hyperpencils in general.

Chapter 4
Compactness

4.1 Two Compactness Theorems for Spaces of Embedded Spheres

In this chapter we prove a pair of compactness results for the moduli spaces of holomorphic spheres that arise in our main applications. Assume (M, ω) is a symplectic 4-manifold with an ω-tame almost complex structure J, and fix an integer $m \geqslant 0$ together with a set of pairwise distinct points $p_1, \ldots, p_m \in M$. As in Sect. 2.1.4, for any given $A \in H_2(M)$ we shall denote by

$$\mathcal{M}_{0,m}(A; J; p_1, \ldots, p_m)$$

the moduli space of unparametrized J-holomorphic spheres homologous to A with m marked points satisfying the constraint

$$\mathrm{ev}(u) = (p_1, \ldots, p_m),$$

where $\mathrm{ev} : \mathcal{M}_{0,m}(A; J) \to M^m$ is the natural evaluation map (see (2.3)). In other words, the maps $u : S^2 \to M$ representing curves in $\mathcal{M}_{0,m}(A; J; p_1, \ldots, p_m)$ each have a set of distinguished points $\zeta_1, \ldots, \zeta_m \in S^2$ such that

$$u(\zeta_1) = p_1, \quad u(\zeta_2) = p_2, \quad \ldots \quad u(\zeta_m) = p_m. \tag{4.1}$$

Recall that the virtual dimension of $\mathcal{M}_{0,m}(J; p_1, \ldots, p_m)$ is, by (2.7),

$$\text{vir-dim}\, \mathcal{M}_{0,m}(A; J; p_1, \ldots, p_m) = -2 + c_1(A) - 2m. \tag{4.2}$$

© Springer International Publishing AG, part of Springer Nature 2018
C. Wendl, *Holomorphic Curves in Low Dimensions*, Lecture Notes
in Mathematics 2216, https://doi.org/10.1007/978-3-319-91371-1_4

The union of these moduli spaces for all homology classes will be denoted by

$$\mathcal{M}_{0,m}(J; p_1, \ldots, p_m) = \bigcup_{A \in H_2(M)} \mathcal{M}_{0,m}(A; J; p_1, \ldots, p_m).$$

To obtain reasonable compactness results, we will need to impose a genericity condition on the almost complex structure. Specifically, we shall write

$$J \in \mathcal{J}^{\text{reg}}(\omega; p_1, \ldots, p_m)$$

if J has the property that for every $A \in H_2(M)$ and every ordered set $(p_{i_1}, \ldots, p_{i_r})$ of distinct points in $\{p_1, \ldots, p_m\}$, we have

$$\text{vir-dim} \, \mathcal{M}_{0,r}(A; J; p_{i_1}, \ldots, p_{i_r}) \geqslant 0$$

whenever $\mathcal{M}_{0,r}(A; J; p_{i_1}, \ldots, p_{i_r})$ contains a somewhere injective curve. By the transversality results in Sect. 2.1.4, this condition can be achieved for all J outside some meager subset of $\mathcal{J}_\tau(M, \omega)$; in fact, since the virtual dimension is always even, it also suffices to have J belonging to a generic 1-parameter family (cf. Remark 2.20).

Our results concern the following special open subsets of the above moduli space of constrained holomorphic spheres.

Definition 4.1. Let $\mathcal{M}^0_{emb}(J; p_1, \ldots, p_m) \subset \mathcal{M}_{0,m}(J; p_1, \ldots, p_m)$ denote the subset consisting of curves $u : S^2 \to M$ that are embedded and satisfy

$$[u] \cdot [u] = m - 1.$$

In the case $m = 0$ we'll denote this space simply by $\mathcal{M}^0_{emb}(J)$. Similarly, define

$$\mathcal{M}^2_{emb}(J; p_1, \ldots, p_m) \subset \mathcal{M}_{0,m}(J; p_1, \ldots, p_m)$$

or $\mathcal{M}^2_{emb}(J) \subset \mathcal{M}_0(J)$ respectively to consist of curves u that are embedded and satisfy

$$[u] \cdot [u] = m.$$

Observe that for $m = 0$, the images of curves $\mathcal{M}^0_{emb}(J)$ are exceptional spheres. The curves we will use in Chap. 6 to trace out the fibers of Lefschetz fibrations or pencils will be elements of $\mathcal{M}^2_{emb}(J; p_1, \ldots, p_m)$, and the singular fibers will be nodal curves with components in spaces of the form $\mathcal{M}^0_{emb}(J; p_{i_1}, \ldots, p_{i_r})$.

The subscripts in the above notation refer to the virtual dimensions of the moduli spaces:

Proposition 4.2. *If* $u \in \mathcal{M}^0_{emb}(J; p_1, \ldots, p_m)$, *then*

$$c_1([u]) = m + 1, \quad and \quad \text{vir-dim}\,\mathcal{M}_{0,m}([u]; J; p_1, \ldots, p_m) = 0.$$

If $u \in \mathcal{M}^2_{emb}(J; p_1, \ldots, p_m)$, *then*

$$c_1([u]) = m + 2, \quad and \quad \text{vir-dim}\,\mathcal{M}_{0,m}([u]; J; p_1, \ldots, p_m) = 2.$$

Proof. Suppose $u \in \mathcal{M}^0_{emb}(J; p_1, \ldots, p_m)$ and $[u] = A \in H_2(M)$. Since u is embedded, the adjunction formula (Theorem 2.51) gives

$$m - 1 = A \cdot A = 2\delta(u) + c_1(A) - \chi(S^2) = c_1(A) - 2,$$

thus $c_1(A) = m + 1$, and vir-dim $\mathcal{M}_{0,m}(A; J; p_1, \ldots, p_m) = 0$ follows by (4.2). Similarly if $u \in \mathcal{M}^2_{emb}(J; p_1, \ldots, p_m)$ and $[u] = A$, adjunction gives

$$m = A \cdot A = c_1(A) - 2,$$

hence $c_1(A) = m + 2$, and this implies vir-dim $\mathcal{M}_{0,m}(A; J; p_1, \ldots, p_m) = 2$. $\square$

The fact that $\mathcal{M}^0_{emb}(J; p_1, \ldots, p_m)$ has virtual dimension 0 implies that for generic J, it is a discrete space. In fact, this is true for *all* J due to automatic transversality, Theorem 2.46, but there is also an easier way to prove that $\mathcal{M}^0_{emb}(J; p_1, \ldots, p_m)$ is discrete, without knowledge of transversality:

Proposition 4.3. *If u and u' are any two distinct elements of* $\mathcal{M}^0_{emb}(J; p_1, \ldots, p_m)$, *then* $[u] \neq [u'] \in H_2(M)$.

Proof. Assume u and u' are inequivalent curves but are homologous. Then they cannot both be covers of the same simple curve, thus they have at most finitely many intersections, and since they are forced to intersect at $p_1, \ldots, p_m$, positivity of intersections implies the contradiction

$$m - 1 = [u] \cdot [u] = [u] \cdot [u'] \geq m.$$

$\square$

By a completely analogous argument, we have:

Proposition 4.4. *If u and u' are two inequivalent but homologous elements of the space* $\mathcal{M}^2_{emb}(J; p_1, \ldots, p_m)$, *then they intersect only at the constraint points* $p_1, \ldots, p_m$, *and those intersections are transverse.* $\square$

The local structure of $\mathcal{M}^2_{emb}(J; p_1, \ldots, p_m)$ was discussed already in Sect. 2.2.3: in particular, the combination of Proposition 2.53 with Proposition

4.4 above implies that $\mathcal{M}^2_{\text{emb}}(J; p_1, \ldots, p_m)$ is a smooth 2-manifold consisting of embedded curves that intersect each other only at the constraint points $p_1, \ldots, p_m$ and foliate an open subset of $M \backslash \{p_1, \ldots, p_m\}$. Our goal now is to understand the *global* structure of $\mathcal{M}^2_{\text{emb}}(J; p_1, \ldots, p_m)$, namely its compactness or lack thereof. As a first step, we must also study the compactness of $\mathcal{M}^0_{\text{emb}}(J; p_1, \ldots, p_m)$.

The proofs of the following two theorems will occupy the bulk of this chapter. For both statements, we fix a C^∞-convergent sequence of symplectic forms

$$\omega_k \to \omega_\infty,$$

together with a corresponding C^∞-convergent sequence of tame almost complex structures

$$J_k \to J_\infty, \qquad J_k \in \mathcal{J}_\tau(M, \omega_k), \qquad k = 1, \ldots, \infty.$$

Theorem 4.5. *Assume* $J_\infty \in \mathcal{J}^{\text{reg}}(\omega_\infty; p_1, \ldots, p_m)$ *and* $u_k \in \mathcal{M}^0_{\text{emb}}(J_k; p_1, \ldots, p_m)$ *is a sequence satisfying a uniform energy bound*

$$\int_{\omega_k} u_k^* \omega_k \leqslant C$$

for some constant $C > 0$. *Then* u_k *has a subsequence converging to an embedded curve* $u_\infty \in \mathcal{M}^0_{\text{emb}}(J_\infty; p_1, \ldots, p_m)$.

This implies in particular that for any fixed generic $J \in \mathcal{J}^{\text{reg}}(\omega; p_1, \ldots, p_m)$, any subset of $\mathcal{M}^0_{\text{emb}}(J; p_1, \ldots, p_m)$ satisfying a given energy bound is finite. The situation for $\mathcal{M}^2_{\text{emb}}(J; p_1, \ldots, p_m)$ is slightly more complicated: we will not have compactness in general, but we can give a very precise description of its compactification, the upshot of which is that the only possible nodal curves look like Lefschetz singular fibers.

Theorem 4.6. *Assume* $J_\infty \in \mathcal{J}^{\text{reg}}(\omega_\infty; p_1, \ldots, p_m)$ *and* $u_k \in \mathcal{M}^2_{\text{emb}}(J_k; p_1, \ldots, p_m)$ *is a sequence satisfying a uniform energy bound*

$$\int_{\omega_k} u_k^* \omega_k \leqslant C$$

for some constant $C > 0$. *Then a subsequence of* u_k *converges to one of the following:*

(1) An embedded curve $u_\infty \in \mathcal{M}^2_{\text{emb}}(J_\infty; p_1, \ldots, p_m)$;
(2) A nodal curve with exactly two smooth components

$$v_+ \in \mathcal{M}^0_{\text{emb}}(J_\infty; p_{i_1}, \ldots, p_{i_r}), \qquad v_- \in \mathcal{M}^0_{\text{emb}}(J_\infty; p_{i_{r+1}}, \ldots, p_{i_m})$$

for some permutation $i_1, \ldots, i_m$ *of* $1, \ldots, m$ *and* $0 \leqslant r \leqslant m$, *where* v_+ *and* v_- *have exactly one intersection, which is transverse and lies in* $M \backslash \{p_1, \ldots, p_m\}$.

Moreover, for any given energy bound $C > 0$, there exist at most finitely many nodal curves of the second type.

Theorem 4.6 says essentially that the closure

$$\overline{\mathcal{M}}_{emb}^2(J_\infty; p_1, \ldots, p_m) \subset \overline{\mathcal{M}}_{0,m}(J_\infty)$$

of $\mathcal{M}_{emb}^2(J_\infty; p_1, \ldots, p_m)$ in the Gromov compactification is much nicer than one could a priori expect—while a sequence of embedded holomorphic curves may in general converge to a nodal curve with many non-embedded (e.g. multiply covered) components, this does not happen in $\overline{\mathcal{M}}_{emb}^2(J_\infty; p_1, \ldots, p_m)$. In order to prove the theorem, we will have to use the adjunction formula in concert with the genericity assumption to rule out all possible nodal limits that are less well behaved.

Given the compactness theorem, Proposition 4.4 can now be extended as follows:

Proposition 4.7. *Given $J_\infty \in \mathcal{J}^{reg}(\omega; p_1, \ldots, p_m)$ as in Theorem 4.6, suppose $u, v \in \overline{\mathcal{M}}_{emb}^2(J_\infty; p_1, \ldots, p_m)$ are two inequivalent (possibly nodal) curves representing the same homology class. Then u and v intersect each other only at the constraint points $p_1, \ldots, p_m$, and those intersections are transverse.*

Proof. The case where neither u nor v has nodes has already been dealt with in Proposition 4.4, using positivity of intersections. Essentially the same argument works if u is a nodal curve and v is smooth, or if both curves are nodal but neither of the components of u is a reparametrization of a component of v. To finish, we claim that if u and v are both nodal curves with components (u^1, u^2) and (v^1, v^2) respectively and u^1 has the same image as v^1, then the two nodal curves are equivalent. Indeed, suppose $u^1 = v^1$. Then both of these curves pass through a certain subset of the constraint points $p_1, \ldots, p_m$, and the points that they miss must be hit by both u^2 and v^2, so it follows that for each $i = 1, 2$, u^i and v^i hit exactly the same sets of constraint points. In particular, if $m = m_1 + m_2$ where u^i hits m_i constraint points for $i = 1, 2$, then unless u^2 and v^2 are equivalent curves, positivity of intersections implies

$$[u^2] \cdot [v^2] \geq m_2.$$

But by Theorem 4.6, $[u^1] \cdot [u^1] = m_1 - 1$, while $[u^1] \cdot [u^2] = [v^1] \cdot [v^2] = 1$, so the assumption $u^1 = v^1$ implies

$$\begin{aligned}
m = [u] \cdot [v] &= [u^1] \cdot [v^1] + [u^2] \cdot [v^1] + [u^1] \cdot [v^2] + [u^2] \cdot [v^2] \\
&= [u^1] \cdot [u^1] + [u^2] \cdot [u^1] + [v^1] \cdot [v^2] + [u^2] \cdot [v^2] \\
&\geq (m_1 - 1) + 1 + 1 + m_2 = m + 1,
\end{aligned}$$

a contradiction. $\qquad\square$

4.2   Index Counting in Dimension Four

We will give complete proofs of Theorems 4.5 and 4.6 over the next three sections, beginning with the unconstrained cases ($m = 0$) and then treating the general case in Sect. 4.4. One fundamental piece of intuition behind both results is the notion that, *generically*, nodal degenerations are a "codimension two phenomenon," hence they should not happen at all in a 0-dimensional space, and should happen at most finitely many times in a 2-dimensional space. The same intuition underlies the construction of the Gromov-Witten invariants (see Sect. 7.2), though it has some technical complications in general, as the "codimension two" claim is not strictly true unless one can achieve transversality for all curves, including the multiple covers. As explained in [MS12], this problem can be circumvented in any symplectic manifold that satisfies a technical condition known as *semipositivity*, as multiply covered curves can then be confined to subsets whose dimension is sufficiently small. As luck would have it, *all* symplectic manifolds of dimension four and six are semipositive, but in dimension four, one can also proceed in a somewhat simpler manner by deriving index relations between nodal or multiply covered curves and their simple components. We will derive such relations in this section as preparation for proving the compactness theorems.

Suppose M is a $2n$-dimensional almost complex manifold with a C^∞-convergent sequence of almost complex structures $J_k \to J_\infty$, and we have a sequence $u_k \in \mathcal{M}_g(A; J_k)$ with

$$u_k \to u_\infty \in \overline{\mathcal{M}}_g(A; J_\infty),$$

where the stable nodal curve $u_\infty := [(S, j, u_\infty, \varnothing, \Delta)]$ has connected components

$$\left\{ u_\infty^i := [(S_i, j, u_\infty^i, \varnothing)] \in \mathcal{M}_{g_i}(A_i; J_\infty) \right\}_{i=1,\dots,V}.$$

Here, $\sum_{i=1}^{V} A_i = A$, and some of the smooth curves u_∞^i may be constant; the latter is the case if and only if $A_i = 0$. For $i = 1, \dots, V$, let N_i denote the number of nodal points on the component u_∞^i, i.e. $N_i = |S_i \cap \Delta|$, so stability (see Definition 2.34) implies

$$\chi(S_i) - N_i < 0 \quad \text{whenever} \quad A_i = 0. \tag{4.3}$$

Recall that nodal points always come in pairs, so one can define the integer $E := \frac{1}{2} \sum_{i=1}^{V} N_i$. Then u_∞ determines a graph with $V \geqslant 1$ vertices corresponding to the connected components $u_\infty^1, \dots, u_\infty^V$ and $E \geqslant 0$ edges corresponding to the nodes. We can deduce a relation between the numbers $g_1, \dots, g_V$ and g by observing that if Σ denotes a surface diffeomorphic to the domains of the curves u_k, then Σ can be constructed by gluing together the components S_i, each with N_i disks removed, hence

$$\chi(\Sigma) = \sum_{i=1}^{V} [\chi(S_i) - N_i] = \sum_{i=1}^{V} \chi(S_i) - 2E. \qquad (4.4)$$

Let us define the *index* of a limiting nodal curve such that it matches the indices of the curves in the sequence, i.e.

$$\text{ind}(u_\infty) := \text{ind}(u_k) = (n-3)\chi(\Sigma) + 2c_1(A).$$

Since $\text{ind}(u_\infty^i) = (n-3)\chi(\Sigma_i) + 2c_1(A_i)$, combining (4.4) with the relation $\sum_i A_i = A$ then gives

$$\text{ind}(u_\infty) = \sum_{i=1}^{V} \left[\text{ind}(u_\infty^i) - (n-3)N_i \right].$$

Observe that whenever u_∞^i is a constant component, we have $\text{ind}(u_\infty^i) - (n-3)N_i = (n-3)[\chi(S_i) - N_i]$, where the term in brackets is negative due to the stability condition (4.3). In particular, in dimension four, we plug in $n = 2$ and find the appealing relation

$$\text{ind}(u_\infty) = \sum_{i=1}^{V} \left[\text{ind}(u_\infty^i) + N_i \right],$$

where each term in the sum corresponding to a constant component must be *strictly positive*. Notice that unless u_∞ has no nodes at all, there is always a contribution of $N_i \geq 1$ from at least one nonconstant component, and strictly more than this unless there are also constant components, which also contribute positively in the sum. This proves:

Proposition 4.8. *Suppose (M, J) is an almost complex 4-manifold and u_∞ is a nonconstant nodal J-holomorphic curve in M with connected components $u_\infty^1, \ldots, u_\infty^V$, each component u_∞^i having N_i nodal points. Then*

$$\text{ind}(u_\infty) \geq \sum_{\{i \,|\, u_\infty^i \neq \text{const}\}} \left[\text{ind}(u_\infty^i) + N_i \right],$$

with equality if and only if there are no constant components. In particular, we have

$$\text{ind}(u_\infty) \geq 2 + \sum_{\{i \,|\, u_\infty^i \neq \text{const}\}} \text{ind}(u_\infty^i)$$

unless u_∞ has no nodes. ☐

One can see why such a relation might be useful when $\operatorname{ind}(u_k)$ is small: if transversality can be achieved for all the components u_∞^i, then $\operatorname{ind}(u_\infty^i) \geqslant 0$ will imply $\operatorname{ind}(u_\infty) \geqslant 2$ if u_∞ has nodes. So this rules out nodal degenerations entirely if $\operatorname{ind}(u_k) = 0$, and it places severe constraints on the possible nodal curves if $\operatorname{ind}(u_k) = 2$.

Of course one cannot generally assume that transversality holds for all the nonconstant components u_∞^i: some of these might in principle be multiply covered and have negative index, even if J is generic. Recall however that every curve covers an underlying simple curve, which certainly will have nonnegative index for generic J, so one can still derive constraints on $\operatorname{ind}(u_\infty^i)$ by understanding the index relations between a curve and its multiple covers.

Suppose again that (M, J) is a $2n$-dimensional almost complex manifold, $u : (\Sigma, j) \to (M, J)$ is a closed J-holomorphic curve with $[u] = A \in H_2(M)$, and $\varphi : (\widetilde{\Sigma}, \widetilde{j}) \to (\Sigma, j)$ is a holomorphic map of degree $k \geqslant 1$; recall that φ is necessarily a biholomorphic diffeomorphism if $k = 1$, and it is a branched cover if $k > 1$, with a branch point wherever $d\varphi = 0$. This gives rise to another J-holomorphic curve

$$\widetilde{u} := u \circ \varphi : (\widetilde{\Sigma}, \widetilde{j}) \to (M, J),$$

with $[\widetilde{u}] =: \widetilde{A} = kA \in H_2(M)$. To relate $\operatorname{ind}(\widetilde{u})$ and $\operatorname{ind}(u)$, we will need the following relation between $\chi(\Sigma)$ and $\chi(\widetilde{\Sigma})$:

Proposition 4.9 (Riemann-Hurwitz Formula). *Suppose $\varphi : (\widetilde{\Sigma}, \widetilde{j}) \to (\Sigma, j)$ is a nonconstant holomorphic map of closed Riemann surfaces, with $k := \deg(\varphi) \geqslant 1$. Let $Z(d\varphi) \in \mathbb{Z}$ denote the sum of the orders of the critical points of φ, defined as the orders of the zeroes of $d\varphi \in \Gamma\left(\operatorname{Hom}_{\mathbb{C}}\left(T\widetilde{\Sigma}, \varphi^* T\Sigma\right)\right)$. Then*

$$-\chi(\widetilde{\Sigma}) + k\chi(\Sigma) = Z(d\varphi) \geqslant 0.$$

Proof. The integer $Z(d\varphi)$ is an algebraic count of the zeroes of a nontrivial holomorphic section of a holomorphic line bundle, namely

$$\operatorname{Hom}_{\mathbb{C}}\left(T\widetilde{\Sigma}, \varphi^* T\Sigma\right) = \left(T\widetilde{\Sigma}\right)^* \otimes \varphi^* T\Sigma,$$

hence it computes the first Chern number of this bundle:

$$Z(d\varphi) = c_1\left((T\widetilde{\Sigma})^* \otimes \varphi^* T\Sigma\right) = -c_1\left(T\widetilde{\Sigma}\right) + c_1\left(\varphi^* T\Sigma\right)$$

$$= -\chi(\widetilde{\Sigma}) + \deg(\varphi) \cdot c_1(T\Sigma) = -\chi(\widetilde{\Sigma}) + k\chi(\Sigma).$$

The observation that $Z(d\varphi)$ is nonnegative follows immediately from the fact that nontrivial holomorphic $\mathbb{C}$-valued functions on domains in $\mathbb{C}$ can only have isolated and positive zeroes. $\square$

Exercise 4.10. Use the Riemann-Hurwitz formula to give a new proof of Proposition 2.7, that multiply covered J-holomorphic spheres are always covers of other spheres, never curves with higher genus.

Writing $\operatorname{ind}(\tilde{u}) = (n-3)\chi(\widetilde{\Sigma}) + 2c_1(\widetilde{A})$ and $\operatorname{ind}(u) = (n-3)\chi(\Sigma) + 2c_1(A)$, then plugging in $\widetilde{A} = kA$ and Proposition 4.9, we find

$$\operatorname{ind}(\tilde{u}) = k\operatorname{ind}(u) - (n-3)Z(d\varphi).$$

This is most useful when dim $M = 4$, as then $n = 2$ implies:

Proposition 4.11. *Suppose u is a closed nonconstant J-holomorphic curve in an almost complex manifold of dimension four, $\tilde{u} = u \circ \varphi$ is a k-fold branched cover of u for some $k \geqslant 1$, and $Z(d\varphi)$ is the algebraic count of branch points. Then*

$$\operatorname{ind}(\tilde{u}) = k\operatorname{ind}(u) + Z(d\varphi).$$

In particular, $\operatorname{ind}(\tilde{u}) \geqslant k\operatorname{ind}(u)$, *with equality if and only if the cover has no branch points.* □

We now combine the above results to prove a lemma that will play a key role in the unconstrained (i.e. $m = 0$) cases of Theorems 4.5 and 4.6. The crucial assumption we need is that all somewhere injective J-holomorphic curves have nonnegative index—this holds for generic choices of J, and by Remark 2.20, it even holds for all J in a generic 1-parameter family.

Lemma 4.12. *Suppose (M, J) is an almost complex 4-manifold admitting no somewhere injective J-holomorphic curves with negative index, and $u_\infty \in \overline{\mathcal{M}}_0(A; J)$ is a nonconstant stable nodal J-holomorphic sphere. If $\operatorname{ind}(u_\infty) = 0$, then u_∞ is a smooth (i.e. non-nodal) and simple curve. If $\operatorname{ind}(u_\infty) = 2$, then it is one of the following:*

- *A smooth simple curve;*
- *A smooth branched double cover of a simple J-holomorphic sphere with index 0;*
- *A nodal curve with exactly two components connected by a single node, where both components are simple J-holomorphic spheres with index 0.*

Proof. By Proposition 4.11, we may assume that *all* (not only simple) J-holomorphic curves in M have nonnegative index. It then follows immediately from Proposition 4.8 that if $\operatorname{ind}(u_\infty) = 0$ then u_∞ has no nodes, i.e. it is a smooth J-holomorphic sphere $u_\infty : S^2 \to M$. If $u_\infty = v \circ \varphi$ for some simple curve $v : \Sigma \to M$ and branched k-fold cover $\varphi : S^2 \to \Sigma$, then Proposition 4.11 implies that φ has no branch points, $Z(d\varphi) = 0$. But by Proposition 2.7 or Exercise 4.10,

Σ must also be a sphere, and the Riemann-Hurwitz formula (Proposition 4.9) then gives

$$Z(d\varphi) = -\chi(S^2) + k\chi(S^2) = 2k - 2,$$

hence $0 = Z(d\varphi) = 2k - 2$ implies $k = 1$, meaning u_∞ is a simple curve.

Suppose now that $\mathrm{ind}(u_\infty) = 2$ and u_∞ has no nodes and is a k-fold cover $v \circ \varphi$ of a simple curve v. By Proposition 2.7 or Exercise 4.10, v must also be a sphere, so the Riemann-Hurwitz formula gives $Z(d\varphi) = 2k - 2$, and Proposition 4.11 then gives

$$2 = \mathrm{ind}(u_\infty) = k\,\mathrm{ind}(v) + 2(k - 1) \geqslant 2(k - 1)$$

since $\mathrm{ind}(v) \geqslant 0$. This implies that either $k = 2$ and $\mathrm{ind}(v) = 0$, or u_∞ is simple.

Finally, suppose $\mathrm{ind}(u_\infty) = 2$ and u_∞ has nodes; denote its connected components by $u_\infty^1, \ldots, u_\infty^V$. These components must all be spheres since u_∞ has arithmetic genus 0. By Exercise 4.13 below, at least two of the nodal points must lie on nonconstant components, so Proposition 4.8 then implies that all components are nonconstant and have index 0, and there can be at most (and therefore exactly) two of them, each with exactly one nodal point, forming a single nodal pair. By the same argument used above for the case $\mathrm{ind}(u_\infty) = 0$, the index 0 components cannot be multiple covers. □

Exercise 4.13. Suppose $u := [(S, j, u, \Theta, \Delta)] \in \overline{\mathcal{M}}_{0,m}(A; J)$ is a nonconstant stable nodal J-holomorphic curve with arithmetic genus 0 and m marked points, such that no ghost bubble (i.e. constant component) in u has more than one of the marked points.

  (a) Show that if u has a ghost bubble, then it also has at least two nonconstant connected components.
  (b) Show that if u has a marked point $\zeta \in \Theta$ lying on a constant component, then it also has at least two distinct nodal points $z, z' \in \Delta$ that lie on nonconstant components and satisfy $u(z) = u(z') = u(\zeta)$.

Hint: the stability condition is crucial here. Think about the graph with vertices corresponding to connected components of S and edges corresponding to nodes. Since the arithmetic genus is 0, this graph must be a tree, i.e. it cannot have any nontrivial loops.

Exercise 4.14. Find an alternative proof of Lemma 4.12 by relating the first Chern numbers of the nodal curve and its connected components. *Note: You will probably find the alternative argument simpler, but it is very specific to the situation at hand, while the index counting arguments we've discussed in this section are more widely applicable, e.g. we will apply them to the constrained case in Sect. 4.4, and to the definition of Gromov-Witten invariants in Sect. 7.2.*

Remark 4.15. Though we have not used any intersection theory in this section, we have nonetheless made use of the assumption $n = 2$ several times. In higher dimensions, such index relations typically do not work out so favorably, though with a bit more effort one can still carry out similar arguments in the semipositive case, see [MS12].

4.3   Proof of the Compactness Theorems When $m = 0$

Consider $J_k \rightarrow J_\infty$ as in the statements of Theorems 4.5 and 4.6, and assume the sequence u_k belongs to $\mathcal{M}_{emb}^0(J_k)$ or $\mathcal{M}_{emb}^2(J_k)$ and satisfies a uniform energy bound. By Gromov compactness, we may without loss of generality replace u_k by a subsequence that converges to a stable nodal J_∞-holomorphic sphere $u_\infty \in \overline{\mathcal{M}}_0(J_\infty)$. After taking a further subsequence, we may also assume

$$[u_\infty] = [u_k] \in H_2(M) \quad \text{for all } k. \tag{4.5}$$

The preparations of the previous section make the proof of Theorem 4.5 almost immediate: Lemma 4.12 implies namely that if $\text{ind}(u_\infty) = \text{ind}(u_k) = 0$, then u_∞ is a simple J_∞-holomorphic sphere. In light of (4.5), the adjunction formula (Theorem 2.51) now implies that since u_k is embedded, u_∞ is as well, and

$$[u_\infty] \cdot [u_\infty] = [u_k] \cdot [u_k] = -1,$$

so $u_\infty \in \mathcal{M}_{emb}^0(J_\infty)$ as claimed.

For Theorem 4.6, we instead assume $u_k \in \mathcal{M}_{emb}^2(J_k)$, so $[u_k] \cdot [u_k] = 0$ and $\text{ind}(u_k) = 2$. Lemma 4.12 now allows the following possibilities:

(1) u_∞ is a smooth simple J_∞-holomorphic sphere;
(2) u_∞ is a branched double cover of a simple J_∞-holomorphic sphere v with $\text{ind}(v) = 0$;
(3) u_∞ is a nodal curve with a single node connecting two components u^1, u^2, both of which are simple J_∞-holomorphic spheres with index 0.

In the first case, applying the adjunction formula as in the previous paragraph shows that $u_\infty \in \mathcal{M}_{emb}^2(J_\infty)$. We claim now that the second case can never happen. If it does, then since v is simple, it satisfies the adjunction formula

$$[v] \cdot [v] = 2\delta(v) + c_1([v]) - 2 = 2\delta(v) - 1,$$

where we've plugged in $c_1([v]) = 1$ since $\text{ind}(v) = -\chi(S^2) + 2c_1([v]) = 0$. Its homological self-intersection number is therefore odd. But since $[u_\infty] = 2[v]$, we then have

$$0 = [u_k] \cdot [u_k] = [u_\infty] \cdot [u_\infty] = 4[v] \cdot [v] \quad \Rightarrow \quad [v] \cdot [v] = 0,$$

implying that zero is an odd number.

To understand the third case, we still must distinguish two possibilities: either u^1 and u^2 are the same curve up to parametrization, or they are distinct curves that intersect each other at most finitely many times. In the first case, we can call them both v and repeat the homological adjunction calculation above with $[u_\infty] = 2[v]$, leading to the same contradiction. Thus u^1 and u^2 must be distinct curves, and since they necessarily intersect each other at a node, positivity of intersections (Theorem 2.49) implies

$$[u^1] \cdot [u^2] \geq 1.$$

They also each satisfy the adjunction formula since they are simple curves, so using $[u_\infty] = [u^1] + [u^2]$, we find

$$
\begin{aligned}
0 = [u_\infty] \cdot [u_\infty] &= [u^1] \cdot [u^1] + [u^2] \cdot [u^2] + 2[u^1] \cdot [u^2] \\
&= 2\delta(u^1) + c_1([u^1]) - 2 + 2\delta(u^2) + c_1([u^2]) - 2 + 2[u^1] \cdot [u^2] \\
&= 2\delta(u^1) + 2\delta(u^2) + 2([u^1] \cdot [u^2] - 1),
\end{aligned}
$$

where we've plugged in $c_1([u^i]) = 1$ since $\mathrm{ind}(u^i) = -\chi(S^2) + 2c_1([u^i]) = 0$ for $i = 1, 2$. Each term on the right hand side is nonnegative, therefore they are all 0, so we have $\delta(u^1) = \delta(u^2) = 0$ implying that both components are embedded, and $[u^1] \cdot [u^2] = 1$, meaning that their obvious intersection is the only one, and is transverse. Finally, applying the adjunction formula again to u^i for $i = 1, 2$ gives

$$[u^i] \cdot [u^i] = 2\delta(u^i) + c_1([u^i]) - 2 = -1,$$

thus $u^i \in \mathcal{M}^0_{emb}(J_\infty)$, and the compactness statement for $\mathcal{M}^0_{emb}(J_\infty)$ then implies that there are finitely many such curves satisfying the given energy bound. This completes the proof of Theorem 4.6 in the $m = 0$ case.

4.4 Proof of the Compactness Theorems with Constraints

We will prove both theorems for $m > 0$ using essentially the same arguments as in the previous two sections. Where the general argument seems more complicated, it is mostly a matter of extra bookkeeping: the point is to keep track of those terms that must always be nonnegative and show in the end that they must all be zero.

As in the statements of the theorems, we assume $p_1, \ldots, p_m \in M$ are pairwise distinct points and the limiting almost complex structure $J_\infty = \lim_k J_k$ belongs to the subset $\mathcal{J}^{reg}(\omega_\infty; p_1, \ldots, p_m)$, which will prevent the existence of somewhere injective J_∞-holomorphic curves lying in constrained moduli spaces of negative virtual dimension, with any subset of $\{p_1, \ldots, p_m\}$ taken as constraints. Given a sequence u_k of curves in $\mathcal{M}^0_{emb}(J_k; p_1, \ldots, p_m)$ or $\mathcal{M}^2_{emb}(J_k; p_1, \ldots, p_m)$ that satisfy a uniform energy bound, we can apply Gromov compactness to replace u_k by a subsequence of curves representing a fixed homology class $[u_k] \in H_2(M)$ and converging to a stable nodal curve $u_\infty \in \overline{\mathcal{M}}_{0,m}(J_\infty)$.

We will first use the index relations of Sect. 4.2 to deduce as much as possible about u_∞. Recall from Sect. 2.1.4 that whenever $p_1, \ldots, p_m$ are distinct points in an almost complex 4-manifold (M, J), the index of a constrained J-holomorphic curve $u \in \mathcal{M}_{g,m}(J; p_1, \ldots, p_m)$ is related to the virtual dimension of the constrained moduli space by

$$\text{vir-dim } \mathcal{M}_{g,m}([u]; J; p_1, \ldots, p_m) = \text{ind}(u) - 2m.$$

In the situation at hand, it will be convenient to consider the nodal curve $\widehat{u}_\infty \in \overline{\mathcal{M}}_0(J_\infty)$ defined by deleting the marked points from $u_\infty \in \overline{\mathcal{M}}_{0,m}(J_\infty)$ and then stabilizing, i.e. we operate on u_∞ with the natural map

$$\overline{\mathcal{M}}_{0,m}(J_\infty) \to \overline{\mathcal{M}}_0(J_\infty).$$

This map is defined by deleting all marked points and then "collapsing" any ghost bubbles (i.e. constant spherical components) that become unstable as a result, so $\widehat{u}_\infty$ has all the same nonconstant components as u_∞ but retains only those constant components for which the stability condition is satisfied (see Fig. 4.1). Notice that u_∞ and $\widehat{u}_\infty$ have the same arithmetic genus and represent the same homology class, so $\text{ind}(u_\infty) = \text{ind}(\widehat{u}_\infty)$.

Suppose $\widehat{u}_\infty$ has connected components $u_\infty^1, \ldots, u_\infty^V$, and N_i denotes the number of nodal points on u_∞^i for $i = 1, \ldots, V$. In light of the marked point constraints on u_∞, each of the points $p_1, \ldots, p_m$ is in the image of some (not necessarily unique) nonconstant component of $\widehat{u}_\infty$; note that marked points of u_∞ can appear on

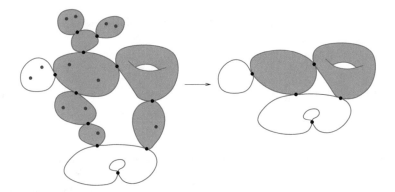

Fig. 4.1 An example illustrating the map $\overline{\mathcal{M}}_{g,m}(J) \to \overline{\mathcal{M}}_g(J)$, which deletes all marked points and then stabilizes the result by collapsing non-stable ghost bubbles. In this scenario with $g = 3$ and $m = 12$, we have a nodal curve with eight constant components (shown shaded in the picture), six of which must be collapsed in order to define a stable curve without marked points.

constant components, but whenever this happens, there is also a nonconstant component that maps a nodal point to the same constraint point. We can therefore choose integers $m_i \geq 0$ with $\sum_{i=1}^{V} m_i = m$ and $m_i = 0$ whenever u_∞^i is constant, such that each nonconstant component u_∞^i lifts to a curve in $\mathcal{M}_{0,m_i}(J_\infty)$ that maps its m_i marked points to distinct points in $\{p_1, \ldots, p_m\}$. In this case, u_∞^i can be regarded as belonging to a moduli space of J_∞-holomorphic spheres with m_i constrained marked points, which has virtual dimension $\mathrm{ind}(u_\infty^i) - 2m_i$. By Proposition 4.8, these virtual dimensions are related to vir-dim $\mathcal{M}_{0,m}([u_\infty]; J_\infty; p_1, \ldots, p_m) = \mathrm{ind}(u_\infty) - 2m \in \{0, 2\}$ by

$$\mathrm{ind}(u_\infty) - 2m \geq \sum_{\{i \,|\, u_\infty^i \neq \mathrm{const}\}} \left[\mathrm{ind}(u_\infty^i) - 2m_i + N_i \right], \tag{4.6}$$

with equality if and only if $\hat{u}_\infty$ has no constant components. If u_∞^i is a nonconstant component, then $u_\infty^i = v^i \circ \varphi_i$ for some simple J_∞-holomorphic sphere v^i and holomorphic map $\varphi_i : S^2 \to S^2$ of degree $k_i \geq 1$, and since the constraint points are all distinct, φ_i maps the m_i chosen marked points of u_∞^i to distinct points in the domain of v^i. We can therefore also regard v^i as a curve with m_i constrained marked points, so the genericity assumption $J_\infty \in \mathcal{J}^{\mathrm{reg}}(\omega_\infty; p_1, \ldots, p_m)$ implies

$$\mathrm{ind}(v^i) - 2m_i \geq 0.$$

By Proposition 4.11 and the Riemann-Hurwitz formula (Proposition 4.9), it follows that

$$\begin{aligned}
\mathrm{ind}(u_\infty^i) - 2m_i &= k_i \, \mathrm{ind}(v^i) + Z(d\varphi_i) - 2m_i \\
&= k_i \left[\mathrm{ind}(v^i) - 2m_i \right] + 2(k_i - 1) + 2m_i(k_i - 1) \\
&= k_i \left[\mathrm{ind}(v^i) - 2m_i \right] + 2(m_i + 1)(k_i - 1).
\end{aligned} \tag{4.7}$$

Combining (4.6) and (4.7), we now have

$$\mathrm{ind}(u_\infty) - 2m \geq \sum_{\{i \,|\, u_\infty^i \neq \mathrm{const}\}} \left(k_i \left[\mathrm{ind}(v^i) - 2m_i \right] + 2(m_i + 1)(k_i - 1) + N_i \right), \tag{4.8}$$

where the summands on the right hand side are sums of three nonnegative terms, and equality is achieved if and only if $\hat{u}_\infty$ has no constant components. If $u_k \in \mathcal{M}_{\mathrm{emb}}^0(J_k; p_1, \ldots, p_m)$, then $\mathrm{ind}(u_\infty) - 2m = 0$ and we conclude that $\hat{u}_\infty$ has no nodes and is a simple curve. This does not immediately imply the same for u_∞, which may in principle be a nodal curve with only one nonconstant component and additional ghost bubbles that become unstable when marked points are removed. Observe however that since all the constraint points are distinct, no ghost bubble can have more than one marked point. It then follows from Exercise 4.13 that u_∞ does not have any ghost bubbles, so we've proved:

Lemma 4.16. *If $u_k \in \mathcal{M}^0_{emb}(J_k; p_1, \ldots, p_m)$, then u_∞ is a smooth simple J_∞-holomorphic sphere.* □

Since curves $u_k \in \mathcal{M}^0_{emb}(J_k; p_1, \ldots, p_m)$ are embedded, the adjunction formula implies that the simple curve u_∞ of Lemma 4.16 is also embedded, and in this case

$$[u_\infty] \cdot [u_\infty] = [u_k] \cdot [u_k] = m - 1,$$

so $u_\infty \in \mathcal{M}^0_{emb}(J_\infty; p_1, \ldots, p_m)$ and the proof of Theorem 4.5 is complete.

When $u_k \in \mathcal{M}^2_{emb}(J_\infty; p_1, \ldots, p_m)$, the left hand side of (4.8) is 2. If $\hat{u}_\infty$ has nodes, then Exercise 4.13 implies $\sum_i N_i \geq 2$, thus equality in (4.8) is achieved, there are no constant components, there are exactly two nonconstant components u^1_∞ and u^2_∞ connected by a single pair of nodal points, and they satisfy

$$k_1 = k_2 = 1, \qquad \text{ind}(u^1_\infty) - 2m_1 = \text{ind}(u^2_\infty) - 2m_2 = 0. \qquad (4.9)$$

Putting back the original marked points, u_∞ may in principle have additional ghost bubbles, but if this happens, then since each ghost bubble can have at most one marked point, Exercise 4.13 implies that u^1_∞ and u^2_∞ each map their nodal points to one of the constraint points $p_1, \ldots, p_m$. At least one of these constraints was not accounted for when we used genericity to deduce $\text{ind}(u^i_\infty) - 2m_i \geq 0$; putting in the extra constraint, we find a component u^i_∞ that actually satisfies

$$\text{ind}(u^i_\infty) - 2(m_i + 1) \geq 0.$$

This turns (4.8) into a *strict* inequality and thus gives a contradiction. It follows that u_∞ also has exactly two components u^1_∞ and u^2_∞, both nonconstant and simple. The m marked points are distributed among the two components at points separate from the node, so we can say u^1_∞ and u^2_∞ have m'_1 and m'_2 constrained marked points respectively, where $m'_1 + m'_2 = m$. Since both are simple and $J \in \mathcal{J}^{reg}(\omega_\infty; p_1, \ldots, p_m)$, we have $\text{ind}(u^i_\infty) - 2m'_i \geq 0$ for $i = 1, 2$, which implies in light of (4.9) that $m_1 - m'_1$ and $m_2 - m'_2$ are both nonnegative. But since $m_1 + m_2 = m'_1 + m'_2 = m$, this implies $m_1 = m'_1$ and $m_2 = m'_2$.

Suppose next that $\hat{u}_\infty$ has no nodes. Applying Exercise 4.13 again with the fact that no ghost bubble of u_∞ can have more than one marked point, this implies that u_∞ is also a smooth curve with no nodes. If $u_\infty = v \circ \varphi$ for a simple curve v and a holomorphic map $\varphi : S^2 \to S^2$ of degree $k \geq 1$, then (4.8) now reduces to

$$2 = \text{ind}(u_\infty) - 2m = k\left[\text{ind}(v) - 2m\right] + 2(m + 1)(k - 1),$$

so either $k = 1$, or $k = 2$ and $m = 0$. We've proved:

Lemma 4.17. *If $u_k \in \mathcal{M}^2_{\mathrm{emb}}(J_k; p_1, \ldots, p_m)$, then u_∞ is one of the following:*

(1) A smooth simple curve;

(2) A smooth branched double cover of a simple J-holomorphic sphere with index 0 (arising only in the case $m = 0$);

(3) A nodal curve with exactly two components connected by a single node, where the two components u^1 and u^2 are simple J-holomorphic spheres with m_1 and m_2 marked points respectively, such that $\mathrm{ind}(u^i) = 2m_i$ for $i = 1, 2$.

$\square$

The second possibility was ruled out already in our proof for the $m = 0$ case, see Sect. 4.3. If the first possibility occurs, then the adjunction formula implies as usual that u_∞ is embedded, so we conclude $u_\infty \in \mathcal{M}^2_{\mathrm{emb}}(J_\infty; p_1, \ldots, p_m)$.

In the third case, we can first rule out the possibility $u^1 = u^2$. Indeed, since both curves satisfy $\mathrm{ind}(u^i) = 2m_i$, our genericity condition implies that neither can pass through any of the constraint points $p_1, \ldots, p_m$ anywhere other than at its m_i marked points. Since the constraint points are all distinct, this permits $u^1 = u^2$ only if $m_1 = m_2 = 0$, in which case it was already ruled out in Sect. 4.3. Now assuming $u^1 \neq u^2$, note that since $\mathrm{ind}(u^i) = -\chi(S^2) + 2c_1([u^i]) = 2m_i$, we have $c_1([u^i]) = m_i + 1$ for $i = 1, 2$. Let us compute $[u_\infty] \cdot [u_\infty] - m$, applying the adjunction formula to each of u^1 and u^2:

$$0 = [u_\infty] \cdot [u_\infty] - m = ([u^1] \cdot [u^1] - m_1) + ([u^2] \cdot [u^2] - m_2) + 2[u^1] \cdot [u^2]$$

$$= 2\delta(u^1) + \left(c_1([u^1]) - m_1\right) - 2 + 2\delta(u^2) + \left(c_1([u^2]) - m_2\right) - 2 + 2[u^1] \cdot [u^2]$$

$$= 2\delta(u^1) + 2\delta(u^2) + 2\left([u^1] \cdot [u^2] - 1\right).$$

We conclude $\delta(u^1) = \delta(u^2) = 0$ and, since u^1 and u^2 have at least one intersection due to the node, $[u^1] \cdot [u^2] = 1$, implying that both are embedded and they have exactly one intersection, which is transverse. The node is necessarily disjoint from the marked points on both curves, so since both are embedded and disjoint everywhere else, the intersection point lies in $M \setminus \{p_1, \ldots, p_m\}$.

Finally, we observe that in the nodal curve u_∞, the smooth components u^i for $i = 1, 2$ each belong to spaces of the form $\mathcal{M}^0_{\mathrm{emb}}(J_\infty; p_{j_1}, \ldots, p_{j_{m_i}})$ since the adjunction formula now gives

$$[u^i] \cdot [u^i] = 2\delta(u^i) + c_1([u^i]) - 2 = m_i - 1.$$

They also satisfy the energy bound

$$\int (u^i)^* \omega_\infty < \int_{[u_\infty]} \omega_\infty \leqslant \limsup \int u_k^* \omega_k \leqslant C,$$

thus Theorem 4.5 implies that the space of all such curves that can appear in the limit of u_k is compact. By Proposition 4.3 it is also discrete, and thus finite. This completes the proof of Theorem 4.6.

Chapter 5
Exceptional Spheres

5.1 Deforming Pseudoholomorphic (-1)-Curves

In this chapter, we complete the proofs of Theorems B and C regarding exceptional spheres and the symplectic blowdown. The following is the main technical result behind both of these. We shall follow the convention of saying that any given result is true for "generic" J in some particular space of almost complex structures if it is true for all J outside some meager subset of that space (cf. Sect. 2.1.3).

Theorem 5.1. *For any closed symplectic 4-manifold (M, ω) and generic $J \in \mathcal{J}_\tau(M, \omega)$, every exceptional sphere $E \subset (M, \omega)$ is isotopic through symplectically embedded spheres to the image of a unique J-holomorphic sphere.*

Moreover, suppose $\{\omega_s\}_{s\in[0,1]}$ is a smooth 1-parameter family of symplectic structures on M and $J \in \mathcal{J}_\tau(M, \omega_0)$. Then for generic families $\{J_s\}_{s\in[0,1]} \in \mathcal{J}_\tau(M, \{\omega_s\})$ with $J_0 = J$, every embedded J_0-holomorphic sphere u_0 whose image is an exceptional sphere in (M, ω_0) extends to a smooth family $\{u_s\}$ of embedded J_s-holomorphic spheres for $s \in [0, 1]$.

Proof. The uniqueness part of this statement is an immediate consequence of positivity of intersections: if $J \in \mathcal{J}_\tau(M, \omega)$ and u_1 and u_2 are two distinct J-holomorphic curves both homologous to an exceptional sphere $E \subset (M, \omega)$ then we get the contradiction:

$$0 \leqslant [u_1] \cdot [u_2] = [E] \cdot [E] = -1.$$

Note that a curve homologous to E also cannot be a multiple cover, since any homology class satisfying $A \cdot A = -1$ must be primitive.

Now, fix a generic $J \in \mathcal{J}_\tau(M, \omega)$ and suppose $E \subset (M, \omega)$ is an exceptional sphere. We can then choose (by Proposition 2.2) a tame almost complex structure $J_0 \in \mathcal{J}_\tau(M, \omega)$ that preserves the tangent spaces of E, which means E is the image

© Springer International Publishing AG, part of Springer Nature 2018
C. Wendl, *Holomorphic Curves in Low Dimensions*, Lecture Notes
in Mathematics 2216, https://doi.org/10.1007/978-3-319-91371-1_5

of an embedded J_0-holomorphic sphere

$$u_0 : S^2 \to M.$$

One cannot expect J_0 chosen in this way to be generic in any sense (see Remark 5.2 below), but it turns out that u_0 is Fredholm regular anyway. Indeed, since $[E] \cdot [E] = -1$, the normal bundle of E has first Chern number -1 and thus

$$c_1([u_0]) = c_1([E]) = \chi(E) - 1 = 1,$$

so the index of u_0 is

$$\mathrm{ind}(u_0) = -2 + 2c_1([u_0]) = 0.$$

Since this is strictly greater than $2g - 2 = -2$ and u_0 is an immersed curve in a 4-dimensional manifold, u_0 satisfies the criteria for automatic transversality (Theorem 2.44 or 2.46).

Now extend J_0 to a smooth 1-parameter family $\{J_s\}_{s\in[0,1]}$ with $J_1 = J$. The uniqueness argument above implies that for each $s \in [0, 1]$, there is at most one curve in $\mathcal{M}_0([E]; J_s)$. Moreover, the adjunction formula implies that such a curve $u \in \mathcal{M}_0([E]; J_s)$ must be embedded: indeed, it is somewhere injective since E is a primitive class, and since $[E] \cdot [E] = -1$ and $c_1([E]) = 1$, we have

$$-1 = [u] \cdot [u] = 2\delta(u) + c_1([u]) - \chi(S^2) = 2\delta(u) - 1,$$

hence $\delta(u) = 0$. Finally, u must also be Fredholm regular since it is immersed and $\mathrm{ind}(u) = -2 + 2c_1([u]) = 0 > 2g - 2$, so Theorem 2.44 (or 2.46) applies. It follows (by Theorem 2.15) that if we define the parametric moduli space

$$\mathcal{M}_0([E]; \{J_s\}) := \{(s, u) \mid s \in [0, 1], \, u \in \mathcal{M}_0([E]; J_s)\},$$

then the projection

$$\mathcal{M}_0([E]; \{J_s\}) \to [0, 1] : (s, u) \mapsto s \tag{5.1}$$

is a submersion.

We now need to show that $\mathcal{M}_0([E]; \{J_s\})$ is compact: if that is true, then it follows that (5.1) is a diffeomorphism and thus that there exists a (unique) curve $u_1 \in \mathcal{M}_0([E]; J_1)$, which is isotopic to u_0 through a family of embedded curves $u_s \in \mathcal{M}_0([E]; J_s)$. To achieve compactness, we first make a generic perturbation of the family $\{J_s\}_{s\in[0,1]}$ for $s \in (0, 1)$, i.e. keeping J_0 and J_1 fixed. Then for all $s \in (0, 1)$, we may assume there exist no somewhere injective J_s-holomorphic curves v with $\mathrm{ind}(v) < -1$, and since the index is always even (cf. Remark 2.20), this implies in fact that all such curves satisfy $\mathrm{ind}(v) \geqslant 0$. The same is also true for $s = 1$ since we assumed J_1 generic to start with. Now in the notation of Chap. 4, we have $J_s \in \mathcal{J}^{reg}(\omega_s)$ for every $s \in (0, 1]$, so Theorem 4.5 applies and we conclude that $\mathcal{M}_0([E], \{J_s\})$ is compact.

The proof of the second statement is the same: given a family of symplectic forms $\{\omega_s\}_{s \in [0,1]}$ and $J \in \mathcal{J}_\tau(M, \omega_0)$ with a J_0-holomorphic exceptional sphere u_0, for a generic family $\{J_s\} \in \mathcal{J}_\tau(M, \{\omega_s\})$ with $J_0 = J$, the same argument implies that the moduli space $\mathcal{M}_0([u_0]; \{J_s\})$ is smooth and diffeomorphic to $[0, 1]$ via the projection $(s, u) \mapsto s$, and every curve in this space is embedded.

$\square$

Remark 5.2. The following technical point is sometimes misunderstood: given a symplectically embedded surface $S \subset (M, \omega)$, one cannot generally assume that a tame almost complex structure preserving TS can be chosen to be "generic" in the usual sense. One can easily see this from the example $(M, \omega) = (S^2 \times \Sigma_g, \sigma_1 \oplus \sigma_2)$ where Σ_g is a surface with genus $g \geq 0$ and σ_1 and σ_2 are area forms. If J is any almost complex structure for which one of the surfaces $\{point\} \times \Sigma_g$ is the image of an embedded J-holomorphic curve, then this curve has trivial normal bundle, so its first Chern number is $\chi(\Sigma_g)$ and thus its index is

$$-\chi(\Sigma_g) + 2\chi(\Sigma_g) = 2 - 2g.$$

This is fine if $g = 0$, and in this case the curve in question satisfies the criteria for "automatic" transversality (Theorem 2.44). But if $g \geq 2$, then the index is negative, which means that the curve could not exist if J were generic—indeed, it will disappear if J is perturbed generically.

The above remark demonstrates that the use of automatic transversality in our proof of Theorem 5.1 above was crucial: it allowed us to find a generic almost complex structure for which any given exceptional sphere can be regarded as a holomorphic curve, though this is not possible for arbitrary symplectically embedded surfaces. Since it will be useful in several other proofs, let us state a more general lemma on this subject. Notice that all the "generic" subsets of $\mathcal{J}_\tau(M, \omega)$ we've considered, even those which depend on a choice of constraint points $p_1, \ldots, p_m$, are invariant under all symplectomorphisms that fix the constraint points. This follows from the observation that if $\varphi : (M, \omega) \to (M, \omega)$ is such a symplectomorphism and $J \in \mathcal{J}_\tau(M)$, then composition with φ induces an obvious one-to-one correspondence between spaces of J-holomorphic curves and $\varphi^* J$-holomorphic curves satisfying the constraints.

Lemma 5.3. *Asssume (M, ω) is a symplectic 4-manifold, $p_1, \ldots, p_m \in M$ are pairwise distinct points for $m \geq 0$, and $S_1, \ldots, S_k \subset (M, \omega)$ is a collection of pairwise disjoint symplectically embedded spheres, each sphere S_i containing $m_i \geq 0$ of the points $p_1, \ldots, p_m$, such that*

$$[S_i] \cdot [S_i] \geq m_i - 1, \quad for\ i = 1, \ldots, k.$$

Suppose $\mathcal{J}'$ is a dense subset of $\mathcal{J}_\tau(M, \omega)$ that is invariant under symplectomorphisms that fix $p_1, \ldots, p_m$, i.e. for any such symplectomorphism $\varphi : (M, \omega) \to (M, \omega)$, $J \in \mathcal{J}'$ if and only if $\varphi^* J \in \mathcal{J}'$. Then there exists a $J \in \mathcal{J}'$ with $J(T S_i) = T S_i$ for every $i = 1, \ldots, k$.

Proof. To simplify the notation, we consider only the case where $k = 1$, so $S \subset (M, \omega)$ is a single symplectically embedded sphere passing through the points $p_1, \ldots, p_{m'}$ but not through $p_{m'+1}, \ldots, p_m$, where $m' \geqslant 0$ and $[S] \cdot [S] \geqslant m' - 1$. We can then choose $J_0 \in \mathcal{J}_\tau(M, \omega)$ such that S is the image of an embedded J_0-holomorphic sphere $u_0 \in \mathcal{M}_{0,m'}(J_0; p_1, \ldots, p_{m'})$. The normal bundle of this curve has first Chern number equal to $[S] \cdot [S]$, thus

$$c_1([u_0]) = \chi(S^2) + [S] \cdot [S] \geqslant 1 + m',$$

and we have

$$\mathrm{ind}(u_0) = -\chi(S^2) + 2c_1([u_0]) \geqslant 2m'.$$

Then u_0 satisfies the hypotheses for automatic transversality (Theorem 2.46) and is thus Fredholm regular for the constrained problem. It follows that for any $J \in \mathcal{J}_\tau(M, \omega)$ in some C^∞-small neighborhood of J_0, there exists a curve $u \in \mathcal{M}_{0,m'}(J; p_1, \ldots, p_{m'})$ that is correspondingly C^∞-close to u_0. Extend J_0 to a smooth path of almost complex structures $\{J_s\}_{s \in [0,1]}$ that are C^∞-close to J_0 in this sense, with $J := J_1 \in \mathcal{J}'$; the latter can be arranged since $\mathcal{J}' \subset \mathcal{J}_\tau(M, \omega)$ is dense. This extends u_0 to a smooth family of constrained pseudoholomorphic spheres $\{u_s \in \mathcal{M}_{0,m'}(J_s; p_1, \ldots, p_m)\}_{s \in [0,1]}$, which we may assume without loss of generality are all disjoint from $p_{m'+1}, \ldots, p_m$. The images S_s of the curves u_s are therefore a smooth family of symplectically embedded spheres that pass through $p_1, \ldots, p_{m'}$ but miss $p_{m'+1}, \ldots, p_m$, so there exists a smooth family of symplectomorphisms $\varphi_s : (M, \omega) \to (M, \omega)$ fixing $p_1, \ldots, p_m$ with $\varphi_0 \equiv \mathrm{Id}$ and $\varphi_s(S) = S_s$. Now $\varphi_1^* J \in \mathcal{J}'$ is an almost complex structure with the desired properties.

The case of multiple disjoint spheres can be handled in the same way: the key point is that J_0 can be chosen to make all of them simultaneously holomorphic and (automatically) Fredholm regular.   □

5.2 Proofs of Theorems B and C

We now restate and prove Theorem B.

Theorem 5.4. *Suppose M is a closed connected 4-manifold with a smooth 1-parameter family of symplectic structures $\{\omega_s\}_{s \in [0,1]}$, and $E_1, \ldots, E_k \subset M$ is a collection of pairwise disjoint exceptional spheres in (M, ω_0). Then there are*

smooth 1-parameter families of embedded spheres $E_1^s, \ldots, E_k^s \subset M$ *for* $s \in [0, 1]$
such that

- $E_i^0 = E_i$ *for* $i = 1, \ldots, k$;
- *For every* $s \in [0, 1]$, $E_i^s \cap E_j^s = \varnothing$ *for* $i \neq j$;
- *For every* $s \in [0, 1]$ *and* $i = 1, \ldots, k$, E_i^s *is symplectically embedded in* (M, ω_s).

In particular, (M, ω_0) *is minimal if and only if* (M, ω_1) *is minimal.*

Proof. Since the spheres $E_1, \ldots, E_k \subset (M, \omega_0)$ are all symplectically embedded and pairwise disjoint, we can choose $J_0 \in \mathcal{J}_\tau(M, \omega_0)$ such that each is the image of an embedded J_0-holomorphic sphere

$$u_i \in \mathcal{M}_0([E_i]; J_0), \qquad i = 1, \ldots, k.$$

Now extend J_0 to a smooth 1-parameter family $\{J_s\} \in \mathcal{J}_\tau(M, \{\omega_s\})$ which is generic for $s \in (0, 1]$. By Theorem 5.1, the curves u_i now extend to smooth families $u_i^s \in \mathcal{M}_0([E_i]; J_s)$ for $s \in [0, 1]$, and since $[E_i] \cdot [E_j] = 0$ for $i \neq j$, positivity of intersections guarantees that the images of u_i^s and u_j^s for $i \neq j$ are disjoint for all $s \in [0, 1]$. The desired families of symplectically embedded spheres can therefore be defined as the images of the curves u_i^s. $\square$

Finally, Theorem C claimed the following.

Theorem 5.5. *Suppose* (M, ω) *is a closed symplectic 4-manifold and* $E_1, \ldots, E_k \subset M$ *is a maximal collection of pairwise disjoint exceptional spheres. Then the manifold* (M_0, ω_0) *obtained by blowing down* (M, ω) *at all of these spheres is minimal.*

Proof. By assumption, (M, ω) is the symplectic blowup of (M_0, ω_0) along some collection of pairwise disjoint symplectic embeddings

$$\coprod_{i=1}^k (B_{R_i + \epsilon}^4, \omega_{\text{st}}) \hookrightarrow (M_0, \omega_0),$$

where $R_i > 0$ is the weight of the blowup producing the sphere E_i (cf. Definition 3.7). Now for every $\tau \in (0, 1]$, define (M^τ, ω^τ) as the symplectic blowup of (M_0, ω_0) along the same collection of Darboux balls but restricted to shrunken domains of radius $\tau \cdot (R_i + \epsilon)$ and with weights τR_i. By Theorem 3.8, these blowups are all symplectically deformation equivalent to (M, ω), and they all contain $E_1, \ldots, E_k$ as exceptional spheres, with symplectic areas that vary with τ.

If (M_0, ω_0) is not minimal, then it contains an exceptional sphere E. Since being symplectically embedded is an open condition, we can assume after a small perturbation that E does not intersect the centers of any of the Darboux balls above. Then E is also disjoint from the balls of radius $\tau \cdot (R_i + \epsilon)$ for any $\tau > 0$ sufficiently small, and it thus survives the blowup with weights τR_i to define an exceptional sphere in (M^τ, ω^τ) that is disjoint from $E_1, \ldots, E_k$. Now by Theorem 5.4, the pairwise disjoint collection of exceptional spheres $E_1, \ldots, E_k, E \subset (M^\tau, \omega^\tau)$ is isotopic to a similar collection $E_1', \ldots, E_k', E' \subset (M, \omega)$. Using Lemma 5.3, choose a generic $J \in \mathcal{J}_\tau(M, \omega)$ for which $E_1, \ldots, E_k$ are all J-holomorphic. Then by Theorem 5.1, E' is symplectically isotopic to a unique J-holomorphic exceptional sphere, and positivity of intersections implies that this sphere is disjoint from each of the J-holomorphic spheres E_i since $[E'] \cdot [E_i] = [E] \cdot [E_i] = 0$. This contradicts the assumption that $E_1, \ldots, E_k$ is a maximal collection.   □

Chapter 6
Rational and Ruled Surfaces

In this chapter we shall prove Theorems A, D, E and F and sketch the proof of Theorem G. The easiest path is to start with the last two and then prove the others as corollaries.

6.1 Proofs of Theorems F and G

Theorem F stated the following.

Theorem 6.1. *Suppose* (M, ω) *is a closed and connected symplectic 4-manifold that contains a symplectically embedded 2-sphere* $S \subset (M, \omega)$ *with*

$$m := [S] \cdot [S] \geqslant 0.$$

Then for any choice of pairwise distinct points $p_1, \ldots, p_m \in S$, (M, ω) *admits a symplectic Lefschetz pencil with base points* $p_1, \ldots, p_m$ *(or a symplectic Lefschetz fibration if* $m = 0$*), in which* S *is a smooth fiber and no singular fiber contains more than one critical point. Moreover, the set of singular fibers of this pencil (or fibration) is empty if and only if* $m \in \{0, 1\}$ *and* $(M \backslash S, \omega)$ *is minimal.*

Proof. Given $S \subset (M, \omega)$ as stated in the theorem, choose any set of pairwise distinct points

$$p_1, \ldots, p_m \in S.$$

Lemma 5.3 then provides an almost complex structure $J \in \mathcal{J}_\tau(M, \omega)$ such that S is the image of an embedded J-holomorphic sphere u_0 and J also satisfies the genericity criterion of Theorem 4.6 for the chosen points $p_1, \ldots, p_m$. Now u_0 can

© Springer International Publishing AG, part of Springer Nature 2018
C. Wendl, *Holomorphic Curves in Low Dimensions*, Lecture Notes
in Mathematics 2216, https://doi.org/10.1007/978-3-319-91371-1_6

naturally be regarded as an element of the space of embedded constrained curves

$$u_0 \in \mathcal{M}^2_{emb}(J; p_1, \ldots, p_m)$$

that we studied in Chap. 4.

Let $\mathcal{M}_S(J)$ denote the connected component of $\mathcal{M}^2_{emb}(J; p_1, \ldots, p_m)$ containing u_0. Since the curves in this space are all homologous, they all have the same energy

$$E(u) = \int u^*\omega = \langle [\omega], [S] \rangle.$$

Thus by Theorem 4.6, $\mathcal{M}_S(J)$ is compact except for finitely many nodal curves, each of which consists of two embedded curves intersecting once transversely, and Proposition 4.7 guarantees that these nodal curves are disjoint from the curves in $\mathcal{M}_S(J)$ and from each other except for their forced intersections at the points $p_1, \ldots, p_m$. Let

$$\Xi \subset M \backslash \{p_1, \ldots, p_m\}$$

denote the set of points separate from $p_1, \ldots, p_m$ that are in the image of one of these nodal curves. This is a union of finitely many embedded connected surfaces. Now let

$$M_0 \subset M \backslash (\{p_1, \ldots, p_m\} \cup \Xi)$$

denote the set of all points separate from $p_1, \ldots, p_m$ and Ξ that are in the image of any curve in $\mathcal{M}_S(J)$. This is an open subset of $M \backslash (\{p_1, \ldots, p_m\} \cup \Xi)$ due to the implicit function theorem (Proposition 2.53). It is also closed by the compactness result for $\mathcal{M}_S(J)$, i.e. Theorem 4.6: indeed, if $p \in M \backslash (\{p_1, \ldots, p_m\} \cup \Xi)$ is in the closure of M_0, then there exists a sequence $u_k \in \mathcal{M}_S(J)$ whose images contain points in $M \backslash (\{p_1, \ldots, p_m\} \cup \Xi)$ converging to p, and this sequence cannot have any subsequence converging to a nodal curve in Ξ, thus it has a subsequence converging to a curve in $\mathcal{M}_S(J)$, implying $p \in M_0$. But since Ξ is a subset of codimension 2, the space $M \backslash (\{p_1, \ldots, p_m\} \cup \Xi)$ is connected, and we conclude

$$M_0 = M \backslash (\{p_1, \ldots, p_m\} \cup \Xi).$$

Now let $\overline{\mathcal{M}}_S(J)$ denote the closure of $\mathcal{M}_S(J)$ in the Gromov compactification: it is obtained from $\mathcal{M}_S(J)$ by adding the finite set of nodal curves whose images form the subset Ξ. The above argument shows that every point in $M \backslash \{p_1, \ldots, p_m\}$ is either in Ξ or in the image of a unique curve in $\mathcal{M}_S(J)$. Since Ξ has finitely many connected components, each of which is the union of two embedded J-holomorphic curves intersecting transversely and positively at a single node, the result is a foliation of M with a finite set of singular points consisting of the constraint points $p_1, \ldots, p_m$ and the nodes in Ξ. Note that while $\mathcal{M}_S(J)$ is naturally a smooth oriented manifold, $\overline{\mathcal{M}}_S(J)$ at this stage is only a compact topological space, with no natural smooth structure, but we can now use the foliation to assign one to it. Indeed, for a given nodal curve $u_0 \in \overline{\mathcal{M}}_S(J)$, one can choose a 2-disk $\mathcal{D} \subset M \backslash \{p_1, \ldots, p_m\}$ whose center intersects u_0 transversely at one of its smooth points, and then use $\mathcal{D}$ to parametrize a neighborhood of u_0 in $\overline{\mathcal{M}}_S(J)$ via the intersections of nearby curves with $\mathcal{D}$. This makes $\overline{\mathcal{M}}_S(J)$ into a closed oriented surface, and there is also a smooth map

$$\pi : M \backslash \{p_1, \ldots, p_m\} \to \overline{\mathcal{M}}_S(J)$$

taking each point p to the *unique* (possibly nodal) curve in $\overline{\mathcal{M}}_S(J)$ with p in its image. Since all curves in $\overline{\mathcal{M}}_S(J)$ are tangent to J-complex subspaces and their intersections at the constraint points are transverse, each constraint point admits a neighborhood in which π can be identified with the standard local model (3.11) for the base point of a Lefschetz pencil. Appendix A shows similarly that each node has a neighborhood identifiable with the standard local model (3.8) for a Lefschetz critical point, hence π is the desired Lefschetz pencil if $m > 0$, or Lefschetz fibration if $m = 0$. Note that if $m > 0$, we can conclude that $\overline{\mathcal{M}}_S(J) \cong \mathbb{CP}^1$, and an explicit diffeomorphism is defined by choosing an isomorphism of $(T_{p_1} M, J)$ to $(\mathbb{C}^2, i)$ and associating to any curve $u \in \overline{\mathcal{M}}_S(J)$ its tangent space at p_1, which is a complex line in $(T_{p_1} M, J)$ and thus defines a point in $\mathbb{CP}^1$.

We now consider under what circumstances this Lefschetz pencil might have no singular fibers. If $m \geqslant 2$, Proposition 3.31 implies that there must be singular fibers. If $m = 0$, then the singular fibers provided by nodal curves in $\overline{\mathcal{M}}_S(J)$ consist of pairs of embedded J-holomorphic spheres disjoint from S that each have self-intersection -1. When $m = 1$, singular fibers have one component with self-intersection -1 and another that has self-intersection 0 and satisfies a marked point constraint. In either case, no singular fibers can exist if $(M \backslash S, \omega)$ is minimal. Conversely, if there is an exceptional sphere $E \subset (M \backslash S, \omega)$, then Theorem 5.1 implies that E is homologous to an embedded J-holomorphic sphere $u_E : S^2 \to M$. Since $[E] \cdot [S] = 0$, positivity of intersections then implies that u_E either is disjoint from all of the curves in $\mathcal{M}_J(S)$ or has identical image to one of them. The latter is impossible since $[E] \cdot [E] = -1$ and $[S] \cdot [S] \geqslant 0$, thus there must be a singular fiber that contains E. $\square$

Exercise 6.2. Show that if $m > 0$, the map $\overline{\mathcal{M}}_S(J) \to \mathbb{CP}^1$ defined above by associating to each curve its tangent space at p_1 really is a diffeomorphism.

Exercise 6.3. The proof above shows not only that a Lefschetz pencil exists, but also that its fibers are all J-holomorphic for a particular choice of almost complex structure $J \in \mathcal{J}_\tau(M, \omega)$. We can now prove Theorem G as follows:

    (a) Given any smooth 1-parameter family of symplectic forms $\{\omega_s\}_{s \in [0,1]}$ with $\omega_0 = \omega$, choose a generic homotopy $\{J_s\}_{s \in [0,1]}$ of ω-tame almost complex structures with $J_0 = J$, and show that the moduli spaces of

J_0-holomorphic curves produced in the proof of Theorem 6.1 extend to smooth families of moduli spaces of J_s-holomorphic curves which are diffeomorphic for all $s \in [0, 1]$. *Hint: we used the same type of argument in the proof of Theorem 5.1. The genericity of $\{J_s\}$ is needed for the compactness results in Chap. 4, but you will also need to apply automatic transversality to ensure that the deformation is unobstructed for all $s \in [0, 1]$.*

(b) Use positivity of intersections to show that for any $J \in \mathcal{J}_\tau(M, \omega)$ and any $m + 1$ distinct points in the above setting, there is at most one J-holomorphic curve homologous to the fiber and passing through all $m + 1$ of the given points.

As was mentioned in the introduction, Theorem G has an especially useful corollary for the case $(M, \omega) = (\mathbb{CP}^2, \omega_{FS})$, but one must first remove the word "generic" from the statement:

Lemma 6.4. *For $(M, \omega) = (\mathbb{CP}^2, \omega_{FS})$ with $[S] \cdot [S] = 1$, Theorem G holds without any genericity assumptions.*

Proof. The relevant moduli spaces of embedded J-holomorphic spheres satisfy automatic transversality, so their smoothness does not depend on genericity. In Exercise 6.3 above, genericity is only needed in order to apply the compactness results of Chap. 4, which require excluding holomorphic spheres in moduli spaces of negative virtual dimension that could potentially appear in nodal curves. For the situation at hand, we have a single fixed marked point constraint $p \in \mathbb{CP}^2$ and must in particular restrict to almost complex structures $J \in \mathcal{J}_\tau(\mathbb{CP}^2, \omega_{FS})$ with the property that any moduli space of the form $\mathcal{M}_0(A; J)$ or $\mathcal{M}_{0,1}(A; J; p)$ for $A \in H_2(\mathbb{CP}^2)$ has nonnegative virtual dimension if it contains a somewhere injective curve. We claim that that is true for *all* $J \in \mathcal{J}_\tau(\mathbb{CP}^2, \omega_{FS})$. This results from the fact that $H_2(\mathbb{CP}^2)$ is especially simple: any $A \in H_2(\mathbb{CP}^2)$ is of the form $A = d[\mathbb{CP}^1]$ for some $d \in \mathbb{Z}$, and any nonconstant curve $u \in \mathcal{M}_{0,m}(A)$ must then satisfy

$$0 < \int u^* \omega_{FS} = \langle [\omega_{FS}], A \rangle = d \langle [\omega_{FS}], [\mathbb{CP}^1] \rangle,$$

which is true if and only if $d > 0$. Since the generator $[\mathbb{CP}^1]$ satisfies $[\mathbb{CP}^1] \cdot [\mathbb{CP}^1] = 1$ and is represented by an embedded sphere, we deduce also

$$c_1([\mathbb{CP}^1]) = \chi(S^2) + 1 = 3,$$

thus whenever $A = d[\mathbb{CP}^1]$ with $d > 0$, we have

$$\text{vir-dim}\, \mathcal{M}_0(A; J) = -\chi(S^2) + 2c_1(d[\mathbb{CP}^1]) = -2 + 6d \geqslant 4,$$

and

$$\text{vir-dim}\, \mathcal{M}_{0,1}(A; J; p) = -\chi(S^2) + 2c_1(d[\mathbb{CP}^1]) - 2 = -2 + 6d - 2 \geqslant 2.$$

With this understood, compactness holds for all $J \in \mathcal{J}_\tau(\mathbb{CP}^2, \omega_{FS})$ and the rest of the proof of Theorem G goes through as before.   □

Since the choice of base point $p \in S \subset \mathbb{CP}^2$ in Theorem 6.1 is arbitrary, application of Theorem G to $(\mathbb{CP}^2, \omega_{FS})$ now implies the following useful result of Gromov:

Corollary 6.5. *For any tame almost complex structure J on $(\mathbb{CP}^2, \omega_{FS})$ and any two distinct points $p_1, p_2 \in \mathbb{CP}^2$, there is a unique J-holomorphic sphere homologous to $[\mathbb{CP}^1] \in H_2(\mathbb{CP}^2)$ passing through p_1 and p_2, and it is embedded.*   □

6.2   Proofs of Theorems A, D and E

We next prove Theorem A:

Theorem 6.6. *Suppose (M, ω) is a closed and connected symplectic 4-manifold containing a symplectically embedded 2-sphere $S \subset M$ with*

$$[S] \cdot [S] \geq 0.$$

Then (M, ω) is either $(\mathbb{CP}^2, c\omega_{FS})$ for some constant $c > 0$ or a blown-up symplectic ruled surface.

Proof. The first step is to show that under the assumptions of the theorem, (M, ω) also contains a symplectically embedded 2-sphere $S' \subset M$ with

$$[S'] \cdot [S'] \in \{0, 1\}.$$

We argue by contradiction: let m denote the smallest integer that occurs as the self-intersection number of a symplectically embedded sphere $S \subset (M, \omega)$, and assume $m \geq 2$. Theorem F then gives a symplectic Lefschetz pencil

$$\pi : M \setminus \{p_1, \ldots, p_m\} \to \mathbb{CP}^1$$

which has S as a fiber. By Proposition 3.31, this pencil must have at least one singular fiber. The singular fibers provided by Theorem F each have exactly two irreducible components, and by Lemma 3.29, each of these components is a symplectically embedded sphere $E \subset M$ with

$$[E] \cdot [E] = -1 + b,$$

where b is the number of base points in E. Since the total number of base points is positive, this means there exists an irreducible component of a singular fiber satisfying $E \cdot E \in \{0, \ldots, m - 1\}$, which contradicts the initial assumption.

Let us now proceed assuming without loss of generality that $S \subset (M, \omega)$ satisfies $[S] \cdot [S] = m \in \{0, 1\}$. If $m = 0$, then Theorem F gives a symplectic Lefschetz fibration

$$\pi : M \to \Sigma$$

over some smooth, oriented closed surface Σ, diffeomorphic to a certain compactified moduli space of embedded holomorphic spheres homologous to S. The singular fibers consist of pairs of exceptional spheres intersecting transversely, so blowing down one component in each singular fiber produces a smooth symplectic fibration by spheres, i.e. the blowdown is a symplectic ruled surface.

If $m = 1$, then Theorem F instead produces a symplectic Lefschetz pencil

$$\pi : M \backslash \{p\} \to \mathbb{CP}^1,$$

in which each singular fiber has one irreducible component that is an exceptional sphere. Blowing down each of these to get rid of all the singular fibers, the resulting symplectic manifold (M', ω') is diffeomorphic to $\mathbb{CP}^2$ by Proposition 3.27. Moreover, since ω' is symplectic on the fibers of the pencil, Theorem 3.33 implies that after identifying M' with $\mathbb{CP}^2$, ω' can be deformed to the standard symplectic structure ω_{FS} through a 1-parameter family of symplectic forms. Now since $H^2_{dR}(\mathbb{CP}^2)$ has only a single generator, one can also rescale ω_{FS} and all the symplectic forms in this deformation to make them cohomologous, and then Moser's stability theorem implies that ω' and $c\omega_{FS}$ are isotopic for a suitable constant $c > 0$. $\square$

An important ingredient in proving Theorem E is the following easy extension of Proposition 2.2.

Exercise 6.7. Suppose (M, ω) is a symplectic 4-manifold and $S_1, S_2 \subset M$ are two symplectically embedded surfaces that intersect each other transversely and positively. Then there exists an ω-tame almost complex structure J preserving both $T S_1$ and $T S_2$.

We now prove Theorem E:

Theorem 6.8. *Suppose (M, ω) is a closed, connected and minimal symplectic 4-manifold containing a pair of symplectically embedded spheres $S_1, S_2 \subset (M, \omega)$ that satisfy $[S_1] \cdot [S_1] = [S_2] \cdot [S_2] = 0$ and have exactly one intersection with each other, which is transverse and positive. Then (M, ω) admits a symplectomorphism to $(S^2 \times S^2, \sigma_1 \oplus \sigma_2)$ identifying S_1 with $\{S^2\} \times \{0\}$ and S_2 with $\{0\} \times S^2$, where σ_1, σ_2 are any two area forms on S^2 such that*

$$\int_{S^2} \sigma_i = \int_{S_i} \omega \qquad for \ i = 1, 2.$$

Proof. By Exercise 6.7, we can choose $J_0 \in \mathcal{J}_\tau(M, \omega)$ such that S_1 and S_2 are both images of embedded J_0-holomorphic spheres u_1 and u_2 respectively. It is not obvious whether such an almost complex structure can be chosen to be generic, but as usual we can get around this using automatic transversality: since u_i for $i = 1, 2$ both have trivial normal bundles, they both have $c_1([u_i]) = 2$ and thus $\operatorname{ind}([u_i]) = -2 + 2c_1([u_i]) = 2 > 2g - 2$, so after perturbing J_0 to a generic $J \in \mathcal{J}_\tau(M, \omega)$ there exist embedded J-holomorphic spheres u_1' and u_2' close to u_1 and u_2 respectively. Repeating the argument of Theorem F for the case $m = 0$, these two curves generate two compact families $\mathcal{M}_{S_1}(J)$ and $\mathcal{M}_{S_2}(J)$ of embedded J-holomorphic spheres homologous to $[S_1]$ and $[S_2]$ respectively. The compactness of these spaces follows from the assumption that (M, ω) is minimal, as any nodal curve would necessarily contain a J-holomorphic exceptional sphere. Thus the Lefschetz pencil of Theorem F becomes in this case a pair of smooth fibrations

$$\pi_1 : M \to \mathcal{M}_{S_1}(J), \qquad \pi_2 : M \to \mathcal{M}_{S_2}(J).$$

By positivity of intersections, $[S_1] \cdot [S_2] = 1$ implies that every fiber of π_1 intersects every fiber of π_2 exactly once transversely: in particular, every point in the image of u_1' intersects a unique curve in $\mathcal{M}_{S_2}(J)$, so this defines a diffeomorphism of $\mathcal{M}_{S_2}(J)$ to S^2, and there is a similar diffeomorphism of $\mathcal{M}_{S_1}(J)$ to S^2. Under these identifications, the pair of fibrations (π_1, π_2) defines a diffeomorphism

$$(\pi_1, \pi_2) : M \to S^2 \times S^2,$$

in which the fibers of π_1 are identified with $S^2 \times \{*\}$ and those of π_2 are identified with $\{*\} \times S^2$.

One minor point is that in our construction of the above fibrations, the original surfaces S_1 and S_2 are not fibers, as they got perturbed when we replaced the original J_0 with the generic J. Now that we know $M \cong S^2 \times S^2$ however, we can redo the argument without worrying about genericity: indeed, we claim that for the original (non-generic) $J_0 \in \mathcal{J}_\tau(M, \omega)$, every somewhere injective J_0-holomorphic sphere $v : S^2 \to M$ satisfies $\operatorname{ind}(v) \geqslant 0$. To see this, observe that under the identification of M with $S^2 \times S^2$ obtained above, $[S_1]$ and $[S_2]$ are generators of $H_2(M)$, so for any nonconstant curve $v \in \mathcal{M}_0(J_0)$ we can write

$$[v] = k[S_1] + m[S_2]$$

for some $k, m \in \mathbb{Z}$. If v is a reparametrization or multiple cover of u_1, then $[v] \cdot [S_1] = k[S_1] \cdot [S_1] = 0$, and if v is anything else, then positivity of intersections implies $[v] \cdot [S_1] = [v] \cdot [u_1] \geqslant 0$. Applying the same argument with u_2, we have $[v] \cdot [S_2] \geqslant 0$ as well, and since $[S_1] \cdot [S_2] = 1$, we conclude that k and m are both nonnegative, and they cannot both be 0 since v is not constant. Now since $c_1([S_1]) = c_1([S_2]) = 2$, we have

$$\text{ind}(v) = -2 + 2c_1(k[S_1] + m[S_2]) = -2 + 4k + 4m > 0.$$

This implies that J_0 satisfies the conditions needed in Theorem 4.6 to prove compactness of $\mathcal{M}_{S_1}(J_0)$ and $\mathcal{M}_{S_2}(J_0)$. Since the curves in these spaces all satisfy automatic transversality, this was the only step where genericity of J was ever needed—we can therefore dispense with genericity and assume the fibers of π_1 and π_2 are J_0-holomorphic curves, so in particular, the original surfaces S_1 and S_2 are both fibers.

It remains only to prove that our given symplectic structure ω is symplectomorphic to a split structure $\sigma_1 \oplus \sigma_2$. Using the identification above, we can now write $M = S^2 \times S^2$ and assume the J_0-holomorphic fibers of π_1 and π_2 are simply $S^2 \times \{*\}$ and $\{*\} \times S^2$. Here J_0 is an almost complex structure on S^2, and our given ω can be regarded as a symplectic form on $S^2 \times S^2$ that tames J_0. Choose area forms σ_1, σ_2 on S^2 such that

$$\int_{S^2} \sigma_1 = \int_{S^2 \times \{*\}} \omega, \qquad \int_{S^2} \sigma_2 = \int_{\{*\} \times S^2} \omega.$$

Then ω and $\sigma_1 \oplus \sigma_2$ are two symplectic forms on $S^2 \times S^2$ representing the same cohomology class in $H^2_{\text{dR}}(S^2 \times S^2)$. The J_0-holomorphic fibers $S^2 \times \{*\}$ and $\{*\} \times S^2$ are also symplectic submanifolds and are symplectically orthogonal to each other with respect to $\sigma_1 \oplus \sigma_2$. From this, it is easy to show that $\sigma_1 \oplus \sigma_2$ also tames J_0. Now for $s \in [0, 1]$, we define

$$\omega_s := s(\sigma_1 \oplus \sigma_2) + (1 - s)\omega,$$

which is a smooth 1-parameter family of cohomologous closed 2-forms, and they are all symplectic and tame J_0 since J_0 is tamed by both $\sigma_1 \oplus \sigma_2$ and ω. Moser's stability theorem (see [MS17]) then implies that ω and $\sigma_1 \oplus \sigma_2$ are isotopic. □

And finally, we prove Theorem D:

Theorem 6.9. *Suppose (M, ω) is a closed, connected and minimal symplectic 4-manifold that contains a symplectically embedded 2-sphere $S \subset (M, \omega)$ with $[S] \cdot [S] \geqslant 0$. One then has the following possibilities:*

(1) If $[S] \cdot [S] = 0$, then (M, ω) admits a symplectomorphism to a symplectic ruled surface such that S is identified with a fiber.

(2) If $[S] \cdot [S] = 1$, then (M, ω) admits a symplectomorphism to $(\mathbb{CP}^2, c\omega_{\text{FS}})$ for some constant $c > 0$, such that S is identified with the sphere at infinity $\mathbb{CP}^1 \subset \mathbb{CP}^2$.

(3) If $[S] \cdot [S] > 1$, then (M, ω) is symplectomorphic to one of the following:

(a) $(\mathbb{CP}^2, c\omega_{\mathrm{FS}})$ *for some constant* $c > 0$;
(b) $(S^2 \times S^2, \sigma_1 \oplus \sigma_2)$ *for some pair of area forms* σ_1, σ_2 *on* S^2.

Proof. The cases $[S] \cdot [S] = 0$ and $[S] \cdot [S] = 1$ were already shown in the proof of Theorem 6.6, so we focus on the case $m := [S] \cdot [S] > 1$. Then Theorem 6.1 provides a symplectic Lefschetz pencil

$$\pi : M \backslash \{p_1, \ldots, p_m\} \to \mathbb{CP}^1$$

which has S as a smooth fiber, and the set of singular fibers is necessarily non-empty. Each singular fiber has exactly two irreducible components S_+ and S_-, and since (M, ω) is minimal, neither of these is an exceptional sphere, and neither can contain all m of the base points. It follows that S_+ and S_- are both symplectically embedded spheres with

$$[S_\pm] \cdot [S_\pm] \in \{0, \ldots, m - 2\}.$$

If either of these has self-intersection 1, then we conclude from Theorem 6.6 that $(M, \omega) \cong (\mathbb{CP}^2, c\omega_{\mathrm{FS}})$. If not, then either they both have self-intersection 0 or one of them has self-intersection greater than 1 but less than $m - 1$. We can thus repeat this argument until one of the following happens:

(1) We find a symplectically embedded sphere $S' \subset (M, \omega)$ with $[S'] \cdot [S'] = 1$ and thus conclude $(M, \omega) \cong (\mathbb{CP}^2, c\omega_{\mathrm{FS}})$.
(2) We find two symplectically embedded spheres $S_+, S_- \subset (M, \omega)$ that intersect each other once transversely and positively and both have self-intersection 0. (See Fig. 3.2 for a picture of this scenario.) In this case, Theorem 6.8 implies that (M, ω) is symplectomorphic to $S^2 \times S^2$ with a split symplectic form.

$\square$

Chapter 7
Uniruled Symplectic 4-Manifolds

The theorems discussed so far show that there is clearly something special about the class of symplectic 4-manifolds that contain symplectically embedded spheres of nonnegative self-intersection. In this chapter, we will see that this class can also be characterized in terms of enumerative symplectic invariants that count J-holomorphic curves, that is, the Gromov-Witten invariants.

7.1 Further Characterizations of Rational or Ruled Surfaces

We shall introduce a simple version of the Gromov-Witten invariants in Sect. 7.2 below. It leads naturally to the notion of uniruled symplectic manifolds: in essence, a symplectic manifold is *symplectically uniruled* if it has some nonzero Gromov-Witten invariant that guarantees the existence of a nontrivial J-holomorphic sphere through every point for generic tame J. Theorem G shows that the latter property is shared by all symplectic 4-manifolds in our special class, and indeed, it will be easy to show that all manifolds in this class are symplectically uniruled (see Theorem 7.33). It is then natural to wonder what else is. The answer turns out to be *nothing*, i.e. in dimension four, the uniruled symplectic manifolds are precisely those which are rational or blown-up ruled surfaces. The proof of this requires a rather non-obvious generalization of Theorem A: as McDuff showed in [McD92], the theorem remains true if its hypothesis is weakened to allow a symplectic sphere $S \subset M$ that is *immersed* but not necessarily embedded, as long as its self-intersections are positive and its first Chern number is large enough.

Definition 7.1. Given a symplectic 4-manifold (M, ω) and an immersion $\iota : S \looparrowright M$ of a surface S, we say that S is **positively symplectically immersed** in (M, ω) if $\iota^*\omega$ is symplectic, all self-intersections of S are transverse and positive, and there are no triple self-intersections $\iota(z_1) = \iota(z_2) = \iota(z_3)$ for pairwise distinct points $z_1, z_2, z_3 \in S$.

© Springer International Publishing AG, part of Springer Nature 2018
C. Wendl, *Holomorphic Curves in Low Dimensions*, Lecture Notes
in Mathematics 2216, https://doi.org/10.1007/978-3-319-91371-1_7

Exercise 7.2. Show that an immersion $\iota : S \looparrowright M$ with only transverse self-intersections and no triple self-intersections is positively symplectically immersed in (M, ω) if and only if there exists an ω-tame almost complex structure J on M and a complex structure j on S such that $\iota : (S, j) \to (M, J)$ is a J-holomorphic curve.

Theorem 7.3. *If (M, ω) is a closed and connected symplectic 4-manifold, the following are equivalent:*

(1) (M, ω) *is a symplectic rational surface or blown-up ruled surface.*

(2) (M, ω) *admits a positively symplectically immersed sphere $S \looparrowright M$ with* $c_1([S]) \geqslant 2$.

(3) *For some $J \in \mathcal{J}_\tau(M, \omega)$, there exists a somewhere injective Fredholm regular J-holomorphic sphere u with* $\mathrm{ind}(u) \geqslant 2$.

(4) *For some $A \in H_2(M)$ satisfying $-2 + 2c_1(A) \geqslant 2$ and a dense subset $\mathcal{J}^{\mathrm{reg}} \subset \mathcal{J}_\tau(M, \omega)$, there exists a somewhere injective Fredholm regular J-holomorphic sphere u with $[u] = A$ for every $J \in \mathcal{J}^{\mathrm{reg}}$.*

(5) (M, ω) *is symplectically uniruled.*

Corollary 7.4. *Each of the properties listed in Theorem 7.3 for symplectic 4-manifolds is invariant under symplectic deformation equivalence and symplectic blowup or blowdown.*

Proof. We have already seen in Theorem G that the class of symplectic 4-manifolds satisfying the first condition is invariant under symplectic deformations; alternatively, we will see in Sect. 7.2 below that this is immediate for the fifth condition, because the Gromov-Witten invariants are deformation invariant. Using the existence of symplectically embedded spheres to characterize the first condition, it is also manifestly satisfied for any blowup of a rational or ruled surface along a Darboux ball that is small enough to avoid intersecting such a sphere, and all blowups along different or larger Darboux balls are symplectically deformation equivalent to this one. Lemma 7.5 below shows in turn that the conditions are invariant under symplectic blowdown.   □

Before completing the last step in the above argument, let us comment on the idea. Corollary 7.4 was first claimed in [McD90, Theorem 1.2] and later given a correct proof in [McD92]; the proof given in [McD90] had an important gap, though both the argument and its gap are illuminating. One would ideally like to argue as follows: if (M, ω) contains a symplectically embedded sphere S with $[S] \cdot [S] \geqslant 0$ and it also contains an exceptional sphere E, then one can choose an almost complex structure J making S a J-holomorphic sphere $u : S^2 \to M$ and perform the blowdown along E so that the blowdown map $\beta : M \to \check{M}$ is pseudoholomorphic, thus producing a $\check{J}$-holomorphic sphere $\check{u} = \beta \circ u : S^2 \to \check{M}$ in the blowdown. We've seen in Exercise 3.6 that this operation generally increases the first Chern number of the normal bundle of our curve, thus we expect to see $[\check{u}] \cdot [\check{u}] \geqslant [S] \cdot [S] \geqslant 0$ so that the theorems of Chap. 1 still apply to the blowdown $\check{M}$. The trouble, however, is that $\check{u}$ might not be embedded: in fact if

$[S] \cdot [E] \geqslant 2$, then $\check{u}$ will definitely pass through the point $\beta(E) \in \check{M}$ multiple times. This is where the more general conditions in Theorem 7.3 allowing for non-embedded curves become essential, and it is the reason why Corollary 7.4 has been delayed until the present chapter rather than being stated among the main results in Chap. 1.

Lemma 7.5. *If (M, ω) satisfies the fourth condition in Theorem 7.3 and contains a collection of pairwise disjoint exceptional spheres $E_1, \ldots, E_m \subset (M, \omega)$, then its blowdown along $E_1 \amalg \ldots \amalg E_m$ satisfies the third condition in Theorem 7.3.*

Proof. If there exists a simple and regular J-holomorphic sphere u in $M \backslash (E_1 \amalg \ldots \amalg E_m)$ with $\mathrm{ind}(u) \geqslant 2$ for some $J \in \mathcal{J}_\tau(M, \omega)$, then u will still exist after blowing down $E_1, \ldots, E_m$, and we are done. Let us therefore assume that all such spheres intersect at least one of the exceptional spheres E_i for $i = 1, \ldots, m$. Pick symplectomorphisms identifying disjoint neighborhoods $\mathcal{U}_{E_i} \subset M$ of E_i with neighborhoods of the zero-section in the tautological line bundle $\widetilde{\mathbb{C}}^2$, each with a standard symplectic form ω_R for some $R > 0$ (see Sect. 3.2). We can then choose a tame almost complex structure $J_0 \in \mathcal{J}_\tau(M, \omega)$ that is generic outside $\mathcal{U}_E := \mathcal{U}_{E_1} \cup \ldots \cup \mathcal{U}_{E_m}$ and matches the standard (integrable) complex structure of $\widetilde{\mathbb{C}}^2$ in $\mathcal{U}_E$. By assumption, there exists a sequence $J_k \in \mathcal{J}_\tau(M, \omega)$ with $J_k \to J_0$ and a sequence u_k of somewhere injective J_k-holomorphic spheres in a fixed homology class $A \in H_2(M)$ with $\mathrm{ind}(u_k) = -2 + 2c_1(A) \geqslant 2$. The latter means $c_1(A) \geqslant 2$, so by the adjunction formula,

$$A \cdot A = 2\delta(u_k) + c_1(A) - 2 \geqslant 0,$$

implying that A cannot be a multiple of any $[E_i]$. Applying Gromov compactness, a subsequence of u_k then converges to a nodal curve in $\overline{\mathcal{M}}_0(A; J_0)$ whose components cannot all be covers of the spheres E_i, so in particular, at least one such component covers a simple J_0-holomorphic sphere $u : S^2 \to M$ that is not contained in $\mathcal{U}_E$, and we can choose this component such that u has nontrivial intersection with $E_1 \cup \ldots \cup E_m$. Since J_0 is generic outside $\mathcal{U}_E$, we are free to assume that u is Fredholm regular with $\mathrm{ind}(u) \geqslant 0$, hence

$$c_1([u]) \geqslant 1.$$

After possibly perturbing u within its moduli space, we can also assume via Corollaries 2.30 and 2.32 that u is immersed and transverse to each E_i. Note that by positivity of intersections,

$$\sum_{i=1}^{m} [u] \cdot [E_i] \geqslant 1.$$

We can now perform the complex blowdown operation on (M, J_0) along $E_1 \amalg \ldots \amalg E_m$ and, by Theorem 3.14, find a compatible symplectic structure $\check{\omega}$ on the blowdown $(\check{M}, \check{J})$ such that $(\check{M}, \check{\omega})$ is the symplectic blowdown of (M, ω) along $E_1 \sqcup \ldots \sqcup E_m$, and every J-holomorphic curve in M yields a $\check{J}$-holomorphic curve in $\check{M}$ via composition with the blowdown map. In particular, by Exercise 3.6, u gives rise to an immersed $\check{J}$-holomorphic sphere $\check{u}$ in $\check{M}$ with

$$c_1([\check{u}]) = c_1(N_{\check{u}}) + 2 = c_1(N_u) + 2 + \sum_{i=1}^{m} [u] \cdot [E_i] = c_1([u]) + \sum_{i=1}^{m} [u] \cdot [E_i] \geqslant 2.$$

The latter gives $\operatorname{ind}(\check{u}) = -2 + 2c_1([\check{u}]) \geqslant 2$, and since $\check{u}$ is also immersed, it satisfies the automatic transversality criterion of Corollary 2.45 and is therefore Fredholm regular.   □

Two symplectic 4-manifolds are said to be **birationally equivalent** whenever they are related to each other by a finite sequence of symplectic blowup and blowdown operations and symplectic deformations. Corollary 7.4 thus implies that the class of closed symplectic 4-manifolds that contain symplectic spheres of nonnegative self-intersection is closed under birational equivalence.

Recall that a symplectic 4-manifold is called a **rational surface** whenever it is birationally equivalent to $(\mathbb{CP}^2, \omega_{FS})$. We can now characterize rational surfaces in terms of Lefschetz pencils as follows.

Theorem 7.6. *A closed and connected symplectic 4-manifold (M, ω) is a rational surface if and only if it admits a symplectic Lefschetz pencil with fibers of genus zero.*

Proof. By Corollary 7.4, everything birationally equivalent to $\mathbb{CP}^2$ necessarily contains a symplectically embedded sphere of nonnegative self-intersection and therefore admits a symplectic Lefschetz pencil or fibration $\pi : M \backslash M_{\text{base}} \to \Sigma$ with genus zero fibers. Given this, we need to show that the condition of being birationally equivalent to $\mathbb{CP}^2$ is equivalent to Σ being a sphere. If indeed $\Sigma \cong \mathbb{CP}^1$, then after blowing up all base points and then blowing down an irreducible component of every singular fiber, we obtain a symplectic ruled surface $\pi : M' \to S^2$ that is birationally equivalent to (M, ω). By the classification of ruled surfaces (see Remark 1.13), M' is symplectically deformation equivalent to either $S^2 \times S^2$ with a product symplectic structure or $\mathbb{CP}^2 \# \overline{\mathbb{CP}}^2$, viewed as the blowup of $(\mathbb{CP}^2, \omega_{FS})$. Both are birationally equivalent to $(\mathbb{CP}^2, \omega_{FS})$; note that in the case of $S^2 \times S^2$, this follows from the proof of Proposition 3.45, which describes a sequence of one blowup and two blowdowns leading from $S^2 \times S^2$ to $\mathbb{CP}^2$.

Conversely, we claim that if (M, ω) admits a symplectic Lefschetz fibration $\pi : M \to \Sigma$ with Σ a surface of positive genus, then (M, ω) is not a rational surface. Assume the contrary, that there is a sequence of deformations, blowup and blowdown operations leading from (M, ω) to $(\mathbb{CP}^2, \omega_{FS})$. Each deformation and blowup operation clearly preserves the property of admitting a symplectic Lefschetz fibration over Σ, as in particular the blowups can all be performed along small balls centered at regular points, thus adding critical points to the Lefschetz fibration, and the blown-up Lefschetz fibration survives arbitrary symplectic deformations due to Theorem G. For blowdowns, we observe that for a generic choice of almost complex structure J such that all fibers of $\pi : M \to \Sigma$ are J-holomorphic and all exceptional spheres have unique J-holomorphic representatives, each J-holomorphic exceptional sphere must be contained in a fiber. Indeed, any embedded J-holomorphic sphere $u : S^2 \to M$ that does not have this property must intersect every fiber positively, so that the map $\pi \circ u : S^2 \to \Sigma$ has positive degree, which is impossible since Σ has a contractible universal cover and thus $\pi_2(\Sigma) = 0$. It follows that all blowdown operations on (M, ω) can be realized by blowing down irreducible components of singular fibers, producing a new Lefschetz fibration on the blowdown that still has base Σ. After following a finite sequence of such operations, we would therefore obtain a symplectic Lefschetz fibration $\pi : \mathbb{CP}^2 \to \Sigma$ with genus zero fibers, and it is easy to show that $\mathbb{CP}^2$ does not admit any such structure, e.g. the fiber would need to represent a nontrivial homology class with self-intersection zero, and there is no such class in $H_2(\mathbb{CP}^2)$. □

Exercise 7.7. Show that up to symplectic deformation equivalence, $\mathbb{CP}^2 \# \overline{\mathbb{CP}}^2$ is the only symplectic ruled surface that is not minimal.

Exercise 7.8. Show that two blown-up symplectic ruled surfaces are birationally equivalent if and only if they admit symplectic Lefschetz fibrations with genus zero fibers over bases of the same genus. *Hint: see Example 3.46.*

Another easy consequence of Theorem 7.3 is that the minimal blowdown of any closed symplectic 4-manifold is essentially unique unless it is rational or ruled. We saw in Proposition 3.45 and Example 3.46 that the caveat for the rational or ruled case is necessary.

Corollary 7.9. *Suppose (M_1, ω_1) and (M_2, ω_2) are two minimal symplectic 4-manifolds with symplectomorphic blowups*

$$(\widetilde{M}, \widetilde{\omega}) := (\widetilde{M}_1, \widetilde{\omega}_1) \cong (\widetilde{M}_2, \widetilde{\omega}_2),$$

such that $(\widetilde{M}, \widetilde{\omega})$ is not a rational surface or blown-up ruled surface. Then (M_1, ω_1) and (M_2, ω_2) are symplectomorphic.

Proof. The hypotheses mean that $(\widetilde{M}, \widetilde{\omega})$ contains two maximal collections of pairwise disjoint exceptional spheres $E_1^1, \ldots, E_k^1 \subset \widetilde{M}$ and $E_1^2, \ldots, E_\ell^2 \subset \widetilde{M}$ such that blowing $(\widetilde{M}, \widetilde{\omega})$ down along $E_1^1 \amalg \ldots \amalg E_k^1$ or $E_1^2 \amalg \ldots \amalg E_\ell^2$ gives (M_1, ω_1) or (M_2, ω_2) respectively. Recall from Theorem 3.11 that the symplectomorphism type of a blowdown is determined by the symplectic isotopy class of the union of exceptional spheres being blown down. By Theorem 5.1, we can choose a generic tame almost complex structure $\widetilde{J}$ on $(\widetilde{M}, \widetilde{\omega})$ and assume after suitable symplectic isotopies that all of the exceptional spheres $E_j^i \subset \widetilde{M}$ are images of embedded $\widetilde{J}$-holomorphic curves. Positivity of intersections then implies

$$[E_i^1] \cdot [E_j^2] \geqslant 0$$

for all i, j, except in cases where $E_i^1 = E_j^2$. Moreover, by Corollary 2.32, we can assume after a further generic perturbation of $\tilde{J}$ that E_i^1 and E_j^2 are always transverse unless they are identical.

Now if (M_1, ω_1) and (M_2, ω_2) are not symplectomorphic, it means $E_1^1 \amalg \ldots \amalg E_k^1 \neq E_1^2 \amalg \ldots \amalg E_\ell^2$, so after reordering $E_1^1, \ldots, E_k^1$, we can assume $E_1^1 \neq E_j^2$ for all $j = 1, \ldots, \ell$. This implies $[E_1^1] \cdot [E_j^2] > 0$ for some j, as otherwise E_1^1 would need to be disjoint from every E_j^2, contradicting the assumption that the latter is a maximal collection. So without loss of generality, assume $[E_1^1] \cdot [E_1^2] > 0$. There are now at least two distinct ways to see that $(\widetilde{M}, \widetilde{\omega})$ satisfies one of the conditions in Theorem 7.3: we shall describe one way explicitly and outline the other in Exercise 7.10 below.

The first approach is to blow down E_1^1. More precisely, we start by modifying $\tilde{J}$ near E_1^1 to match the particular integrable model needed in Theorem 3.14; note that after this modification, $\tilde{J}$ is still sufficiently generic for the purposes of Theorem 5.1 because every $\tilde{J}$-holomorphic curve other than E_1^1 and its multiple covers (which necessarily have positive index since $c_1([E_1^1]) = 1$) passes through regions outside a neighborhood of E_1^1, in which arbitrary small perturbations of $\tilde{J}$ are allowed (cf. Remark 2.17). We can then use Theorem 3.14 to construct a symplectic blowdown (M, ω) of $(\widetilde{M}, \widetilde{\omega})$ along E_1^1 that carries a compatible almost complex structure J for which there is a natural pseudoholomorphic blowdown map $\beta : (\widetilde{M}, \tilde{J}) \to (M, J)$. Since E_1^1 and E_1^2 are transverse, Exercise 3.6 then implies that the image of E_1^2 under β becomes an immersed J-holomorphic sphere $u : S^2 \to M$ with $c_1([u]) = c_1([E_1^2]) + [E_1^2] \cdot [E_1^1] = 1 + [E_1^2] \cdot [E_1^1] \geq 2$, so $\mathrm{ind}(u) = -2 + 2c_1([u]) \geq 2$ and u is Fredholm regular by Corollary 2.45, thus establishing the third condition in Theorem 7.3 for (M, ω).   □

Exercise 7.10. For an alternative version of the last step in the above proof, suppose $S_1, S_2 \subset (M, \omega)$ is a pair of symplectically embedded surfaces that intersect each other transversely and positively in a nonempty set. Show that given $p \in S_1 \cap S_2$, there exists a positively symplectically immersed surface $S \looparrowright M$ that can be formed by deleting small disks from S_1 and S_2 near p and then gluing in an annulus to attach them to each other. In particular, if S_1 and S_2 are both spheres, this produces a positively symplectically immersed sphere S with $c_1([S]) = c_1([S_1]) + c_1([S_2])$. *Hint: choose local coordinates near p in which $S_1 \cup S_2$ looks like the standard local model of a singular fiber of a Lefschetz fibration, holomorphic with respect to an integrable almost complex structure tamed by ω. Then perturb the singular fiber to a regular fiber; note that since the symplectic condition is open, one need not do anything fancy to connect the perturbed singularity with $S_1 \cup S_2$ outside a neighborhood of p.*

To begin the proof of Theorem 7.3, observe that the implication (4) $\Rightarrow$ (3) is obvious, and (1) $\Rightarrow$ (4) is an immediate consequence of Theorem G. The next easiest step is (2) $\Leftrightarrow$ (3), which follows from the results in Sects. 2.1.4 and 2.1.5

on moduli spaces with constrained marked points, plus the automatic transversality criterion of Sect. 2.2.1:

Lemma 7.11. *The second and third conditions in the statement of Theorem 7.3 are equivalent.*

Proof. If $S \hookrightarrow (M, \omega)$ is a positively symplectically immersed sphere with $c_1([S]) \geq 2$, then using Exercise 7.2, one can choose $J \in \mathcal{J}_\tau(M, \omega)$ such that S is the image of an immersed and somewhere injective J-holomorphic curve $u : S^2 \to M$ with $\operatorname{ind}(u) = -\chi(S^2) + 2c_1([u]) \geq 2$. This curve satisfies the automatic transversality criterion of Corollary 2.45, so it is Fredholm regular. Conversely, if a Fredholm regular J-holomorphic sphere $u : S^2 \to M$ with $\operatorname{ind}(u) \geq 2$ exists, then by Corollaries 2.26, 2.30 and 2.32, we can assume after possibly a generic perturbation of J and a small perturbation of u within the moduli space of holomorphic curves that u is immersed, with only transverse self-intersections and no triple self-intersections. The image of u is then a positively immersed symplectic sphere with $c_1([u]) \geq 2$ since $\operatorname{ind}(u) \geq 2$. □

The rest of the proof of Theorem 7.3 will take a bit more work. We shall prove $(1) \Rightarrow (5) \Rightarrow (2)$ in the next section, on Gromov-Witten invariants and the uniruled condition. The hard part will then be dealt with in Sect. 7.3, which covers the main result of [McD92], giving a proof of $(2) \Rightarrow (1)$ and an independent proof of $(2) \Rightarrow (4)$ (without passing through (1)).

7.2 Gromov-Witten Invariants

7.2.1 The Invariants in General

We shall begin this section with a heuristic discussion of the Gromov-Witten invariants and uniruled symplectic manifolds in general, and then give a rigorous definition of a slightly simplified version of the invariants for symplectic 4-manifolds, proving the implications $(1) \Rightarrow (5) \Rightarrow (2)$ in Theorem 7.3 along the way.

In their simplest form, the Gromov-Witten invariants associate to any closed symplectic $2n$-manifold (M, ω), given integers $g, m \geq 0$ and a homology class $A \in H_2(M)$, a symplectically deformation-invariant homomorphism

$$\mathrm{GW}_{g,m,A}^{(M,\omega)} : H^*(M; \mathbb{Q})^{\otimes m} \to \mathbb{Q}, \tag{7.1}$$

defined in principle by counting J-holomorphic curves that satisfy constraints at their marked points. Let us regard $\mathrm{GW}_{g,m,A}^{(M,\omega)}$ as a multilinear map and write

$$\mathrm{GW}_{g,m,A}^{(M,\omega)}(\alpha_1, \ldots, \alpha_m) \in \mathbb{Q} \quad \text{for} \quad \alpha_1, \ldots, \alpha_m \in H^*(M; \mathbb{Q})$$

instead of $\mathrm{GW}_{g,m,A}^{(M,\omega)}(\alpha_1 \otimes \ldots \otimes \alpha_m)$. Heuristically, the number $\mathrm{GW}_{g,m,A}^{(M,\omega)}$ $(\alpha_1, \ldots, \alpha_m) \in \mathbb{Q}$ is intended to be the answer to the following question:

For generic $J \in \mathcal{J}_\tau(M, \omega)$, given smooth submanifolds $\bar{\alpha}_i \subset M$ Poincaré dual to $\alpha_i \in H^*(M)$ for $i = 1, \ldots, m$, how many curves $[(\Sigma, j, u, (\zeta_1, \ldots, \zeta_m))] \in \mathcal{M}_{g,m}(A; J)$ exist subject to the constraints

$$u(\zeta_i) \in \bar{\alpha}_i \text{ for } i = 1, \ldots, m?$$

Recalling the evaluation map

$$\mathrm{ev} = (\mathrm{ev}_1, \ldots, \mathrm{ev}_m) : \overline{\mathcal{M}}_{g,m}(A; J) \to M^m,$$

the set of curves satisfying the constraints described above is precisely $\mathrm{ev}^{-1}(\bar{\alpha}_1 \times \ldots \bar{\alpha}_m)$. Thus if we adopt the convenient (though usually fictitious) assumption that $\overline{\mathcal{M}}_{g,m}(A; J)$ is a smooth, closed and oriented manifold of dimension equal to $d :=$ vir-dim $\mathcal{M}_{g,m}(A; J)$, and denote its fundamental class by

$$[\overline{\mathcal{M}}_{g,m}(A; J)]^{\mathrm{vir}} \in H_d(\overline{\mathcal{M}}_{g,m}(A; J)),$$

we obtain a homology class $[\mathrm{ev}] := \mathrm{ev}_*[\overline{\mathcal{M}}_{g,m}(A; J)]^{\mathrm{vir}} \in H_d(M^m)$ and can define the desired count of constrained curves as a homological intersection number

$$\mathrm{GW}_{g,m,A}^{(M,\omega)}(\alpha_1, \ldots, \alpha_m) = [\mathrm{ev}] \cdot (\mathrm{PD}(\alpha_1) \times \ldots \times \mathrm{PD}(\alpha_m)), \qquad (7.2)$$

where PD denotes the Poincaré duality isomorphism. Implicit in this expression is that the intersection number and hence the Gromov-Witten invariant is zero unless the two homology cycles are of complementary dimension, which means

$$\text{vir-dim}\, \mathcal{M}_{g,m}(A; J) = \sum_{i=1}^{m} \deg(\alpha_i).$$

The fact that the intersection number generally turns out to be in $\mathbb{Q}$ rather than $\mathbb{Z}$, even if the α_i are all assumed to be integral classes, is related to the fact that $\overline{\mathcal{M}}_{g,m}(A; J)$ is not actually a manifold in general—even if transversality can be established for multiply covered curves, the moduli space will then have orbifold singularities whenever those curves have nontrivial automorphisms, so intersections must be counted with rational weights. It is common to rewrite (7.2) as an integral, interpreted as the evaluation of a product of pulled back cohomology classes on the fundamental class $[\overline{\mathcal{M}}_{g,m}(A; J)]^{\mathrm{vir}}$,

$$\mathrm{GW}_{g,m,A}^{(M,\omega)}(\alpha_1, \ldots, \alpha_m) = \int_{[\overline{\mathcal{M}}_{g,m}(A;J)]^{\mathrm{vir}}} \mathrm{ev}_1^* \alpha_1 \cup \ldots \cup \mathrm{ev}_m^* \alpha_m.$$

For a given $m \geqslant 0$, homomorphisms of the form (7.1) are called m-**point Gromov-Witten invariants**. They reduce to rational numbers in the case $m = 0$, i.e. the 0-*point invariant* $\mathrm{GW}_{g,0,A}^{(M,\omega)} \in \mathbb{Q}$ is defined when vir-dim $\mathcal{M}_g(A; J) = 0$ and interpreted as a count of (isolated) curves in $\overline{\mathcal{M}}_g(A; J)$.

Most of the time, the failure of transversality for multiple covers prevents $\overline{\mathcal{M}}_{g,m}(A; J)$ from being anything nearly as nice as an orbifold, so one must resort to more abstract perturbations of the nonlinear Cauchy-Riemann equation in order to either define a "virtual" fundamental class $[\overline{\mathcal{M}}_{g,m}(A; J)]^{\mathrm{vir}}$ or otherwise give a rigorous interpretation of the intersection number in (7.2). In the following, we will address this issue only in dimension four and in the case vir-dim $\mathcal{M}_g(J; A) > 0$, for which a relatively straightforward solution is available. For some recent approaches to the general case, see for instance [HWZa, CM07, Ger, Par16].

If $2g + m \geqslant 3$, then one can define a more general version of the Gromov-Witten invariants by imposing an additional constraint on the complex structures of the domains of the curves. Let $\overline{\mathcal{M}}_{g,m}$ denote the compactified moduli space of stable nodal Riemann surfaces of genus g with m marked points, i.e. it is the same as $\overline{\mathcal{M}}_{g,m}(A; J)$ if M is taken to be a one point space. Elements of $\overline{\mathcal{M}}_{g,m}$ are thus equivalence classes of tuples $(S, j, (\zeta_1, \ldots, \zeta_m), \Delta)$, where stability (cf. Definition 2.34) means that every connected component of S after removing Δ and all marked points has negative Euler characteristic, which is impossible unless $2g + m \geqslant 3$. In general, $\overline{\mathcal{M}}_{g,m}$ is a smooth orbifold of real dimension

$$\dim \overline{\mathcal{M}}_{g,m} = 6g - 6 + 2m,$$

and it is a manifold if $g = 0$. We then consider the natural **forgetful map**

$$\Phi : \overline{\mathcal{M}}_{g,m}(A; J) \to \overline{\mathcal{M}}_{g,m},$$

sending each element of $\overline{\mathcal{M}}_{g,m}(A; J)$ represented by a stable nodal J-holomorphic curve $(S, j, u, (\zeta_1, \ldots, \zeta_m), \Delta)$ to the equivalence class of its domain $(S, j, (\zeta_1, \ldots, \zeta_m), \Delta)$ in $\overline{\mathcal{M}}_{g,m}$. (See Remark 7.12 below on the subtleties involved in this definition.) This gives rise to a homomorphism

$$\mathrm{GW}_{g,m,A}^{(M,\omega)} : H^*(M; \mathbb{Q})^{\otimes m} \otimes H_*(\overline{\mathcal{M}}_{g,m}; \mathbb{Q}) \to \mathbb{Q},$$

whose action on $\alpha_1 \otimes \ldots \otimes \alpha_m \otimes \beta$ we shall denote by

$$\mathrm{GW}_{g,m,A}^{(M,\omega)}(\alpha_1, \ldots, \alpha_m; \beta) \in \mathbb{Q} \text{ for } \alpha_1, \ldots, \alpha_m \in H^*(M; \mathbb{Q}), \ \beta \in H_*(\overline{\mathcal{M}}_{g,m}; \mathbb{Q}),$$

such that if the α_i are Poincaré dual to submanifolds $\bar{\alpha}_i \subset M$ and β is represented by a submanifold $\bar{\beta} \subset \overline{\mathcal{M}}_{g,m}$, then $\mathrm{GW}_{g,m,A}^{(M,\omega)}(\alpha_1, \ldots, \alpha_m; \beta)$ is a count of elements in $(\mathrm{ev}, \Phi)^{-1}(\bar{\alpha}_1 \times \ldots \times \bar{\alpha}_m \times \bar{\beta})$. Given a well-defined virtual fundamental class for

$\overline{\mathcal{M}}_{g,m}(A; J)$, this can again be interpreted as a homological intersection number

$$\text{GW}_{g,m,A}^{(M,\omega)}(\alpha_1, \ldots, \alpha_m; \beta) = [(\text{ev}, \Phi)] \cdot (\text{PD}(\alpha_1) \times \ldots \times \text{PD}(\alpha_m) \times \beta) \qquad (7.3)$$

or an integral

$$\text{GW}_{g,m,A}^{(M,\omega)}(\alpha_1, \ldots, \alpha_m; \beta) = \int_{[\overline{\mathcal{M}}_{g,m}(A;J)]^{\text{vir}}} \text{ev}_1^* \alpha_1 \cup \ldots \cup \text{ev}_m^* \alpha_m \cup \Phi^* \text{PD}(\beta),$$

and dimensional considerations now dictate that $\text{GW}_{g,m,A}^{(M,\omega)}(\alpha_1, \ldots, \alpha_m; \beta)$ vanishes unless

$$\text{vir-dim} \, \mathcal{M}_{g,m}(A; J) = \sum_{i=1}^{m} \deg(\alpha_i) + \dim \mathcal{M}_{g,m} - \deg(\beta). \qquad (7.4)$$

A natural special case is to define β as the fundamental class $[\overline{\mathcal{M}}_{g,m}] \in H_{6g-6+2m}(\overline{\mathcal{M}}_{g,m})$, represented by $\bar{\beta} = \overline{\mathcal{M}}_{g,m}$, which amounts to not imposing any constraint at all on the forgetful map, hence

$$\text{GW}_{g,m,A}^{(M,\omega)}(\alpha_1, \ldots, \alpha_m; [\overline{\mathcal{M}}_{g,m}]) = \text{GW}_{g,m,A}^{(M,\omega)}(\alpha_1, \ldots, \alpha_m).$$

Alternatively, choosing $\beta = [\text{pt}] \in H_0(\overline{\mathcal{M}}_{g,m})$, i.e. the homology class of a point, means fixing a domain $(\Sigma, j, (\zeta_1, \ldots, \zeta_m))$ and counting J-holomorphic maps $u : (\Sigma, j) \to (M, J)$ with both the complex structure j and the marked points $(\zeta_1, \ldots, \zeta_m)$ fixed in place. The homological invariance of intersection numbers implies that the resulting count of curves will not depend on which fixed domain is chosen, but notice that by (7.4), the indices of the curves being counted are now larger than in the case $\beta = [\overline{\mathcal{M}}_{g,m}]$.

Remark 7.12. The description of the forgetful map above ignores one important detail: Definition 2.34 allows $(S, j, u, (\zeta_1, \ldots, \zeta_m), \Delta)$ to have spherical components on which u is nonconstant but there are fewer than three marked or nodal points, in which case the domain $(S, j, (\zeta_1, \ldots, \zeta_m), \Delta)$ is not stable and thus does not represent an element of $\overline{\mathcal{M}}_{g,m}$. However, every nodal Riemann surface with $2g + m \geqslant 3$ has a well-defined **stabilization**, obtained by collapsing the unwanted spherical components, i.e. they are eliminated from S along with their nodal points, and any orphaned marked point on such a component is then placed in the position of the corresponding orphaned nodal point on the adjacent component. Thus for a general stable nodal curve $u \in \overline{\mathcal{M}}_{g,m}(A; J)$, $\Phi(u) \in \overline{\mathcal{M}}_{g,m}$ is defined as the stabilization of the domain of u.

7.2.2 The Uniruled Condition

In complex algebraic geometry, a proper variety is called *uniruled* if it contains rational curves through every point, see e.g. [Deb01]. The following definition plays this role in the symplectic category.

Definition 7.13. Let $[\mathrm{pt}] \in H_0(M)$ denote the homology class of a point. We say that (M, ω) is **symplectically uniruled** if there exist $A \in H_2(M)$, an integer $m \geqslant 3$ and classes $\alpha_2, \ldots, \alpha_m \in H^*(M; \mathbb{Q})$, $\beta \in H_*(\overline{\mathcal{M}}_{0,m}; \mathbb{Q})$ such that

$$\mathrm{GW}_{0,m,A}^{(M,\omega)}(\mathrm{PD}[\mathrm{pt}], \alpha_2, \ldots, \alpha_m; \beta) \neq 0.$$

Proposition 7.14. *If (M, ω) is symplectically uniruled, then for every $J \in \mathcal{J}_\tau(M, \omega)$ and every $p \in M$, there exists a nonconstant J-holomorphic sphere passing through p.*

At the moment, we are only in a position to justify this result heuristically since we have not given a rigorous definition for the intersection number (7.3), but the idea is simple enough. Let us assume that $\mathrm{GW}_{0,m,A}^{(M,\omega)}(\mathrm{PD}[\mathrm{pt}], \alpha_2, \ldots, \alpha_m; \beta) \neq 0$ and that each of the homology classes $\mathrm{PD}(\alpha_i) \in H_*(M)$ and $\beta \in H_*(\overline{\mathcal{M}}_{0,m})$ can be represented by smooth submanifolds $\bar{\alpha}_i \subset M$ and $\bar{\beta} \subset \overline{\mathcal{M}}_{0,m}$; note that by a theorem of Thom [Tho54], the latter is true for all integral homology classes after multiplication by a natural number, so there is no loss of generality. Now for generic and arbitrarily small perturbations of the nonlinear Cauchy-Riemann equation[1] defined by $J \in \mathcal{J}_\tau(M, \omega)$, we can assume (ev, Φ) is transverse to the submanifold $\{p\} \times \bar{\alpha}_2 \times \ldots \times \bar{\alpha}_m \times \bar{\beta} \subset M^m \times \overline{\mathcal{M}}_{0,m}$, and the nontriviality of $\mathrm{GW}_{0,m,A}^{(M,\omega)}(\mathrm{PD}[\mathrm{pt}], \alpha_2, \ldots, \alpha_m; \beta)$ implies the existence of at least one solution $u \in (\mathrm{ev}, \Phi)^{-1}(\{p\} \times \bar{\alpha}_2 \times \ldots \times \bar{\alpha}_m \times \bar{\beta})$. In particular, given a sequence of such generic perturbations converging to the standard Cauchy-Riemann equation for J, there exists a corresponding sequence of solutions u_k, which are all homologous to A and thus have bounded energy, so one can use a version of Gromov compactness to extract a subsequence that converges to an element u_∞ in $\overline{\mathcal{M}}_{0,m}(A; J)$ satisfying $\mathrm{ev}_1(u_\infty) = p$. This limit may be a nodal curve, but it has at least one nonconstant smooth component that is a smooth J-holomorphic sphere passing through p.

The next lemma, in conjunction with Proposition 7.14, furnishes the implication $(5) \Rightarrow (2)$ in Theorem 7.3.

---

[1] We are being intentionally vague here about the meaning of the words "perturbations of the nonlinear Cauchy-Riemann equation". This can mean various things depending on the context and the precise definition of the intersection number (7.3) that one adopts, e.g. in the approaches of [MS12, CM07], the usual equation $\bar{\partial}_J u = 0$ is generalized to allow generic dependence of J on points in the domain of u, while in [RT97, Ger], one instead introduces a generic nonzero term (inhomogeneous perturbation) on the right hand side of the equation. A more abstract functional-analytic approach is taken in [HWZa].

Lemma 7.15. *Suppose* (M, ω) *is a closed symplectic manifold of dimension* $2n \geqslant 4$ *with a point* $p \in M$ *such that for every* $J \in \mathcal{J}_\tau(M, \omega)$, *there exists a nonconstant J-holomorphic sphere through* p. *Then for generic* $J \in \mathcal{J}_\tau(M, \omega)$, *there also exists an immersed J-holomorphic sphere* $u : (S^2, i) \looparrowright (M, J)$ *that satisfies* $c_1([u]) \geqslant 2$ *and has no triple self-intersections or tangential self-intersections.*

Proof. Since every holomorphic sphere covers one that is somewhere injective, we can add the latter to the hypotheses of the lemma without loss of generality. Then if $u : (S^2, i) \to (M, J)$ is somewhere injective and passes through p, we can add a marked point and regard it as an element of the moduli space $\mathcal{M}_{0,1}^*(A; J; p)$ of somewhere injective curves in $\mathrm{ev}^{-1}(p) \subset \mathcal{M}_{0,1}(A; J)$, for $A := [u] \in H_2(M)$. This constrained moduli space has virtual dimension

$$\text{vir-dim}\,\mathcal{M}_{0,1}(A; J; p) = \mathrm{ind}(u)+2-2n = 2(n-3)+2c_1(A)+2-2n = 2c_1(A)-4,$$

which must be nonnegative for generic J, thus $c_1([u]) \geqslant 2$. Corollaries 2.26, 2.30, and 2.32 then combine to furnish an immersed curve $u' \in \mathcal{M}_0^*(A; J)$ close to u that has no triple points or tangential double points.   □

Remark 7.16. One could include the case $1 \leqslant m < 3$ in Definition 7.13 and allow the conditions $\mathrm{GW}_{0,1,A}^{(M,\omega)}(\mathrm{PD}[\mathrm{pt}]) \neq 0$ or $\mathrm{GW}_{0,2,A}^{(M,\omega)}(\mathrm{PD}[\mathrm{pt}], \alpha) \neq 0$, but this would not add any generality. The reason is that any nontrivial Gromov-Witten invariant with $m < 3$ can be related to one with $m \geqslant 3$ using Exercise 7.17 below.

Exercise 7.17. Deduce from the heuristic description of the Gromov-Witten invariants as counts of constrained J-holomorphic curves that for any $\alpha_1, \ldots, \alpha_m \in H^*(M, \mathbb{Q})$ with $\alpha_m \in H^2(M; \mathbb{Q})$,

$$\mathrm{GW}_{g,m,A}^{(M,\omega)}(\alpha_1, \ldots, \alpha_m) = \left(\int_A \alpha_m \right) \cdot \mathrm{GW}_{g,m-1,A}^{(M,\omega)}(\alpha_1, \ldots, \alpha_{m-1}).$$

In the terminology of Kontsevich-Manin [KM94], this is a special case of the so-called *divisor axiom*. *Hint: if* $\alpha_m \in H^2(M)$ *is Poincaré dual to a smooth submanifold* $\bar{\alpha}_m \subset M$ *that is transverse to some curve* $u \in \mathcal{M}_g(A; J)$, *then* $\int_A \alpha_m$ *is the signed count of intersections between* u *and* $\bar{\alpha}_m$.

7.2.3 Pseudocycles and the Four-Dimensional Case

Let us now give a rigorous definition of the m-point invariant

$$\mathrm{GW}_{g,m,A}^{(M,\omega)}(\alpha_1, \ldots, \alpha_m) \in \mathbb{Q}$$

under the assumptions

$$\dim M = 4,$$
$$\text{vir-dim}\,\mathcal{M}_g(A; J) = 2c_1(A) + 2g - 2 > 0 \quad \text{or} \quad g = 0.$$

(7.5)

The condition on vir-dim $\mathcal{M}_g(A; J)$ for $g > 0$ implies that $\text{GW}_{g,m,A}^{(M,\omega)}$ can be nontrivial only if $m \geqslant 1$, so we are now excluding 0-point invariants from discussion except in the genus zero case (see Remark 7.31). We are also excluding all choices of $\beta \in H_*(\overline{\mathcal{M}}_{g,m})$ other than the fundamental class $[\overline{\mathcal{M}}_{g,m}]$.

In turns out that under the assumptions (7.5), we can restrict our attention to the smooth manifold $\mathcal{M}_g^*(A; J)$ of somewhere injective curves, and one version of the definition of $\text{GW}_{g,m,A}^{(M,\omega)}$ can then be stated as follows.

Theorem 7.18. *Assume (M, ω) is a closed symplectic 4-manifold, $A \in H_2(M)$, $g \geqslant 0$ and $m \geqslant 1$ are integers, and $\alpha_1, \ldots, \alpha_m \in H^*(M)$ are integral cohomology classes Poincaré dual to smooth submanifolds $\bar{\alpha}_1, \ldots, \bar{\alpha}_m \subset M$ such that*

$$2c_1(A) + 2g - 2 = \sum_{i=1}^{m} (\deg(\alpha_i) - 2)$$

and either the latter expression is positive or $g = 0$. Then for generic $J \in \mathcal{J}_\tau(M, \omega)$, the map $\text{ev}\big|_{\mathcal{M}_{g,m}^*(A;J)} : \mathcal{M}_{g,m}^*(A; J) \to M^m$ has only finitely many intersections with $\bar{\alpha}_1 \times \ldots \times \bar{\alpha}_m$, all of them transverse, and the signed count of these intersections

$$\text{GW}_{g,m,A}^{(M,\omega)}(\alpha_1, \ldots, \alpha_m) := \left(\text{ev}\big|_{\mathcal{M}_{g,m}^*(A;J)}\right) \cdot (\bar{\alpha}_1 \times \ldots \times \bar{\alpha}_m) \in \mathbb{Z}$$

depends only on the cohomology classes $\alpha_1, \ldots, \alpha_m$ and the symplectic deformation class of ω.

There is a slight abuse of notation in the above theorem: where "·" usually denotes the homological intersection product, here it is simply a signed count of intersections which has no immediate homological interpretation since the manifold $\mathcal{M}_{g,m}^*(A; J)$ is noncompact. It may seem surprising at first that one can define a finite intersection count in this way, but the reasons why it works are not hard to imagine if you remember the index counting relations in Sect. 4.2. The following lemma forces intersections of $\text{ev} : \overline{\mathcal{M}}_{g,m}(A; J) \to M^m$ with $\bar{\alpha}_1 \times \ldots \times \bar{\alpha}_m$ to stay away from $\overline{\mathcal{M}}_{g,m}(A; J)\backslash\mathcal{M}_{g,m}^*(A; J)$ for dimensional reasons, and thus remain in a compact subset of $\mathcal{M}_{g,m}^*(A; J)$.

Lemma 7.19. *Given a closed symplectic 4-manifold (M, ω), integers $g, m \geqslant 0$ and $A \in H_2(M)$ such that assumptions (7.5) are satisfied, along with a collection of smooth submanifolds $\bar{\alpha}_1, \ldots, \bar{\alpha}_m \subset M$, there exists a comeager subset $\mathcal{J}^{\text{reg}} \subset \mathcal{J}_\tau(M, \omega)$ such that for all $J \in \mathcal{J}^{\text{reg}}$, the image of the map*

$$\overline{\mathcal{M}}_{g,m}(A; J)\backslash\mathcal{M}_{g,m}^*(A; J) \xrightarrow{\text{ev}} M^m$$

is contained in a countable union of sets of the form $f_i(X_i) \subset M^m$, where the X_i are smooth manifolds with

$$\dim X_i \leqslant \dim \mathcal{M}_{g,m}^*(A; J) - 2$$

and $f_i : X_i \to M^m$ are smooth maps transverse to $\bar{\alpha}_1 \times \ldots \times \bar{\alpha}_m$.

Proof. By choosing J to lie in a countable intersection of certain comeager subsets of $\mathcal{J}_\tau(M, \omega)$, we can assume that the spaces $\mathcal{M}_{h,k}^*(B; J)$ are smooth manifolds of the correct dimension for every $h, k \geqslant 0$ and that all evaluation maps ev : $\mathcal{M}_{h,k}^*(B; J) \to M^k$ are transverse to all submanifolds of M^k formed via Cartesian products of the submanifolds $\bar{\alpha}_1, \ldots, \bar{\alpha}_m \subset M$.

Now recall from Sect. 4.4 the natural map

$$\pi : \overline{\mathcal{M}}_{g,m}(A; J) \to \overline{\mathcal{M}}_g(A; J)$$

which forgets all the marked points and collapses any resulting unstable ghost bubbles (see in particular Fig. 4.1). If $u_k \in \mathcal{M}_{g,m}^*(A; J)$ is a sequence converging to some $u_\infty \in \overline{\mathcal{M}}_{g,m}(A; J) \backslash \mathcal{M}_{g,m}^*(A; J)$, then the sequence $\hat{u}_k := \pi(u_k) \in \mathcal{M}_g^*(A; J)$ converges likewise to $\hat{u}_\infty := \pi(u_\infty) \in \overline{\mathcal{M}}_g(A; J)$, which has the same nonconstant components as u_∞ but may have fewer ghost bubbles. We consider three cases.

Case 1 Suppose $\hat{u}_\infty \in \mathcal{M}_g^*(A; J)$. Then u_∞ consists of a single somewhere injective component u_∞^1 of genus g plus a nonempty set of ghost bubbles. Assume u_∞^1 has $N > 0$ nodal points, each of which attaches it to a tree of ghost bubbles. Stability dictates that each tree of ghost bubbles has at least two of the marked points, so the number of marked points remaining on u_∞^1 is at most $m - 2N$. The position of $\text{ev}(u_\infty)$ in M^m is thus determined by three pieces of data: (1) the curve u_∞^1, which lives in a smooth moduli space of dimension $\text{ind}(u_\infty^1) = \text{ind}(u_\infty)$; (2) the positions of at most $m - 2N$ marked points on u_∞^1; (3) the positions of the N nodal points on u_∞^1, each of which determines the image of every marked point on the attached tree of ghost bubbles. All these degrees of freedom add up to something less than or equal to

$$\text{ind}(u_\infty) + 2(m - 2N) + 2N = \text{ind}(u_\infty) + 2(m - N)$$
$$\leqslant \text{ind}(u_k) + 2m - 2 = \dim \mathcal{M}_{g,m}^*(A; J) - 2.$$

Case 2 Suppose $\hat{u}_\infty \in \overline{\mathcal{M}}_g(A; J) \backslash \mathcal{M}_g(A; J)$, so $\hat{u}_\infty$ is a nodal curve. Call its nonconstant components $\hat{u}_\infty^1, \ldots, \hat{u}_\infty^V$ and assume each $\hat{u}_\infty^i$ is a k_i-fold cover of a simple curve $\hat{v}_\infty^i$ for some $k_i \in \mathbb{N}$. Then the combination of Propositions 4.8 and 4.11 implies

$$\sum_{i=1}^{V} \text{ind}(\hat{v}_\infty^i) \leqslant \text{ind}(u_k) - 2.$$

The position of $ev(u_\infty)$ in M^m is now determined by (1) the curves $\hat{v}_\infty^i$, which contribute $\sum_i \mathrm{ind}(\hat{v}_\infty^i)$ degrees of freedom; (2) the positions of at most m points on the curves $\hat{v}_\infty^i$, which may be a mixture of marked points and nodal points attached to constant components of u_∞ that contain more marked points. These degrees of freedom add up to something bounded above by

$$\sum_{i=1}^{V} \mathrm{ind}(\hat{v}_\infty^i) + 2m \leqslant \mathrm{ind}(u_k) + 2m - 2 = \dim \mathcal{M}_{g,m}^*(A; J) - 2.$$

Case 3 Suppose $\hat{u}_\infty \in \mathcal{M}_g(A; J)$ is of the form $\hat{v}_\infty \circ \varphi$ for a simple curve $\hat{v}_\infty$ and a k-fold branched cover φ with $k \geqslant 2$. Then Proposition 4.11 gives $k\,\mathrm{ind}(\hat{v}_\infty) = \mathrm{ind}(\hat{u}_\infty) - Z(d\varphi)$. If $\mathrm{ind}(\hat{u}_\infty) = \mathrm{ind}(u_\infty) = 2c_1(A) + 2g - 2 > 0$, then since the Fredholm index is always even, we conclude

$$\mathrm{ind}(\hat{v}_\infty) \leqslant \mathrm{ind}(u_k) - 2,$$

and this is true regardless for $g = 0$ because $\hat{v}_\infty$ then must also have genus zero and the Riemann-Hurwitz formula then implies $Z(d\varphi) = 2k - 2 \geqslant 2$. Now $ev(u_\infty)$ is determined the curve $\hat{v}_\infty$ and the positions of at most m points on this curve, which again may be a mixture of marked points and nodal points attached to constant components of u_∞, so the total number of degrees of freedom is bounded above by

$$\mathrm{ind}(\hat{v}_\infty) + 2m \leqslant \mathrm{ind}(u_k) + 2m - 2 = \dim \mathcal{M}_{g,m}^*(A; J) - 2.$$

$\square$

To prove that the count in Theorem 7.18 is not only finite but also invariant, one can use the fact that any symplectic deformation $\{\omega_s\}_{s\in[0,1]}$ can be accompanied by a generic homotopy of tame almost complex structures $\{J_s\}_{s\in[0,1]}$, so that the extension of ev to the parametric moduli space $\mathcal{M}_{g,m}^*(A; \{J_s\})$ defines a cobordism from $\mathcal{M}_{g,m}^*(A; J_0)$ to $\mathcal{M}_{g,m}^*(A; J_1)$. This cobordism is generally noncompact, but by a similar dimensional argument as in the above lemma, it will contain $ev^{-1}(\bar{\alpha}_1 \times \ldots \times \bar{\alpha}_m)$ as a compact 1-dimensional cobordism between the finite intersection sets defined via J_0 and J_1. The fact that these intersection sets have the same signed count is then a basic principle of differential topology in the spirit of [Mil97], cf. Fig. 2.1 in Chap. 2.

One can similarly use a cobordism argument to prove that $\mathrm{GW}_{g,m,A}^{(M,\omega)}(\alpha_1, \ldots, \alpha_m)$ is independent of the choices of submanifolds $\bar{\alpha}_i \subset M$ representing $\mathrm{PD}(\alpha_i)$, and since we know from [Tho54] that every homology class $\mathrm{PD}(\alpha) \in H_*(M)$ can be written as $c[\bar{\alpha}]$ for some $c \in \mathbb{Q}$ and a closed oriented submanifold $\bar{\alpha} \subset M$, this uniquely determines a $\mathbb{Q}$-multilinear function $\mathrm{GW}_{g,m,A}^{(M,\omega)} : H^*(M; \mathbb{Q}) \to \mathbb{Q}$. However, appealing to Thom's theorem in this way gives a slightly less direct definition than we would like, and it obscures an interesting detail: whenever the $\alpha_i \in H^*(M)$ are all integral classes and (7.5) holds, we will see that the numbers

$GW_{g,m,A}^{(M,\omega)}(\alpha_1, \ldots, \alpha_m)$ are also *integers*. This is a special property of the invariants in dimension four, and does not hold for Gromov-Witten invariants more generally.[2]

To formalize the dimension-counting trick behind Theorem 7.18, we will show that under assumptions (7.5), the intersection number (7.2) can be given a precise meaning in terms of *pseudocycles*, a notion first introduced by McDuff and Salamon in their presentation of the genus zero Gromov-Witten invariants in the semipositive case [MS94]. Intuitively, a pseudocycle in M is the next best thing to a homology class of the form $[f] := f_*[V] \in H_*(M)$ for a closed oriented manifold V and smooth map $f : V \to M$. The idea is to relax the assumption about V being compact, but impose weaker conditions so that intersection numbers are still well defined for dimensional reasons. We introduce the following notation: for a smooth map $f : V \to M$ defined on a (possibly noncompact) manifold V, its **omega-limit set** is

$$\Omega_f := \left\{ \lim f(x_n) \,\middle|\, \text{sequences } x_n \in V \text{ with no limit points} \right\} \subset M.$$

Definition 7.20. A d-dimensional **pseudocycle** in a smooth manifold M is a smooth map $f : V \to M$, whose domain V is a smooth oriented d-dimensional manifold without boundary, such that $f(V) \subset M$ has compact closure and $\Omega_f \subset M$ is contained in a countable union of images of smooth maps defined on manifolds of dimension at most $d - 2$.

A **bordism** between two d-dimensional pseudocycles $f_+ : V_+ \to M$ and $f_- : V_- \to M$ is a smooth map $f : V \to M$, where V is a smooth oriented $(d + 1)$-dimensional manifold with boundary $\partial V = -V_- \amalg V_+$, such that $f|_{V_\pm} = f_\pm$, $f(V) \subset M$ has compact closure, and $\Omega_f \subset M$ is contained in a countable union of images of smooth maps defined on manifolds of dimension at most $d - 1$.

It is straightforward to show that the existence of bordisms between pseudocycles defines an equivalence relation, and the resulting bordism classes of d-dimensional pseudocycles define an abelian group which we will denote by

$$H_d^{\Psi}(M) = \{d\text{-dimensional pseudocycles in } M\} \,/\, \text{bordism}.$$

The identity element in this group is represented by the *empty* pseudocycle (with $V = \varnothing$), and addition is defined via disjoint unions.

Exercise 7.21. Show that for any pseudocycle $f : V \to M$, the inverse of $[f] \in H_d^{\Psi}(M)$ is represented by the same map $f : -V \to M$ with the orientation of its domain reversed.

---

[2] For slightly different reasons, the genus 0 invariants $GW_{0,m}^{(M,\omega)}(\alpha_1, \ldots, \alpha_m; \beta)$ for $m \geq 3$ are also integers whenever $\alpha_1, \ldots, \alpha_m$ and β are all integral classes and (M, ω) is *semipositive*, see [MS12]. The semipositivity condition is always satisfied when $\dim M \leq 6$.

Example 7.22. Since smooth manifolds of negative dimension are empty by definition, the definition of a zero-dimensional pseudocycle $f : V \to M$ requires V to be a compact oriented 0-manifold, i.e. a finite set of points with signs. Similarly, bordism between 0-dimensional pseudocycles $f_+ : V_+ \to M$ and $f_- : V_- \to M$ reduces to the usual notion of bordism between maps, meaning a smooth map $f : V \to M$ where V is a compact oriented 1-manifold with boundary $\partial V = -V_- \amalg V_+$ and $f|_{V_\pm} = f_\pm$. It follows via standard arguments as in [Mil97] that the signed count of points in domains defines a natural homomorphism

$$H_0^\Psi(M) \to \mathbb{Z}, \tag{7.6}$$

which is an isomorphism if M is connected. More generally, there is a natural isomorphism between $H_0^\Psi(M)$ and $H_0(M)$.

Example 7.23. Any smooth map $f : V \to M$ defined on a closed and oriented manifold V is a pseudocycle of dimension $\dim V$. Notice that if $f_\pm : V_\pm \to M$ are two pseudocycles of this form that are bordant in the classical sense, meaning $f_\pm = f|_{V_\pm}$ for some smooth map $f : V \to M$ defined on a compact oriented manifold V with boundary $\partial V = -V_- \amalg V_+$, then they induce the same homology class $(f_+)_*[V_+] = (f_-)_*[V_-] \in H_*(M)$.

It turns out that $H_d^\Psi(M)$ and $H_d(M)$ are naturally isomorphic for every $d \geqslant 0$ (see [Zin08]), though this is a much stronger result than we will need. More important for our purposes is Proposition 7.28 below, which states that every integral homology class can be represented by a bordism class of pseudocycles in a way that respects the homological intersection product. We must first define the corresponding intersection pairing on pseudocycles.

Definition 7.24. Two pseudocycles $f_1 : V_1 \to M$ and $f_2 : V_2 \to M$ are called **strongly transverse** if for each $i = 1, 2$, there exists a smooth map $f_i^\Omega : V_i^\Omega \to M$ with $\Omega_{f_i} \subset f_i^\Omega(V_i^\Omega)$, where V_i^Ω is a countable disjoint union of smooth manifolds of dimension at most $\dim V_i - 2$, such that

$$f_1 \pitchfork f_2, \quad f_1 \pitchfork f_2^\Omega, \quad f_1^\Omega \pitchfork f_2 \quad \text{and} \quad f_1^\Omega \pitchfork f_2^\Omega.$$

Standard perturbation arguments (e.g. using the Sard-Smale theorem) can be used to show that generic perturbations of any pseudocycle make it strongly transverse to any other pseudocycle. Notice that whenever $f : V \to M$ is a pseudocycle and $\varphi : M \to M$ is a diffeomorphism, $\varphi \circ f : V \to M$ is also a pseudocycle, and moreover, the bordism class of $\varphi \circ f$ depends only on the diffeotopy class of φ.

Lemma 7.25 (cf. [MS12, Lemma 6.5.5]). *Given a pair of pseudocycles $f_1 : V_1 \to M$ and $f_2 : V_2 \to M$, fix an open subset $\mathcal{U} \subset M$ with compact closure such that $\mathcal{U}$ contains the closure of $f_1(V_1) \cap f_2(V_2)$, and let $\mathrm{Diff}(M, M \backslash \mathcal{U})$ denote the group of smooth diffeomorphisms that are the identity outside of $\mathcal{U}$, with its natural C^∞-*

topology. Then there exists a comeager subset $\mathrm{Diff}^{\mathrm{reg}} \subset \mathrm{Diff}(M, M \backslash \mathcal{U})$ *such that for all* $\varphi \in \mathrm{Diff}^{\mathrm{reg}}$, $\varphi \circ f_1 : V_1 \to M$ *is strongly transverse to* $f_2 : V_2 \to M$.   □

Definition 7.26. Given a strongly transverse pair of pseudocycles $f_i : V_i \to M$ of dimensions $d_i \geqslant 0$ for $i = 1, 2$ in an n-dimensional manifold M, we define their intersection product

$$f_1 \cdot f_2$$

as the map $f : V \to M$, where the domain is the $(d_1 + d_2 - n)$-dimenisonal manifold $V = (f_1, f_2)^{-1}(\Delta) \subset V_1 \times V_2$ for the diagonal $\Delta \subset M \times M$, and $f(x_1, x_2) = f_1(x_1) = f_2(x_2)$.

Exercise 7.27. Verify that the intersection product $f_1 \cdot f_2$ of two strongly transverse pseudocycles is also a pseudocycle, and that the bordism class of $f_1 \cdot f_2$ depends only on the bordism classes of f_1 and f_2.

The exercise implies that the intersection product on pseudocycles descends to a homomorphism

$$H_{d_1}^{\Psi}(M) \otimes H_{d_2}^{\Psi}(M) \to H_{d_1 + d_2 - \dim M}^{\Psi}(M) : [f_1] \otimes [f_2] \mapsto [f_1] \cdot [f_2],$$

which is well defined independently of any transversality assumptions in light of Lemma 7.25. In the important special case $d_1 + d_2 = \dim M$, if M is connected, the natural isomorphism $H_0^{\Psi}(M) = \mathbb{Z}$ turns this into an integer-valued intersection number

$$[f_1] \cdot [f_2] \in \mathbb{Z},$$

which can be computed as the signed count of intersections between any strongly transverse representatives $f_1 : V_1 \to M$ and $f_2 : V_2 \to M$; here the dimensional conditions on Ω_{f_1} and Ω_{f_2} ensure that these counts are finite and bordism-invariant.

Proposition 7.28. *For every integer* $d \geqslant 0$, *there exists a natural homomorphism*

$$\Psi : H_d(M) \to H_d^{\Psi}(M) \tag{7.7}$$

with the following properties:

 (1) *If* $f : V \to M$ *is a pseudocycle defined on a closed oriented manifold* V, *then* $\Psi(f_*[V]) = [f]$.
 (2) *For any* $A, B \in H_*(M)$, $\Psi(A) \cdot \Psi(B) = \Psi(A \cdot B)$.

Notice that by the first condition, Ψ is the natural isomorphism $H_0(M) \to H_0^{\Psi}(M)$ in dimension zero, so for classes of complementary dimension, the second condition implies an equality of integer-valued intersection numbers

$$A \cdot B = \Psi(A) \cdot \Psi(B) \in \mathbb{Z}.$$

Proof of Proposition 7.28. Given $A \in H_d(M)$, the idea is roughly to represent A by a map from a simplicial complex to M, then remove the codimension 2 skeleton to turn the simplicial complex into a smooth (but generally noncompact) manifold. We shall work with *smooth* singular homology, i.e. the chain complex defining $H_*(M)$ is generated by smooth maps from simplices into M. The equivalence between this and the usual *continuous* singular homology on smooth manifolds is a standard result, see e.g. [Lee03, Theorem 16.6].

Pick a cycle $\sum_{i=1}^{N} c_i f_i$ in the smooth singular chain complex representing $A \in H_d(M)$; here $c_i \in \mathbb{Z}$ and each f_i is a smooth map $\Delta^d \xrightarrow{f_i} M$ defined on the standard d-simplex Δ^d. We can assume without loss of generality that $c_i = \pm 1$ for all i. Now for each i, define V_i to be the (noncompact if $d \geq 2$) oriented manifold with boundary consisting of the union of the interior of Δ^d with the interiors of its $(d-1)$-dimensional boundary faces, with the orientation assigned according to the coefficient $c_i = \pm 1$. Denote the disjoint union of the continuous maps $f_i : V_i \to M$ by

$$f : \coprod_{i=1}^{N} V_i \to M.$$

Each boundary component of each V_i is now naturally identified with the interior of Δ^{d-1} but inherits an orientation dependent on c_i. The fact that $\sum_i c_i f_i$ is a cycle then implies a cancelation property: since each singular simplex in $\sum_i c_i \partial f_i$ must be canceled by another one, there exists a (not necessarily unique) orientation-reversing diffeomorphism

$$\varphi : \coprod_{i=1}^{N} \partial V_i \to \coprod_{i=1}^{N} \partial V_i, \tag{7.8}$$

mapping each connected boundary component of $\coprod_{i=1}^{N} V_i$ to a different one, such that $f \circ \varphi = f$. Indeed, φ can be defined on each component of ∂V_i by choosing a canceling boundary component of some ∂V_j and using the natural identification of both with the interior of Δ^{d-1}. We obtain from this a topological space

$$V := \coprod_{i=1}^{N} V_i \Big/ (x \sim \varphi(x)),$$

which becomes a smooth d-dimensional manifold if we choose suitable collars near the components of each ∂V_i before gluing, and we have a well-defined continuous map $f : V \to M$ which is smooth except at the glued boundary faces. Using standard approximation results such as [Hir94, Theorem 2.6], one can now perturb f to a smooth map $f' : V \to M$ that is arbitrarily close to f in the strong C^0-topology; in particular, we can find such a perturbation so that for some metric on M, every sequence $x_k \in V$ without limit points satisfies $\text{dist}(f'(x_k), f(x_k)) \to 0$. It follows that limit points $f'(x_n)$ for such sequences are the same as for $f(x_n)$ and thus are contained in the image of some f_i (a smooth map) restricted to a k-dimensional face of Δ^d with $k \leqslant d - 2$. This proves that $f' : V \to M$ is a pseudocycle.

One can use almost the same trick to show that homologous cycles give rise to bordant pseudocycles. Suppose $\sum_i c_i^\pm f_i^\pm$ are two d-dimensional cycles giving rise to pseudocycles $f^\pm : V^\pm \to M$ constructed above, and $\sum_j c_j f_j$ is a $(d + 1)$-dimensional chain whose boundary is $\sum_i c_i^+ f_i^+ - \sum_i c_i^- f_i^-$. Then repeating the procedure above produces a smooth $(d + 1)$-manifold V and a smooth map $f : V \to M$ such that $f(V) \subset M$ has compact closure, Ω_f is contained in the images of smooth maps defined on simplices of dimension at most $d - 1$,

$$\partial V = -\widetilde{V}^- \amalg \widetilde{V}^+$$

and $f|_{\widetilde{V}^\pm} = f^\pm|_{\widetilde{V}^\pm}$, where $\widetilde{V}^\pm \subset V^\pm$ are the open subsets obtained by removing all the $(d - 1)$-dimensional boundary faces of simplices, i.e. each is a disjoint union of the interiors of the d-dimensional simplices in $\sum_i c_i^\pm f_i^\pm$. The following trick to replace the missing pieces of $V^\pm$ in ∂V was suggested by McDuff and implemented in [Zin08]: attach collars of the form $[-1, 0) \times V^-$ and $(0, 1] \times V^+$ to ∂V, glued in the obvious way along $\widetilde{V}^-$ and $\widetilde{V}^+$ respectively. The enlarged object can be given a smooth manifold structure, with f extended over the collars and then approximated by a smooth map such that it now defines a bordism between the pseudocycles f^+ and f^-. Note that this construction also explains why, up to bordism, the pseudocycle constructed in the previous paragraph does not depend on the choice of the diffeomorphism in (7.8).

Having defined a map $\Psi : H_d(M) \to H_d^\Psi(M)$, it is clearly a homomorphism since addition on both sides can be viewed as a disjoint union construction. Next suppose $A = f_*[V]$ where $f : V \to M$ is smooth and V is a closed oriented d-manifold. Picking a triangulation of V gives it the structure of a simplicial complex, and choosing an ordering of its vertices then determines a singular cycle $\sum_i c_i g_i$ that represents the fundamental class $[V] \in H_d(V)$, where each c_i is ± 1 and each g_i is a diffeomorphism from the standard simplex Δ^d to one of the simplices in the complex on V. The class $f_*[V]$ is now represented by $\sum_i c_i (f \circ g_i)$, and applying our previous construction to this singular cycle produces a manifold $\widetilde{V}$ that is naturally identified with the complement of the $(d - 2)$-skeleton in V, with a pseudocycle of the form $\widetilde{f} := f|_{\widetilde{V}} : \widetilde{V} \to M$. One can again use McDuff's collar trick to construct a bordism between the pseudocycles f and $\widetilde{f}$; the domain of this bordism is the union of two collars $[-1, 0] \times \widetilde{V}$ and $(0, 1] \times V$, glued together in the obvious way. This proves $\Psi(f_*[V]) = [f]$.

With the first property established, observe finally that the relation $\Psi(A) \cdot \Psi(B) = \Psi(A \cdot B)$ becomes a standard fact from smooth intersection theory (see e.g. [Bre93, §VI.11]) whenever A and B are both represented by smooth maps from closed manifolds. But by Thom's theorem [Tho54], every homology class has an integer multiple that has this property, so the formula follows in general via bilinearity. □

With this topological language in place, Lemma 7.19 and the subsequent discussion of cobordisms defined via parametric moduli spaces can be rephrased as follows.

Theorem 7.29. *For generic* $J \in \mathcal{J}_\tau(M, \omega)$ *in any closed symplectic 4-manifold* (M, ω), *the map*

$$\mathcal{M}^*_{g,m}(A; J) \xrightarrow{\text{ev}} M^m$$

is a pseudocycle for every $g, m \geqslant 0$ *and* $A \in H_2(M)$ *such that either* vir-dim $\mathcal{M}_g(A; J) > 0$ *or* $g = 0$. *Moreover, for any smooth family of symplectic forms* $\{\omega_s\}_{s \in [0,1]}$ *with* $\omega_0 = \omega$ *and generic* $J_1 \in \mathcal{J}_\tau(M, \omega_1)$ *for which*

$$\mathcal{M}^*_{g,m}(A; J_1) \xrightarrow{\text{ev}} M^m$$

is also a pseudocycle, the two pseudocycles belong to the same bordism class. □

Definition 7.30. For any closed symplectic 4-manifold (M, ω) with integers $g, m \geqslant 0$ and $A \in H_2(M)$ satisfying either $2c_1(A) + 2g - 2 > 0$ or $g = 0$, the m-point Gromov-Witten invariant

$$\mathrm{GW}^{(M,\omega)}_{g,m,A} : H^*(M)^{\otimes m} \to \mathbb{Z}$$

is defined by

$$\mathrm{GW}^{(M,\omega)}_{g,m,A}(\alpha_1, \ldots, \alpha_m) = [\mathrm{ev}] \cdot \Psi(\mathrm{PD}(\alpha_1) \times \ldots \times \mathrm{PD}(\alpha_m)) \in \mathbb{Z}$$

whenever $2c_1(A) + 2g - 2 = \sum_{i=1}^m (\deg(\alpha_i) - 2)$, and is otherwise 0. Here $[\mathrm{ev}] \in H^\Psi_*(M^m)$ denotes the bordism class of the pseudocycle in Theorem 7.29, and $\Psi : H_*(M^m) \to H^\Psi_*(M^m)$ is the homomorphism from Proposition 7.28.

It follows immediately from Theorem 7.29 and the existence of the intersection product on $H^\Psi_*(M^m)$ that $\mathrm{GW}^{(M,\omega)}_{g,m,A}(\alpha_1, \ldots, \alpha_m)$ is independent of the choice of generic $J \in \mathcal{J}_\tau(M, \omega)$ and is invariant under symplectic deformations, and moreover, Proposition 7.28 implies that it can be computed in precisely the way indicated by Theorem 7.18 whenever the classes $\alpha_1, \ldots, \alpha_m$ are all Poincaré dual to smooth submanifolds. This completes the proof of Theorem 7.18.

Remark 7.31. It turns out that the pseudocycle condition is also satisfied in the case vir-dim $\mathcal{M}_g(J; A) = 0$ with $g > 0$, meaning that $\mathcal{M}^*_g(J; A)$ is a finite set of isolated regular curves. This does not follow from any of the standard technical results we have discussed, as one must exclude the possibility that an infinite sequence of isolated simple curves in $\mathcal{M}^*_g(J; A)$ converges to a multiple cover. Our index relations show that if this happens, the cover must be *unbranched*, and one can then appeal to a much more recent result of [GW17, Wene] stating that unbranched covers of closed J-holomorphic curves are also Fredholm regular for generic J, and thus isolated in the case vir-dim $\mathcal{M}_g(J; A) = 0$. In spite of this, Theorem 7.18 remains false in this case, as counting only the simple curves $\mathcal{M}_g^*(J; A)$ will give a number that is not independent of the choice of J or invariant under deformations. To produce an actual invariant, the multiple covers must also be counted, and the resulting counts are then in $\mathbb{Q}$ instead of $\mathbb{Z}$.

7.2.4 Rational/Ruled Implies Uniruled

As a warmup, the following computation is now an immediate consequence of Theorem 5.1, with the sign ($+1$ rather than -1) provided by Proposition 2.47:

Theorem 7.32. *For any exceptional sphere E in a closed symplectic 4-manifold* (M, ω),

$$\mathrm{GW}_{0,0,[E]}^{(M,\omega)} = 1.$$

□

We next prove that $(1) \Rightarrow (5)$ in Theorem 7.3.

Theorem 7.33. *Suppose (M, ω) is the total space of a symplectic Lefschetz pencil or fibration $\pi : M \backslash M_{\text{base}} \to \Sigma$ whose fibers are embedded spheres. Then if $[F] \in H_2(M)$ is the homology class of the fiber,*

$$\mathrm{GW}_{0,m+1,[F]}^{(M,\omega)}(\mathrm{PD}[\mathrm{pt}], \dots, \mathrm{PD}[\mathrm{pt}]) = 1,$$

where $m \geq 0$ is the number of points in M_{base}. In particular, (M, ω) is symplectically uniruled.

Proof. Let $p_1, \dots, p_m$ denote the points in M_{base}, and pick another point $p_0 \in M \backslash M_{\text{base}}$. By Theorem G, for generic J we can assume after a symplectic isotopy that the fibers of π are all J-holomorphic, and they are the only J-holomorphic curves that pass through all base points and are homologous to $[F]$. Exactly one of these curves also passes through p_0, so it follows that there is exactly one curve u homologous to $[F]$ with $m + 1$ marked points satisfying $\mathrm{ev}(u) = (p_0, \dots, p_m)$. By Proposition 2.47, that curve is counted with positive sign. □

7.3 Positively Immersed Symplectic Spheres

We now turn to the implication $(2) \Rightarrow (1)$ in Theorem 7.3. This follows from the main result in [McD92], which can be restated as follows.

Theorem 7.34 (McDuff [McD92]). *Suppose (M, ω) is a closed symplectic 4-manifold admitting a positively symplectically immersed sphere $S \looparrowright (M, \omega)$ with $c_1([S]) \geq 2$. Then it also contains a symplectically embedded sphere with nonnegative self-intersection number.*

Remark 7.35. If one interprets Theorem A to mean that nonnegative symplectically embedded spheres are relatively rare in symplectic 4-manifolds, then we learn from Theorem 7.34 that their *non-embedded* cousins with $c_1([S]) \geq 2$ are even rarer: it is easy to show in fact that the non-embedded case of this theorem can only occur if (M, ω) is a symplectic rational surface. To see this, note that S can be parametrized by an immersed J-holomorphic sphere $u_S : S^2 \looparrowright M$ for suitable $J \in \mathcal{J}_\tau(M, \omega)$, and it is automatically transverse by Corollary 2.45, so we are free to assume J is generic. Then if the theorem is true, Theorem G provides a Lefschetz pencil/fibration $\pi : M \backslash M_{\text{base}} \to \Sigma$ whose fibers are all embedded J-holomorphic spheres. The curve u_S cannot be an irreducible component of any of these fibers since it is not embedded, and it is also not a multiple cover, thus it has strictly positive intersection with every fiber, implying that the map $\pi \circ u_S : S^2 \to \Sigma$ has positive degree. This is only possible if $\Sigma \cong S^2$, so by Theorem 7.6, (M, ω) is a rational surface.

If you have doubts as to whether the non-embedded case of Theorem 7.34 can actually occur at all, see Example 7.38 below.

7.3.1 A Brief Word from Seiberg-Witten Theory

A few historical remarks are in order before we delve into the details on Theorem 7.34. While McDuff's proof in [McD92] was based only on Gromov's pseudoholomorphic curve theory and standard techniques from algebraic topology, the theorem was superseded a few years later by developments from gauge theory. The following result is closely related to Liu's Theorem 1.21 stated in the introduction, and is yet another consequence of Taubes's "SW=Gr" theorem relating the Seiberg-Witten invariants of symplectic 4-manifolds to certain counts of embedded J-holomorphic curves [Tau95, Tau96a, Tau96b].

Theorem 7.36. *Suppose (M, ω) is a closed symplectic 4-manifold containing a symplectically embedded closed surface $\Sigma \subset M$ with $c_1([\Sigma]) \geq 1$ such that Σ is not an exceptional sphere. Then (M, ω) also contains a symplectically embedded sphere with nonnegative self-intersection number.*

To derive Theorem 7.34 from this statement, one only need observe that a positively immersed symplectic sphere can always have its positive double points "resolved" as in Exercise 7.10, producing a symplectically embedded surface which might have higher genus but will always have the same first Chern number.

We will outline a proof below that is based on [MS96, Corollary 1.5], taking a few results from Taubes-Seiberg-Witten theory (notably Theorem 1.21 in the introduction) as black boxes. We will also need the following elementary lemma, which is popular in the study of oriented 4-manifolds with $b_2^+ = 1$.

Proposition 7.37 ("Light Cone Lemma"). *Let $Q(\ ,\ )$ denote the indefinite inner product on $\mathbb{R}^n$ for which the standard basis $(e_1, \ldots, e_n)$ is orthonormal with $Q(e_1, e_1) = 1$ and $Q(e_j, e_j) = -1$ for $j = 2, \ldots, n$. Then the set*

$$\mathcal{P} = \left\{ \mathbf{v} \in \mathbb{R}^n \backslash \{0\} \mid Q(\mathbf{v}, \mathbf{v}) \geqslant 0 \right\}$$

has two connected components, and any $\mathbf{v}, \mathbf{w} \in \mathcal{P}$ in the same connected component satisfy $Q(\mathbf{v}, \mathbf{w}) \geqslant 0$, with equality if and only if both vectors belong to the boundary of $\mathcal{P}$ and are colinear.

Proof. An element $\mathbf{v} = (v_+, \mathbf{v}_-) \in \mathbb{R} \times \mathbb{R}^{n-1}$ belongs to $\mathcal{P}$ if and only if $|\mathbf{v}_-| \leqslant |v_+|$, so the two connected components are distinguished by the sign of v_+. Now if $\mathbf{w} = (w_+, \mathbf{w}_-) \in \mathcal{P}$ satisfies $v_+ w_+ > 0$, the Cauchy-Schwarz inequality gives

$$Q(\mathbf{v}, \mathbf{w}) = v_+ w_+ - \langle \mathbf{v}_-, \mathbf{w}_- \rangle \geqslant v_+ w_+ - |\mathbf{v}_-||\mathbf{w}_-| \geqslant v_+ w_+ - |v_+||w_+| = 0.$$

Equality is achieved if and only if $\langle \mathbf{v}_-, \mathbf{w}_- \rangle = |\mathbf{v}_-||\mathbf{w}_-|$ and $|v_+| - |\mathbf{v}_-| = |w_+| - |\mathbf{w}_-| = 0$; the former implies that $\mathbf{v}_-$ and $\mathbf{w}_-$ are colinear, and the latter then fixes v_+ and w_+ so that $\mathbf{v}$ and $\mathbf{w}$ are also colinear and belong to the boundary of $\mathcal{P}$. □

Sketch of a Proof of Theorem 7.36. We can assume without loss of generality that Σ is connected. Since it is symplectically embedded, we can choose $J \in \mathcal{J}_\tau(M, \omega)$ so that it is the image of a J-holomorphic curve u, whose index is then

$$\text{ind}(u) = -\chi(\Sigma) + 2c_1([u]) = 2g - 2 + 2c_1([u]) > 2g - 2,$$

where g is the genus of Σ. This establishes the criterion of Theorem 2.44 for automatic transversality, hence u is Fredholm regular and will survive sufficiently small perturbations of J, so that we are free to assume J is generic.

If (M, ω) is not minimal, choose a maximal collection of pairwise disjoint exceptional spheres $E_1, \ldots, E_k \subset M$; note that by assumption, Σ is not one of them. We can assume they are all J-holomorphic by Theorem 5.1, and by Corollary 2.32, we are also free to assume that they are all transverse to Σ. Then blowing down (M, ω) along $E_1 \amalg \ldots \amalg E_k$ using Theorem 3.14 produces a symplectic manifold $(\check{M}, \check{\omega})$ with tame almost complex structure $\check{J}$ and a pseudoholomorphic blowdown map $\beta : (M, J) \to (\check{M}, \check{J})$, so that $\check{u} := \beta \circ u$ is an immersed $\check{J}$-holomorphic curve with $c_1(N_{\check{u}}) \geqslant c_1(N_u)$ by Exercise 3.6. This makes the image of $\check{u}$ a positively symplectically immersed surface $\check{\Sigma}$ in $(\check{M}, \check{\omega})$ with $c_1([\check{\Sigma}]) \geq c_1([\Sigma]) \geq 1$. One can now use Exercise 7.10 to resolve the double points of this immersion, producing a symplectically embedded surface $\check{\Sigma}' \subset (\check{M}, \check{\omega})$, possibly with larger genus than the original, but satisfying $c_1([\check{\Sigma}']) = c_1([\check{\Sigma}]) \geq 1$.

The previous paragraph reduces the theorem to the case where (M, ω) is minimal, so assume this from now on. Observe also that since Σ is not an exceptional sphere, the adjunction formula gives

$$[\Sigma] \cdot [\Sigma] = c_1([\Sigma]) - \chi(\Sigma) \geq 0. \tag{7.9}$$

We claim next that $b_2^+(M) = 1$. Note that $b_2^+(M) \geq 1$ automatically since (M, ω) is symplectic, so arguing by contradiction, assume $b_2^+(M) > 1$. Then using the notation of [MS17, §13.3], the following fundamental results of Taubes apply:

- $\mathrm{SW}(M, \mathfrak{o}_\omega, \Gamma_{\omega, \mathrm{PD}(A)}) = \mathrm{Gr}(M, \omega, A)$ for all $A \in H_2(M)$;
- $\mathrm{SW}(M, \mathfrak{o}_\omega, \Gamma_{\omega, K-a}) = \pm \mathrm{SW}(M, \mathfrak{o}_\omega, \Gamma_{\omega, a})$ for all $a \in H^2(M)$.

Here $K := -c_1(M, \omega)$ is the canonical class of (M, ω), $\mathrm{SW}(M, \mathfrak{o}, \Gamma) \in \mathbb{Z}$ denotes the Seiberg-Witten invariant associated to a homological orientation $\mathfrak{o}$ and a spin^c structure Γ, $\mathfrak{o}_\omega$ and Γ_ω are the canonical homological orientation and spin^c structure respectively associated to the formal homotopy class of ω, and $\Gamma_{\omega, a}$ is the latter's tensor product with a Hermitian line bundle having first Chern class a. Likewise, $\mathrm{Gr}(M, \omega, A) \in \mathbb{Z}$ is the *Gromov invariant* defined in [Tau96a], which counts a special class of constrained J-holomorphic curves for generic $J \in \mathcal{J}_\tau(M, \omega)$ in the homology class A. The two properties above imply that for the canonical class K,

$$\pm \mathrm{Gr}(M, \omega, \mathrm{PD}(K)) = \pm \mathrm{SW}(M, \mathfrak{o}_\omega, \Gamma_{\omega, K}) = \mathrm{SW}(M, \mathfrak{o}_\omega, \Gamma_\omega) = \mathrm{GW}(M, \omega, 0) = 1,$$

where the last equality is essentially a definition since $\mathrm{GW}(M, \omega, 0)$ is a count of the "empty holomorphic curve," of which there is exactly one. Knowing $\mathrm{Gr}(M, \omega, \mathrm{PD}(K)) \neq 0$, we deduce the existence for generic $J \in \mathcal{J}_\tau(M, \omega)$ of a possibly disconnected J-holomorphic curve v homologous to $\mathrm{PD}(K)$. Since Σ is also J-holomorphic, every connected component of v is either a cover of Σ or intersects it at most finitely many times, always positively, implying in light of (7.9) that

$$0 \leq [v] \cdot [\Sigma] = \mathrm{PD}(K) \cdot [\Sigma] = \langle -c_1(M, \omega), [\Sigma] \rangle = -c_1([\Sigma]),$$

This contradicts the assumption $c_1([\Sigma]) > 0$.

Having established that $b_2^+(M) = 1$, let us abbreviate $\alpha \cdot \beta := \langle \alpha \cup \beta, [M] \rangle$ for $\alpha, \beta \in H^2(M; \mathbb{R})$ and assume

$$K \cdot K \geq 0 \quad \text{and} \quad K \cdot [\omega] \geq 0,$$

since the desired result otherwise follows from Theorem 1.21. The first condition means that either K or $-K$ belongs to the "positive light cone" in $H^2(M; \mathbb{R})$, i.e. the connected component of $\{\alpha \in H^2(M; \mathbb{R}) \mid \alpha \cdot \alpha \geq 0\}$ containing $[\omega]$. It follows then from Proposition 7.37 that $K \cdot [\omega]$ cannot be 0, as $[\omega]$ is not on the boundary of the light cone since $[\omega] \cdot [\omega] > 0$. Thus $K \cdot [\omega] > 0$, implying that K is in the positive light cone. But (7.9) and $\int_\Sigma \omega > 0$ imply that $\text{PD}[\Sigma]$ is also in the positive light cone, so the assumption $K \cdot \text{PD}[\Sigma] = -c_1([\Sigma]) < 0$ now contradicts Proposition 7.37.   □

Results of this kind provide good motivation to study the Seiberg-Witten theory of symplectic 4-manifolds, and we refer the reader to [MS17, §13.3] and [MS96] for a nice overview of what else can be proved with such techniques. In particular, these methods have been greatly successful in extending standard theorems about complex surfaces into the symplectic category. From the author's admittedly subjective point of view, however, holomorphic curve theory has the advantage of often leading to more transparent geometric arguments than are typically possible via gauge theory, and moreover, the theory remains well defined (though weaker) in higher dimensions, where gauge-theoretic techniques are completely unavailable.

7.3.2   Outline of the Proof

We will leave Theorem 7.36 out of the following discussion and present a variation on McDuff's original proof of Theorem 7.34, which is completely independent of Seiberg-Witten theory. The proof below contains a few simplifications in comparison with the original, most of which are also due to McDuff but have not appeared in the literature before. Here is a summary of the argument.

The beginning of the proof will seem familiar: we shall choose an almost complex structure turning the given immersed sphere S into a J-holomorphic curve, which will be regular due to automatic transversality. The condition $c_1([S]) \geq 2$ makes the index of this curve at least 2, so we can then impose enough pointwise constraints to view S as an element in a 2-dimensional moduli space $\mathcal{M}_S(J)$ consisting of somewhere injective curves homologous to S with $m \geq 0$ marked points mapped to fixed positions $p_1, \ldots, p_m \in M$. It will be straightforward to show that the natural compactification $\overline{\mathcal{M}}_S(J)$ of $\mathcal{M}_S(J)$ is obtained by adding a finite collection of nodal curves that each have two simple components of (constrained) index 0.

At this point the argument diverges from the previous chapters, as we can no longer assume that any of the component curves in $\overline{\mathcal{M}}_S(J)$ are embedded or that neighboring curves in this moduli space are disjoint, so they do not give rise to a nice geometric decomposition of M. The key instead is to show that the set of nodal curves in $\overline{\mathcal{M}}_S(J)$ must be nonempty, and one of the components of these nodal curves must be an *embedded* curve with nonnegative self-intersection, see Fig. 7.1. The features that guarantee this result are essentially topological: first, by adding

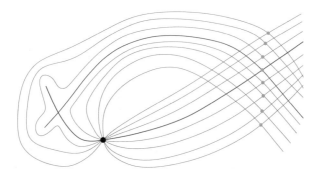

Fig. 7.1 Schematic picture of a 2-dimensional family of immersed holomorphic spheres with one constrained marked point, each with one transverse self-intersection. The family can degenerate to a nodal curve whose smooth components are a pair of embedded spheres, one of which passes through the constraint point and has self-intersection number 0

an extra marked point to the curves in $\overline{\mathcal{M}}_S(J)$, we obtain a 4-dimensional moduli space $\overline{\mathcal{U}}_S(J)$ with two natural maps

$$\pi_m : \overline{\mathcal{U}}_S(J) \to \overline{\mathcal{M}}_S(J) \quad \text{and} \quad \mathrm{ev}_{m+1} : \overline{\mathcal{U}}_S(J) \to M,$$

where the first is defined by forgetting the extra marked point and the second by evaluating it. We will see that $\overline{\mathcal{U}}_S(J)$ can be given the structure of a smooth, closed and oriented 4-manifold, and its topology is of a very specific type: the map π_m is a Lefschetz fibration, which makes $\overline{\mathcal{U}}_S(J)$ topologically a blown-up ruled surface.

The second crucial topological observation is that while the curves in $\overline{\mathcal{U}}_S(J)$ can no longer be assumed to foliate M in any reasonable sense, generically almost all of them are immersed and thus subject to automatic transversality arguments, which we will be able to use to show that

$$\deg(\mathrm{ev}_{m+1}) \geq 1.$$

This and a standard generic homotopy argument establish the implication (2) $\Rightarrow$ (4) in Theorem 7.3, allowing us to reduce to the minimal case via Lemmas 7.5 and 7.11. Now if (M, ω) is minimal but the theorem does not hold, we know that none of the nodal curves in $\overline{\mathcal{M}}_S(J)$ contain an embedded component. Under this assumption, we will use our knowledge of the topology of blown-up ruled surfaces to draw topological conclusions about (M, ω) that are impossible for any symplectic 4-manifold, e.g. that it violates the Hirzebruch signature theorem.

Note that on rational surfaces, it is easy to construct (e.g. via gluing) non-embedded holomorphic spheres with positive index. The following concrete example illustrates one way that such non-embedded spheres can be arranged into a 2-dimensional moduli space that reveals embedded spheres in its compactification.

Example 7.38. Suppose M is $S^2 \times S^2$ with a product symplectic structure and its standard complex structure $J = i \oplus i$. Consider the real 2-dimensional family of holomorphic spheres $u_a : S^2 \to S^2 \times S^2$ defined by

$$u_a(z) = \left(\frac{(z-1)(z-a)}{(z+a)(z-i)}, \frac{(z-1)(z+a)}{(z-a)(z-i)} \right), \qquad a \in \mathbb{C} \setminus \{0, 1, -1, i, -i\}.$$

It is easy to check that all of these curves are simple, as for instance $z = 1$ is an injective point for all of them. Denoting the generators of $H_2(S^2 \times S^2)$ by $[S_1] = [S^2 \times \{\text{const}\}]$ and $[S_2] = [\{\text{const}\} \times S^2]$, we have $[u_a] = 2[S_1] + 2[S_2]$ and thus $[u_a] \cdot [u_a] = c_1([u_a]) = 8$ for all a, so by the adjunction formula, $\delta(u_a) = 1$, implying that none of the u_a are embedded. If we add the six marked points $\zeta_1 = 0$, $\zeta_2 = 1$, $\zeta_3 = \infty$, $\zeta_4 = i$, $\zeta_5 = a$ and $\zeta_6 = -a$, we find that u_a satisfies the constraints

$$u_a(\zeta_1) = (i, i), \quad u_a(\zeta_2) = (0, 0),$$
$$u_a(\zeta_3) = (1, 1), \quad u_a(\zeta_4) = (\infty, \infty),$$
$$u_a(\zeta_5) = (0, \infty), \quad u_a(\zeta_6) = (\infty, 0)$$

and the moduli space of spheres homologous to $2[S_1] + 2[S_2]$ with six marked points satisfying these constraints has virtual dimension 2. In the limit as

$$a \to 0,$$

u_a degenerates into a nodal curve with two embedded components u_0^+ and u_0^- both satisfying

$$[u_0^{\pm}] \cdot [u_0^{\pm}] = 2.$$

The former is the limit of the maps $u_a : S^2 \to S^2 \times S^2$ in $C^{\infty}_{\text{loc}}(S^2 \setminus \{0\})$,

$$u_0^+(z) = \left(\frac{z-1}{z-i}, \frac{z-1}{z-i} \right),$$

which is just a reparametrization of the diagonal curve $z \mapsto (z, z)$. The other component is obtained by reparametrizing u_a and taking the limit in $C^{\infty}_{\text{loc}}(\mathbb{C})$ of $u_a(az)$, giving the curve

$$u_0^-(z) = \left(\frac{z-1}{i(z+1)}, \frac{z+1}{i(z-1)} \right).$$

The original six marked points are split evenly between these two components: u_0^+ gets the points that stay outside a fixed neighborhood of 0, namely $\zeta_2 = 1$, $\zeta_3 = \infty$

and $\zeta_4 = i$, while the reparametrization realizes the other three points on u_0^- as $\zeta_1 = 0$, $\zeta_5 = 1$ and $\zeta_6 = -1$. Since both curves satisfy $c_1([u_0^{\pm}]) = c_1([S_1] + [S_2]) = 4$, they each have unconstrained index 6 and the three constraint points bring their indices down to 0. Observe finally that in spite of the lack of genericity in our choice of J relative to the constraint points, Theorem 2.46 implies that both components of the nodal curve are regular for the constrained problem, so we would necessarily see a small perturbation of the same degeneration (though we could not write it down so explicitly) if J were a generic perturbation of $i \oplus i$.

Exercise 7.39. Show that in the above example, a similar nodal degeneration to the case $a \to 0$ occurs when $a \to \infty$. Show however that the degenerations for the cases $a \to \{1, -1, i, -i\}$ all involve nodal curves that could not exist if J were generic relative to the given constraint points.

7.3.3 The Universal J-Holomorphic Curve

We will make use of the following general construction. For any almost complex manifold (M, J), integers $g, m \geq 0$ and a homology class $A \in H_2(M)$, let

$$\overline{\mathcal{U}}_{g,m}(A; J) = \overline{\mathcal{M}}_{g,m+1}(A; J)$$

and consider the natural map

$$\pi_m : \overline{\mathcal{U}}_{g,m}(A; J) \to \overline{\mathcal{M}}_{g,m}(A; J)$$

defined by forgetting the "extra" marked point (i.e. the last of the $m + 1$) and stabilizing the nodal curve that remains. Here, "stabilizing" means in practice the following: if the extra marked point lies on a ghost bubble with only three marked or nodal points, then removing it produces a curve that is not stable, but there is a uniquely determined element of $\overline{\mathcal{M}}_{g,m}(A; J)$ obtained from this by collapsing the non-stable ghost bubble—if another marked point loses its domain component in this process, it can naturally be placed on a neighboring component in place of the orphaned nodal point (see Fig. 7.2). We denote the evaluation map for the extra marked point by

$$\mathrm{ev}_{m+1} : \overline{\mathcal{U}}_{g,m}(A; J) \to M.$$

Though $\pi_m : \overline{\mathcal{U}}_{g,m}(A; J) \to \overline{\mathcal{M}}_{g,m}(A; J)$ is not generally a fibration, it will be convenient to think of it in this way, and to regard each of its "fibers" as a parametrization of the underlying holomorphic curve. To make this precise, let us associate to any nodal curve $[(S, j, u, (\zeta_1, \ldots, \zeta_m), \Delta)] \in \overline{\mathcal{M}}_{g,m}(A; J)$ the singular Riemann surface

$$\check{S} = S/\sim \quad \text{where} \quad z \sim z' \text{ for each node } \{z, z'\} \in \Delta,$$

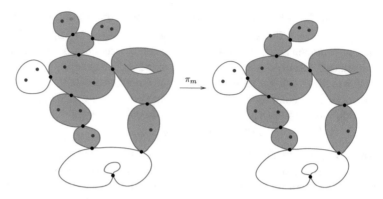

Fig. 7.2 An example illustrating the map $\overline{\mathcal{U}}_{g,m}(J) \to \overline{\mathcal{M}}_{m,g}(J)$ that forgets the extra marked point and collapses the resulting non-stable domain component (Ghost components are the shaded ones in the picture.)

and observe that $u : S \to M$ descends to a well-defined continuous map $\check{S} \to M$. The automorphism group of u, i.e. the set of all equivalences of $(S, j, u, (\zeta_1, \ldots, \zeta_m), \Delta)$ to itself, also acts naturally on $\check{S}$ via homeomorphisms $\varphi : \check{S} \to \check{S}$ satisfying $u \circ \varphi = u$, as every automorphism is required to map nodal pairs to nodal pairs. Note that the stability condition implies that these automorphism groups are always finite. The result is a well-defined continuous map $u : \check{S}/\operatorname{Aut}(u) \to M$.

Proposition 7.40. *For each tuple* $(S, j, u, (\zeta_1, \ldots, \zeta_m), \Delta)$ *representing an element of* $\overline{\mathcal{M}}_{g,m}(A; J)$ *with automorphism group* $\operatorname{Aut}(u)$, *there is a natural homeomorphism between the preimage of this element under* π_m *and the quotient* $\check{S}/\operatorname{Aut}(u)$ *such that* $\operatorname{ev}_{m+1}|_{\pi_m^{-1}(u)} : \pi_m^{-1}(u) \to M$ *is identified with the map* $u : \check{S}/\operatorname{Aut}(u) \to M$.

Proof. The correspondence is mostly straightforward if u is a smooth simple curve, as then any point in $S \setminus \{\zeta_1, \ldots, \zeta_m\}$ can be chosen as the extra marked point to define an element of $\pi_m^{-1}(u)$, and no two of these elements are equivalent since there are no nontrivial automorphisms of S preserving both the map u and all the marked points $\zeta_1, \ldots, \zeta_m$. The only slightly tricky detail is to identify which elements of $\pi_m^{-1}(u)$ correspond to the marked points $\zeta_1, \ldots, \zeta_m$, but the answer is simple: for each $i = 1, \ldots, m$, the nodal curve consisting of $u : S \to M$ plus a ghost bubble with two marked points attached to S at ζ_i gives an element of $\pi_m^{-1}(u)$ corresponding to ζ_i, see Fig. 7.3. If u is not simple and has nontrivial automorphisms, then two distinct choices of location for the extra marked point define equivalent elements of $\overline{\mathcal{U}}_{g,m}(A; J)$ if and only if they represent the same element of $S/\operatorname{Aut}(u)$.

When u is a nodal curve, we only need to supplement the above discussion by explaining which element of $\pi_m^{-1}(u)$ corresponds to each nodal pair $\{z, z'\} \in \Delta$, and the answer is again straightforward: insert a ghost bubble between z and z', then place the extra marked point on the ghost bubble (Fig. 7.4).   □

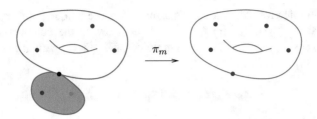

Fig. 7.3 An element of $\pi_m^{-1}(m)$ in which the extra marked point is positioned "in the same place" with one of the original m marked points

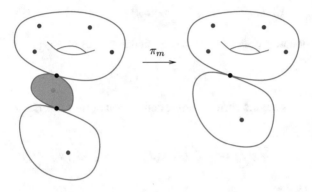

Fig. 7.4 An element of $\pi_m^{-1}(m)$ where u is a nodal curve and the extra marked point is positioned "at a node"

The analogue of π_m as a map on the compactified moduli space of Riemann surfaces $\overline{\mathcal{M}}_{g,m+1} \to \overline{\mathcal{M}}_{g,m}$ is sometimes referred to in the Gromov-Witten literature as the "universal curve," and we shall call the above version the **universal J-holomorphic curve.**

7.3.4   The Moduli Space as a Blown-Up Ruled Surface

For the remainder of this chapter, (M, ω) is a closed and connected symplectic 4-manifold, and $S \hookrightarrow M$ is a positively symplectically immersed sphere with

$$m := c_1([S]) - 2 \geqslant 0.$$

By Exercise 7.2, we can choose $J \in \mathcal{J}_\tau(M, \omega)$ such that S is realized by an immersed J-holomorphic sphere $u_S : S^2 \to M$. Pick a set of pairwise distinct points

$$p_1, \ldots, p_m \subset u_S(S^2)$$

which are all injective points for u_S, meaning there are unique points $\zeta_1, \ldots, \zeta_m \in S^2$ such that $u_S(\zeta_i) = p_i$ for $i = 1, \ldots, m$. Now, in the compactified moduli space $\overline{\mathcal{M}}_{0,m}([S]; J; p_1, \ldots, p_m)$ of J-holomorphic spheres with m marked points constrained at $p_1, \ldots, p_m$, let

$$\overline{\mathcal{M}}_S(J) \subset \overline{\mathcal{M}}_{0,m}([S]; J; p_1, \ldots, p_m)$$

denote the connected component containing the curve u_S with marked points $\zeta_1, \ldots, \zeta_m$. This space has virtual dimension

$$\text{vir-dim}\, \overline{\mathcal{M}}_{0,m}([S]; J; p_1, \ldots, p_m) = \text{ind}(u_S) - 2m = -2 + 2c_1([S]) - 2m = 2,$$

and we will denote the set of smooth (non-nodal) curves in $\overline{\mathcal{M}}_S(J)$ by

$$\mathcal{M}_S(J) \subset \overline{\mathcal{M}}_S(J).$$

We will also consider the corresponding component of the universal J-holomorphic curve,

$$\overline{\mathcal{U}}_S(J) := \pi_m^{-1}\big(\overline{\mathcal{M}}_S(J)\big) \subset \overline{\mathcal{U}}_{0,m}([S]; J),$$

giving rise to a map

$$\pi_m : \overline{\mathcal{U}}_S(J) \to \overline{\mathcal{M}}_S(J)$$

whose fibers (according to Proposition 7.40) are homeomorphic to the nodal domains of elements in $\overline{\mathcal{M}}_S(J)$ modulo automorphisms, with the maps from these domains to M then corresponding to the evaluation map at the extra marked point,

$$\text{ev}_{m+1} : \overline{\mathcal{U}}_S(J) \to M.$$

Note that our definition does not assume the uncompactified space $\mathcal{M}_S(J)$ to be connected, though we will see in a moment that this is true for generic J. Observe that since $\text{ind}(u_S) = 2m + 2$, Theorem 2.46 implies that u_S is Fredholm regular for the problem with fixed point constraints, therefore it will survive any sufficiently small perturbation making J generic, at the cost of starting with a small perturbation of S. We will have more to say about genericity conditions below (see Lemma 7.50), but for now, let us impose the following conditions:

Assumption 7.41 (Genericity).

   (1) *All somewhere injective curves satisfying marked point constraints defined via subsets of $\{p_1, \ldots, p_m\}$ are regular for the constrained problem.*
   (2) *All somewhere injective curves that have constrained index 0 with respect to marked point constraints defined via subsets of $\{p_1, \ldots, p_m\}$ are*

immersed, and any pair of distinct curves of this type are transverse to each other, with intersection points disjoint (in the image) from the self-intersection points of either one.

The first condition was also imposed in Chap. 4, and it implies that

$$\text{vir-dim}\,\mathcal{M}_{0,r}(A; J; p_{i_1}, \ldots, p_{i_r}) \geq 0$$

holds for every $A \in H_2(M)$, $r \geq 0$ and subset $\{p_{i_1}, \ldots, p_{i_r}\} \subset \{p_1, \ldots, p_m\}$ for which the space $\mathcal{M}_{0,r}(A; J; p_{i_1}, \ldots, p_{i_r})$ contains a somewhere injective curve. Corollary 2.23 provides a finite intersection of comeager subsets of $\mathcal{J}_\tau(M, \omega)$ for which this condition holds. Theorem 2.28 and Corollaries 2.30 and 2.32 provide a further finite intersection of comeager subsets for which the second condition holds.

Lemma 7.42. *Given the genericity Assumption 7.41, for any constant $E_0 > 0$, the set of all curves belonging to all moduli spaces of the form $\mathcal{M}_{0,r}^*(A; J; p_{i_1}, \ldots, p_{i_r})$ for $r \geq 0$, arbitrary subsets $\{p_{i_1}, \ldots, p_{i_r}\} \subset \{p_1, \ldots, p_m\}$, and classes $A \in H_2(M)$ satisfying*

$$\langle [\omega], A \rangle \leq E_0 \quad \text{and} \quad -2 + 2c_1(A) - 2r = 0,$$

is finite.

Proof. The proof is almost word-for-word the same as that of Lemma 4.16: the point is to use Gromov compactness and the index relations of Sect. 4.2 to show that any sequence of such curves has a subsequence convergent to a simple curve with no nodes. This set is then a compact 0-dimensional manifold, hence finite. □

Lemma 7.43. *Under the genericity Assumption 7.41, $\overline{\mathcal{M}}_S(J)$ is an oriented 2-dimensional topological manifold containing an at most finite set of isolated nodal curves. Moreover, all curves in $\mathcal{M}_S(J)$ are simple, and each nodal curve in $\overline{\mathcal{M}}_S(J) \backslash \mathcal{M}_S(J)$ has exactly two smooth components, which are distinct simple curves having index 0 with respect to their respective marked point constraints, connected by a node whose image is not any of the points $p_1, \ldots, p_m$.*

Proof. Most of the argument is a repeat of the proof of Theorem 4.6 in Chap. 4, using index relations. As with the previous lemma, the arguments in Lemma 4.17 imply that any sequence $u_k \in \mathcal{M}_S(J)$ has a subsequence converging to a limit $u_\infty \in \overline{\mathcal{M}}_S(J)$ which is one of the following:

(1) A smooth simple curve;
(2) A smooth branched double cover of a simple J-holomorphic sphere with index 0 (arising only in the case $m = 0$);
(3) A nodal curve with exactly two components connected by a single node, where the two components u^1 and u^2 are simple J-holomorphic spheres with m_1 and m_2 marked points respectively, such that $\text{ind}(u^i) = 2m_i$ for $i = 1, 2$.

We need to rule out the second case: our argument for this in the proof of Theorem 4.6 assumed embeddedness, so we need a slightly different argument here, though the idea will be similar. Suppose $u_\infty = v \circ \varphi : S^2 \to M$, where $v : S^2 \to M$ is a simple J-holomorphic curve and $\varphi : S^2 \to S^2$ is a holomorphic branched double cover. Since m must be zero in this case, we have $c_1([u_\infty]) = 2$ and $c_1([v]) = 1$, implying $\text{ind}(v) = 0$, and our genericity assumptions therefore imply that v is immersed. We can then write

$$\delta(v) = \frac{1}{2} \sum_{v(\zeta)=v(\zeta')} i(\zeta, \zeta'),$$

where the sum ranges over the finite set of pairs $(\zeta, \zeta') \in S^2 \times S^2$ outside of the diagonal for which $v(\zeta) = v(\zeta')$, and $i(\zeta, \zeta') \in \mathbb{N}$ denotes the local intersection index for each of these isolated intersections. Suppose (ζ, ζ') is one of these pairs such that ζ and ζ' are both regular values of φ. Then each choice of lift $z \in \varphi^{-1}(\zeta)$ and $z' \in \varphi^{-1}(\zeta')$ gives rise to an isolated self-intersection of u with the same local intersection index $i(z, z') = i(\zeta, \zeta')$, and there are exactly four choices of such pairs of lifts. By positivity of intersections, these isolated self-intersections survive as u_∞ is perturbed to a simple curve u_k for k sufficiently large, hence the double point $v(\zeta, \zeta')$ contributes $4i(\zeta, \zeta')$ to $\delta(u_k)$. Exercise 7.44 below implies that this calculation remains valid even if ζ or ζ' is a critical value of φ, so by positivity of intersections, we conclude

$$\delta(u_k) \geq 4\delta(v) \tag{7.10}$$

for all k sufficiently large. Now let us compare this to what can be deduced from the adjunction formula: since $c_1([v]) = 1$ and $c_1([u_k]) = 2$, we have

$$[v] \cdot [v] = 2\delta(v) - 1, \qquad [u_k] \cdot [u_k] = 2\delta(u_k),$$

so using the fact that $[u_k] = [u_\infty] = 2[v]$,

$$2\delta(u_k) = 4\,(2\delta(v) - 1) = 8\delta(v) - 4,$$

implying $\delta(u_k) = 4\delta(v) - 2$, which contradicts (7.10).

In the third case, we also need to show that the components u^1 and u^2 of our nodal curve are distinct. Indeed, if $u^1 = u^2 =: v$ but $m > 0$, then v appears with two distinct configurations of marked point constraints and has constrained index 0 for both, meaning u^1 and u^2 each have the same positive number of marked points, but v also must pass through *all* of the constraint points and thus violates are genericity assumption. This is a contradiction unless $m = 0$. In the latter case we can adapt the intersection-theoretic argument of the previous paragraph: $[u_\infty] = 2[v]$ and the adjunction formula again imply $\delta(u_k) = 4\delta(v) - 2$, but v also has index zero and is thus immersed, with each of its double points lifting to four isolated double points of u_∞ which survive the perturbation of the nodal curve u_∞ to a smooth simple curve u_k, thus implying the contradiction $\delta(u_k) \geq 4\delta(v)$ for large k.

Since u^1 and u^2 are distinct curves that both pass through exactly as many of the constraint points $p_1, \ldots, p_m$ as their indices will allow, there can be no constraint point that is hit by both of them. This implies that the image of the node is not a constraint point.

Finally, the description of $\overline{\mathcal{M}}_S(J)$ as an oriented topological manifold comes from the combination of two results: away from nodal curves, the usual implicit function theorem (Theorem 2.21) furnishes $\mathcal{M}_S(J)$ with the structure of a smooth oriented 2-dimensional manifold, while neighborhoods of each of the finitely many nodal curves in $\overline{\mathcal{M}}_S(J)$ are homeomorphic to 2-disks by the gluing theorem (Corollary 2.39). Note that the latter requires our second genericity condition, ensuring that the two smooth components in each nodal curve intersect each other transversely at the node. $\square$

Exercise 7.44. Suppose $v_1, v_2 : \mathbb{D}^2 \to \mathbb{C}^2$ are smooth maps with an isolated intersection $v_1(0) = v_2(0)$ that has local intersection index I, and $u_1, u_2 : \mathbb{D}^2 \to \mathbb{C}^2$ are the branched covers $u_1(z) = v_1(z^k)$ and $u_2(z) = v_2(z^\ell)$. Show that the resulting isolated intersection $u_1(0) = u_2(0)$ has local intersection index $k\ell I$.

One consequence of Lemma 7.43 is that every curve in $\overline{\mathcal{M}}_S(J)$ has trivial automorphism group, so we can now extract from Proposition 7.40 a much simpler description of the projection map $\pi_m : \overline{\mathcal{U}}_S(J) \to \overline{\mathcal{M}}_S(J)$. Almost all of its fibers are smooth spheres, except for finitely many which are unions of two spheres intersecting at one point. This should remind you of a Lefschetz fibration, and in fact:

Lemma 7.45. *The spaces $\overline{\mathcal{U}}_S(J)$ and $\overline{\mathcal{M}}_S(J)$ can each be given smooth structures, making them closed manifolds of dimenison 4 and 2 respectively, such that $\pi_m : \overline{\mathcal{U}}_S(J) \to \overline{\mathcal{M}}_S(J)$ is a Lefschetz fibration with fibers of genus zero and one critical point in each singular fiber.*

Proof. Our main task is to show that each element of $\overline{\mathcal{U}}_S(J)$ representing a node under the corresondence in Proposition 7.40 has a neighborhood homeomorphic to a neighborhood of 0 in $\mathbb{C}^2$ such that, for a suitable choice of complex coordinate on a neighborhood of the underlying nodal curve in $\overline{\mathcal{M}}_S(J)$, π_m is identified with the map $(z_1, z_2) \mapsto z_1^2 + z_2^2$. Recall from Exercise 3.21 that after a change of complex coordinates, we can equally well consider the model $\pi(z_1, z_2) := z_1 z_2$. As we saw in Example 3.23, one can parametrize the fibers near 0 in this model as a family of holomorphic annuli that degenerate to a nodal holomorphic curve consisting of two transversely intersecting disks. We shall now use the gluing map from Sect. 2.1.7 to obtain a similar local description of $\pi_m : \overline{\mathcal{U}}_S(J) \to \overline{\mathcal{M}}_S(J)$.

Assume $v_0 \in \overline{\mathcal{M}}_S(J)$ is a nodal curve with smooth components $v_0^+ : (\Sigma^+, j^+) \to (M, J)$ and $v_0^- : (\Sigma^-, j^-) \to (M, J)$ connected by nodal points $z^\pm \in \Sigma^\pm$ with

$$v_0^+(z^+) = v_0^-(z^-) =: p_0 \in M \backslash \{p_1, \ldots, p_m\},$$

so we have an identification of $\pi_m^{-1}(v_0)$ with the singular surface $\check{\Sigma} := (\Sigma^+ \amalg \Sigma^-)/\sim$ in which $z^+ \sim z^-$. Denote the node in $\check{\Sigma}$ by $\check{z}$, and regard it as an element of $\check{\mathcal{U}}_S(J)$. As explained in the proof of Corollary 2.39, the gluing map that describes a neighborhood of v_0 in $\overline{\mathcal{M}}_{0,m}([S]; J)$ specializes to the space of curves constrained by $p_1, \ldots, p_m$ as a smooth map

$$\Psi : [R_0, \infty) \times S^1 \hookrightarrow \mathcal{M}_S(J)$$

for some $R_0 > 0$, such that $\Psi(R, \theta) \to v_0$ in the Gromov topology for every $\theta \in S^1$ as $R \to \infty$. We can be a bit more precise about this by recalling how the gluing map is constructed: first, we fix coordinates identifying punctured neighborhoods of $z^- \in \Sigma^-$ and $z^+ \in \Sigma^+$ biholomorphically with $[0, \infty) \times S^1$ and $(-\infty, 0] \times S^1$ respectively, and for each $(R, \theta) \in [R_0, \infty) \times S^1$, we construct a new Riemann surface $\Sigma_{(R,\theta)}$ by cutting the ends $(R, \infty) \times S^1$ and $(-\infty, -R) \times S^1$ off of Σ^- and Σ^+, then gluing the resulting truncated surfaces together via the map $\{R\} \times S^1 \to \{-R\} \times S^1 : (R, t) \mapsto (-R, t + \theta)$. The Riemann surface $\Sigma_{(R,\theta)}$ serves as the domain of the glued curve

$$v_{(R,\theta)} := \Psi(R, \theta) \in \mathcal{M}_S(J)$$

and is thus a model for the fiber $\pi_m^{-1}(\Psi(R, \theta))$. For our present purposes, it will suffice to understand what is happening on the "neck"

$$Z_{(R,\theta)} = \left([0, R] \times S^1\right) \cup_{(R,t)\sim(-R,t+\theta)} \left([-R, 0] \times S^1\right) \subset \Sigma_{(R,\theta)}.$$

By the construction of the gluing map, the restriction of $v_{(R,\theta)}$ to $Z_{(R,\theta)}$ converges as $R \to \infty$ to $v_- : [0, \infty) \times S^1 \to M$ and $v_+ : (-\infty] \times S^1 \to M$ on the two halves of the increasingly long annulus $Z_{(R,\theta)}$, and to flesh out the implications of this more precisely, let us identify $Z_{(R,\theta)}$ with $[-R, R] \times S^1$ via the biholomorphic map

$$\psi_{(R,\theta)} : [-R, R] \times S^1 \to Z_{(R,\theta)} : (s, t) \mapsto \begin{cases} (s + R, t - \theta) & \text{if } s \leq 0, \\ (s - R, t) & \text{if } s \geq 0. \end{cases}$$

Then writing $\hat{v}_{(R,\theta)} := v_{(R,\theta)} \circ \psi_{(R,\theta)} : [-R, R] \times S^1 \to M$, we have

$$\lim_{R \to \infty} \hat{v}_{(R,\theta)}(s + R, t) = v_0^+(s, t) \qquad \text{for } (s, t) \in (-\infty, 0] \times S^1,$$

$$\lim_{R \to \infty} \hat{v}_{(R,\theta)}(s - R, t) = v_0^-(s, t - \theta) \qquad \text{for } (s, t) \in [0, \infty) \times S^1,$$

with convergence in C_{loc}^∞ on the half-cylinder in both cases. Moreover, given any sequence $(R_k, \theta_k) \in [R_0, \infty) \times S^1$ with $R_k \to \infty$ and a sequence $(s_k, t_k) \in$

$[-R_k, R_k] \times S^1$ with $s_k + R_k \to \infty$ and $R_k - s_k \to \infty$, we have

$$\widehat{v}_{(R_k, \theta_k)}(s_k, t_k) \to p_0.$$

We have intentionally set up these parametrizations to look very similar to the degeneration of the annuli $u_w : [-R, R] \times S^1 \to \mathbb{C}^2$ exhibited in Example 3.23, where $w = re^{2\pi i\theta}$ and

$$R = \frac{1}{4\pi} \cosh^{-1}(\epsilon^2/2r) \tag{7.11}$$

for some small $\epsilon > 0$. And indeed, we can now define a homeomorphism Φ from a neighborhood of $\check{z}$ in $\overline{U}_S(J)$ to a neighborhood of 0 in $\mathbb{C}^2$ as follows. For each $(R, \theta) \in [R_0, \infty) \times S^1$ and $(s, t) \in [-R, R] \times S^1$, denote by (R, θ, s, t) the element of $\pi_m^{-1}(v_{(R,\theta)})$ that has the extra marked point placed at $\psi_{(R,\theta)}(s, t) \in Z_{(R,\theta)} \subset \Sigma_{(R,\theta)}$. We then define Φ outside the singular fiber by

$$\Phi(R, \theta, s, t) = u_w(s, t) \in \mathbb{C}^2,$$

with $w = re^{2\pi i\theta}$ and r defined in terms of R via (7.11). The matching degeneration behavior of the families $\widehat{v}_{(R,\theta)}$ and u_w implies that Φ has a continuous extension defined on the singular fiber as follows: Φ maps the element in $\pi_m^{-1}(v_0)$ with extra marked point at $(s, t) \in (-\infty, 0] \times S^1 \subset \Sigma^+$ to $u_0^+(e^{2\pi(s+it)}) \in \mathbb{C}^2$, while the element with extra marked point at $(s, t) \in [0, \infty) \times S^1 \subset \Sigma^-$ is mapped to $u_0^-(e^{-2\pi(s+it)})$. Defining a complex coordinate chart on $\overline{M}_S(J)$ near v_0 by $\varphi(v_{(R,\theta)}) = re^{2\pi i\theta}$ and $\varphi(v_0) = 0$, we now have $\varphi \circ \pi_m \circ \Phi^{-1}(z_1, z_2) = z_1 z_2$ by construction.   □

Corollary 7.46. *The space $\overline{U}_S(J)$ is homeomorphic to $X \# k\overline{\mathbb{CP}}^2$, where $k \geqslant 0$ is the number of nodal curves in $\overline{M}_S(J)$ and X is a symplectic ruled surface fibering over the closed surface $\overline{M}_S(J)$.*   □

Next, we observe that the above description of the compact moduli space $\overline{M}_S(J)$ is quite stable under generic deformations of J. Indeed, suppose $\{J_s\}_{s \in [0,1]}$ is a generic 1-parameter family of tame almost complex structures with $J_0 = J$, and consider the parametric moduli space

$$\overline{M}_S(\{J_s\}) \subset \overline{M}_{0,m}([S]; \{J_s\}; p_1, \ldots, p_m),$$

defined as the connected component of $\overline{M}_{0,m}([S]; \{J_s\}; p_1, \ldots, p_m)$ that contains the set $\{0\} \times \overline{M}_S(J)$. For each $t \in (0, 1]$, denote

$$\overline{M}_S(J_t) := \left\{ u \in \overline{M}_{0,m}([S]; J_t) \mid (t, u) \in \overline{M}_S(\{J_s\}) \right\},$$

and add an extra marked point to define

$$\overline{\mathcal{U}}_S(\{J_s\}) = \{(s, u) \mid s \in [0, 1] \text{ and } \pi_m(u) \in \overline{\mathcal{M}}_S(J_s)\},$$

along with the obvious continuous maps

$$\pi_m : \overline{\mathcal{U}}_S(\{J_s\}) \to \overline{\mathcal{M}}_S(\{J_s\}),$$

$$\mathrm{ev}_{m+1} : \overline{\mathcal{U}}_S(\{J_s\}) \to M.$$

Lemma 7.47. *For generic homotopies* $\{J_s \in \mathcal{J}_\tau(M, \omega)\}_{s \in [0,1]}$ *where* $J_0 = J$ *and* J_1 *satisfies genericity Assumption 7.41,* $\overline{\mathcal{M}}_S(\{J_s\})$ *is a compact oriented 3-dimensional topological manifold with boundary, defining an oriented cobordism from* $\overline{\mathcal{M}}_S(J)$ *to* $\overline{\mathcal{M}}_S(J_1)$. *Moreover,* $\overline{\mathcal{M}}_S(J_1)$ *also has the structure described in Lemma 7.43, and it contains the same number of nodal curves as* $\overline{\mathcal{M}}_S(J)$.

Proof. We again use the fact that indices of closed holomorphic curves (with or without fixed point constraints) are always even, so adding one parameter for J to depend on does not create any danger of the appearance of index -1 curves. In fact, it does not make possible any of the eventualities that were ruled out via genericity in the proof of Lemma 7.43, as these all involved conditions that have codimension at least 2. In particular, the same argument rules out all nodal curves in $\overline{\mathcal{M}}_S(\{J_s\})$ that are more complicated than those occurring in $\overline{\mathcal{M}}_S(J)$, and their smooth components (with constrained index 0) can still be assumed to be immersed. This last detail implies that all such curves satisfy the hypotheses for automatic transversality in Theorem 2.46, so they are Fredholm regular even at parameter values where J_s is not generic. The nodal curves therefore form a finite collection of 1-parameter families that vary smoothly with $s \in [0, 1]$ and survive the entire homotopy from $s = 0$ to $s = 1$. For the same reason, the gluing theorem also remains valid near the nodal curves even when J_s is not generic, and thus gives a reasonable description of neighborhoods of such curves in $\overline{\mathcal{M}}_S(\{J_s\})$ as topological 3-manifolds. Outside the nodal curves, the arguments of Lemma 7.43 rule out multiple covers so that the usual implicit function theorem as in Theorem 2.22 gives $\mathcal{M}_S(\{J_s\})$ a smooth structure.   □

Another important consequence of automatic transversality as used in the above proof is that the argument of Lemma 7.45 still works for describing the local structure of the space $\overline{\mathcal{U}}_S(\{J_s\})$ near nodes, even in cases where J_s is not generic. Hence:

Corollary 7.48. *In the setting of Lemma 7.47, the space* $\overline{\mathcal{U}}_S(\{J_s\})$ *is a compact oriented 5-dimensional cobordism between* $\overline{\mathcal{U}}_S(J)$ *and*

$$\overline{\mathcal{U}}_S(J_1) := \pi_m^{-1}(\overline{\mathcal{M}}_S(J_1)).$$ □

7.3.5   The Evaluation Map Has Positive Degree

The above results prove that

$$\mathrm{ev}_{m+1} : \overline{\mathcal{U}}_S(J) \to M$$

is a continuous map between closed oriented manifolds of the same dimension, so it has a well-defined mapping degree. Computing this degree requires the following slight modification of Proposition 2.53:

Lemma 7.49. *Suppose $u \in \mathcal{M}_S(J)$ is immersed. Then on some neighborhood of the complement of the marked points in $\pi_m^{-1}(u)$, the map $\mathrm{ev}_{m+1} : \overline{\mathcal{U}}_S(J) \to M$ is a local diffeomorphism and is orientation preserving.*

Proof. The claim that it is a local diffeomorphism follows from Theorem 2.46 since u is immersed and $\mathrm{ind}(u) = 2m + 2$, so automatic transversality holds and remains valid after adding the extra marked point and imposing on it a fixed point constraint. The conclusion that it is orientation preserving then follows from Proposition 2.47.
$\square$

To say more, we need to impose further genericity conditions using the techniques of Sects. 2.1.4 and 2.1.5: in particular, we shall assume from now on that all the results of those sections hold for all choices of submanifolds Z of the form

$$Z = \{(p_{i_1}, \ldots, p_{i_r})\} \subset M^r$$

for arbitrary subsets $\{p_{i_1}, \ldots, p_{i_r}\} \subset \{p_1, \ldots, p_m\}$. Since there are finitely many such subsets, this restricts J to a finite intersection of comeager subsets of $\mathcal{J}_\tau(M, \omega)$, which is also comeager. As with Assumption 7.41, allowing such a perturbation of J implicitly means perturbing the original immersion $S \looparrowright M$ as well.

Now denote by

$$\mathcal{M}_S^{\mathrm{good}}(J) \subset \mathcal{M}_S(J)$$

the open subset consisting of curves $u \in \mathcal{M}_S(J)$ with the following properties:

(1) u is immersed;
(2) u has no tangential self-intersections;
(3) u has no triple self-intersections;
(4) The marked points of u are all injective points;
(5) u has no tangential intersections with any of the finitely many curves in Lemma 7.42 with energy bounded by $E_0 := \langle[\omega], [S]\rangle$.

We denote

$$\mathcal{M}_S^{\mathrm{bad}}(J) = \mathcal{M}_S(J)\backslash\mathcal{M}_S^{\mathrm{good}}(J).$$

The next lemma is a straightforward application of the results in Sects. 2.1.4 and 2.1.5.

Lemma 7.50 (Genericity).

(1) $\mathcal{M}_S^{\mathrm{bad}}(J)$ is a discrete subset of $\mathcal{M}_S(J)$.[3]

(2) Any two distinct curves in $\mathcal{M}_S^{\mathrm{bad}}(J)$ intersect each other transversely.

(3) Critical points, tangential self-intersections or triple self-intersections of curves in $\mathcal{M}_S^{\mathrm{bad}}(J)$ never occur at any of the constraint points $p_1, \ldots, p_m$.

□

Lemma 7.51. The map $\mathrm{ev}_{m+1} : \overline{\mathcal{U}}_S(J) \to M$ has $\deg(\mathrm{ev}_{m+1}) \geqslant 1$, with equality if and only if S is embedded.

Proof. If S is embedded, then the statement reduces to a consequence of the main results from Chap. 1, i.e. $\overline{\mathcal{M}}_S(J)$ forms a foliation of $M \backslash \{p_1, \ldots, p_m\}$ and $\deg(\mathrm{ev}_{m+1})$ is therefore 1. Now assume S is not embedded, and pick a curve $u \in \mathcal{M}_S^{\mathrm{good}}(J)$. The adjunction formula implies that u is not embedded, so we can pick two distinct points $\zeta, \zeta' \in \pi_m^{-1}(u)$ that are separate from the marked points of u and satisfy $\mathrm{ev}_{m+1}(\zeta) = \mathrm{ev}_{m+1}(\zeta') =: q_0$. By Lemma 7.49, ev_{m+1} defines orientation-preserving diffeomorphisms from disjoint neighborhoods of ζ and ζ' in $\overline{\mathcal{U}}_S(J)$ onto a neighborhood of q_0 in M. But by Lemma 7.50, almost every point in M is not in the image of any curve from $\mathcal{M}_S^{\mathrm{bad}}(J)$, thus we can find a point $q \in M$ with this property in the aforementioned neighborhood of q_0. Now Lemma 7.49 implies that every point in $\mathrm{ev}_{m+1}^{-1}(q)$ contributes positively to $\deg(\mathrm{ev}_{m+1})$, and there are at least two such points, hence $\deg(\mathrm{ev}_{m+1}) \geqslant 2$. □

Corollary 7.52. The implication $(2) \Rightarrow (4)$ in Theorem 7.3 holds.

Proof. Given a generic $J_1 \in \mathcal{J}_\tau(M, \omega)$, we can choose a generic homotopy $\{J_s\}_{s \in [0,1]}$ in $\mathcal{J}_\tau(M, \omega)$ from J to J_1 so that Corollary 7.48 gives a compact oriented cobordism between closed oriented 4-manifolds $\overline{\mathcal{U}}_S(J)$ and $\overline{\mathcal{U}}_S(J_1)$. Extending ev_{m+1} in the natural way to this cobordism, its restriction to $\overline{\mathcal{U}}_S(J_1)$ then also has positive degree, implying the existence of somewhere injective J_1-holomorphic spheres homologous to $[S]$. □

For the remainder of this chapter, let

$$N := \deg(\mathrm{ev}_{m+1}).$$

Notice that if $m \geqslant 1$, then for each $i = 1, \ldots, m$, the ith marked point naturally determines a smooth section

$$\Sigma_i \subset \overline{\mathcal{U}}_S(J)$$

---

[3] With a little more effort and intersection theory, one can show that $\mathcal{M}_S^{\mathrm{bad}}(J)$ is actually finite for generic J, but we will not need this.

of the Lefschetz fibration $\pi_m : \overline{\mathcal{U}}_S(J) \to \overline{\mathcal{M}}_S(J)$, such that ev_{m+1} maps Σ_i to a single point p_i. By Lemma 7.50, no curve in $\mathcal{M}_S(J)$ has a non-immersed point at any of its marked points, and this is automatically also true for the nodal curves in $\overline{\mathcal{M}}_S(J)$ since their smooth components are immersed. Thus there exists for each $i = 1, \ldots, m$ a continuous map

$$\sigma_i : \overline{\mathcal{M}}_S(J) \to \mathbb{P}(T_{p_i} M) \cong \mathbb{CP}^1$$

sending each curve $u \in \overline{\mathcal{M}}_S(J)$ to the complex tangent space that it spans at the ith marked point. The next lemma is yet another consequence of automatic transversality.

Lemma 7.53. *The maps σ_i restrict to $\mathcal{M}_S^{\mathrm{good}}(J)$ as orientation-preserving local diffeomorphisms.*

Proof. As in the proof of Theorem 2.46, $T_u \mathcal{M}_S(J)$ can be naturally identified for each $u \in \mathcal{M}_S^{\mathrm{good}}(J)$ with the kernel of the normal Cauchy-Riemann operator $\mathbf{D}_u^N$, defined on a space of sections of the normal bundle N_u of u which are constrained to vanish at the marked points. Since $c_1(N_u) = m$ and zeroes of nontrivial sections $\eta \in \ker \mathbf{D}_u^N = T_u \mathcal{M}_S(J)$ count positively, the zeroes at the marked points are all nondegenerate, and this translates into the statement that the derivative of each σ_i at u is nonsingular. To see that it also preserves orientation, one can check that this is true if $\mathbf{D}_u^N$ happens to be complex linear, and then argue the general case as in Proposition 2.47 by deforming $\mathbf{D}_u^N$ through a family of Cauchy-Riemann type operators to its complex-linear part: the point is that all the operators in this deformation are also surjective and also have the property that nontrivial elements of their kernels have nondegenerate zeroes at the marked points.   □

Lemma 7.54. *For each $i = 1, \ldots, m$, $\deg(\sigma_i) = N$.*

Proof. The discrete subset $\mathcal{M}_S^{\mathrm{bad}}(J)$ can accumulate only near the finite set of nodal curves $\overline{\mathcal{M}}_S(J) \backslash \mathcal{M}_S(J)$, thus if we pick any u belonging to $\mathcal{M}_S^{\mathrm{good}}(J)$ for which $\sigma_i(u)$ lies outside both the finite set of values taken by σ_i on nodal curves and the discrete set of values taken on $\mathcal{M}_S^{\mathrm{bad}}(J)$, then $\sigma_i(u)$ has a neighborhood in $\mathbb{P}(T_{p_i} M)$ consisting of points whose preimages are all in $\mathcal{M}_S^{\mathrm{good}}(J)$ and thus (by the previous lemma) contribute positively to $\deg(\sigma_i)$. This already proves that $\deg(\sigma_i) > 0$. To see that it also matches $\deg(\mathrm{ev}_{m+1})$, one can repeat the argument of Lemma 7.51 using a point q near p_i in the image of a curve near u: since p_i is an injective point of every curve in $\mathcal{M}_S^{\mathrm{good}}(J)$, we can arrange for q to occur as a value of curves in $\overline{\mathcal{M}}_S(J)$ only in some arbitrarily small neighborhood of the ith marked point, and only for curves u' with $\sigma_i(u')$ lying in an arbitrarily small neighborhood of $\sigma_i(u)$. This sets up a bijection between $\mathrm{ev}_{m+1}^{-1}(q)$ and the connected components of $\sigma_i^{-1}(\mathcal{U})$ for some small neighborhood $\mathcal{U} \subset \mathbb{P}(T_{p_i} M)$ of $\sigma_i(u)$.   □

Lemma 7.55. *The sections* $\Sigma_i \subset \overline{\mathcal{U}}_S(J)$ *for* $i = 1, \ldots, m$ *are all disjoint from each other and all satisfy* $[\Sigma_i] \cdot [\Sigma_i] = -N$.

Proof. The fact that they are disjoint is obvious. To compute $[\Sigma_i] \cdot [\Sigma_i]$, we claim that the normal bundle of Σ_i in $\overline{\mathcal{M}}_J(S)$ has first Chern number $-N$. Indeed, this normal bundle is equivalent to the restriction to Σ_i of the vertical subbundle of the Lefschetz fibration $\pi_m : \overline{\mathcal{U}}_S(J) \to \overline{\mathcal{M}}_S(J)$. Since no curve $u \in \overline{\mathcal{M}}_J(S)$ has a non-immersed point at any marked point, the tangent map of ev_{m+1} maps the vertical subspace at $\Sigma_i \cap \pi_m^{-1}(u)$ isomorphically to the subspace $\sigma_i(u)$ in $T_{p_i}M$, so the bundle in question is the pullback via σ_i of the tautological line bundle over $\mathbb{P}(T_{p_i}M) \cong \mathbb{C}\mathbb{P}^1$. The latter has first Chern number -1, so this proves the claim. $\square$

7.3.6 Topology of Ruled Surfaces and the Signature Theorem

The remainder of the proof of Theorem 7.34 consists of essentially topological arguments. Given a topological space X and an integer $k \geqslant 0$, we shall denote by

$$H_k^{\mathrm{free}}(X) := H_k(X)/\mathrm{torsion}, \qquad H_{\mathrm{free}}^k(X) := H^k(X)/\mathrm{torsion}$$

the quotients of the kth homology and cohomology groups of X with integer coefficients by their respective torsion subgroups. Both are free abelian groups of the same rank, which is the k**th Betti number**

$$b_k(M) = \mathrm{rank}\, H_k^{\mathrm{free}}(X) = \mathrm{rank}\, H_{\mathrm{free}}^k(X).$$

Note that the latter is equivalently the real dimension of $H_k(X; \mathbb{R})$ or $H^k(X; \mathbb{R})$. Recall next that if X is a closed oriented and connected 4-manifold, the intersection product defines a nonsingular bilinear pairing on $H_2^{\mathrm{free}}(X)$ which can be identified via Poincaré duality with the cup product pairing

$$H_{\mathrm{free}}^2(X) \times H_{\mathrm{free}}^2(X) \to \mathbb{Z} : (\alpha, \beta) \mapsto \langle \alpha \cup \beta, [X] \rangle. \qquad (7.12)$$

Nonsingularity means in particular that for every primitive element $A \in H_2^{\mathrm{free}}(X)$ or $\alpha \in H_{\mathrm{free}}^2(X)$, there exists $B \in H_2^{\mathrm{free}}(X)$ or $\beta \in H_{\mathrm{free}}^2(X)$ respectively such that $A \cdot B = 1$ or $\langle \alpha \cup \beta, [X] \rangle = 1$, see e.g. [Hat02, Corollary 3.39]. Switching to field coefficients, we can also regard these pairings as nondegenerate quadratic forms on either of the real vector spaces $H_2(X; \mathbb{R})$ or $H^2(X; \mathbb{R})$, and we denote by $b_2^+(X)$ and $b_2^-(X)$ the maximal dimensions of subspaces on which the form is positive- or negative-definite respectively. The nondegeneracy of this form implies

$$b_2(X) = b_2^+(X) + b_2^-(X),$$

and the **signature** of X is defined to be the integer

$$\sigma(X) = b_2^+(X) - b_2^-(X).$$

It will be useful to have a special case of the algebraic classification of nonsingular integral quadratic forms on hand; for more details, see e.g. [CS99, Chapter 15] or [MH73, Chapter II].

Proposition 7.56. *Suppose $Q : \mathbb{Z}^2 \times \mathbb{Z}^2 \to \mathbb{Z}$ is a bilinear form that is nonsingular and indefinite.*[4] *Then there exists a pair of elements e_1, e_2 generating $\mathbb{Z}^2$ such that*

$$\begin{pmatrix} Q(e_1, e_1) & Q(e_1, e_2) \\ Q(e_2, e_1) & Q(e_2, e_2) \end{pmatrix} = \quad either \quad \begin{pmatrix} 1 & 0 \\ 0 & -1 \end{pmatrix} \quad or \quad \begin{pmatrix} 0 & 1 \\ 1 & 0 \end{pmatrix}$$

□

Exercise 7.57. Given a nondegenerate quadratic form Q on $\mathbb{R}^n$ and two elements $\mathbf{v}, \mathbf{w} \in \mathbb{R}^n$ satisfying $Q(\mathbf{v}, \mathbf{v}) > 0$ and $Q(\mathbf{w}, \mathbf{w}) > 0$, show that Q is positive-semidefinite on the span of $\mathbf{v}$ and $\mathbf{w}$ if and only if $[Q(\mathbf{v}, \mathbf{w})]^2 \leq Q(\mathbf{v}, \mathbf{v}) \cdot Q(\mathbf{w}, \mathbf{w})$.

Exercise 7.58 (cf. [Lax07, Chapter 8]). Given a symmetric bilinear form $Q : \mathbb{R}^n \times \mathbb{R}^n \to \mathbb{R}$, denote by b_+ and b_- the maximal dimensions of subspaces on which the quadratic form $\mathbf{v} \mapsto Q(\mathbf{v}, \mathbf{v})$ is positive- or negative-definite respectively, and let $b_0 = \dim \{\mathbf{v} \in \mathbb{R}^n \mid Q(\mathbf{v}, \cdot) = 0\}$. Show that the maximal dimension of subspaces on which the quadratic form is positive- or negative-semidefinite is $b_+ + b_0$ or $b_- + b_0$ respectively, so in particular it is $b_\pm$ whenever the form is nondegenerate.

Let us collect some useful general facts about the cohomology rings of closed 4-manifolds.

Proposition 7.59. *For any closed connected symplectic manifold (X, ω), $b_2^+(X) \geq 1$.*

Proof. The cohomology class of the symplectic form spans a subspace of $H^2(X; \mathbb{R})$ on which the cup product pairing (7.12) is positive-definite. □

Proposition 7.60. *If X is a closed connected oriented 4-manifold with $b_2(X) = 1$, then $\alpha \cup \beta$ is a torsion class for every $\alpha \in H^1(X)$ and $\beta \in H^k(X)$ with $k \in \{1, 2\}$.*

Proof. By assumption rank $H^2_{\text{free}}(X) = 1$, so fix a generator $e \in H^2_{\text{free}}(X)$, and note that by the nonsingularity of the cup product pairing, $e \cup e \neq 0 \in H^4_{\text{free}}(X)$. Then for any $\alpha, \beta \in H^1(X)$, the equivalence class of $\alpha \cup \beta$ in $H^2_{\text{free}}(X)$ is λe for some

---

[4]"Indefinite" means that $Q(e.e)$ is positive for some $e \in \mathbb{Z}^2$ and negative for others. Equivalently, one could stipulate that for the induced real-valued nondegenerate quadratic form on $\mathbb{R}^2$, the maximum dimensions of subspaces on which this form is positive- or negative-definite are each 1.

$\lambda \in \mathbb{Z}$, and since $\alpha \cup \alpha = 0$,

$$0 = (\alpha \cup \beta) \cup (\alpha \cup \beta) = \lambda^2 (e \cup e),$$

implying $\lambda = 0$, hence $\alpha \cup \beta$ is torsion.

If instead $\beta \in H^2(X)$ but $\alpha \cup \beta$ is nontrivial in $H^3_{\text{free}}(X)$, then using the nonsingularity of the cup product again, we can find $\gamma \in H^1(X)$ such that $(\alpha \cup \beta) \cup \gamma \neq 0 \in H^4_{\text{free}}(X)$, but this is impossible since the previous paragraph shows $\alpha \cup \gamma = 0 \in H^2_{\text{free}}(X)$. $\square$

Proposition 7.61. *Suppose X and Y are closed, connected and oriented n-manifolds for some $n \in \mathbb{N}$, and $f : X \to Y$ is a continuous map with $\deg(f) \neq 0$. Fix a field $\mathbb{K}$. Then:*

(1) $f^* : H^*(Y; \mathbb{K}) \to H^*(X; \mathbb{K})$ is injective.
(2) $f_* : H_*(X; \mathbb{K}) \to H_*(Y; \mathbb{K})$ is surjective.
(3) If $n = 4$ and $\deg(f) > 0$, then $b_2^+(X) \geq b_2^+(Y)$ and $b_2^-(X) \geq b_2^-(Y)$.

Proof. By assumption $f_*[X] = D[Y]$ for some integer $D \neq 0$. If $f^*\alpha = 0 \in H^*(X; \mathbb{K})$ for some nonzero $\alpha \in H^*(Y; \mathbb{K})$, the nonsingularity of the cup product implies that we can find $\beta \in H^*(Y; \mathbb{K})$ with $\langle \alpha \cup \beta, [Y] \rangle = 1$, hence

$$0 = \langle f^*\alpha \cup f^*\beta, [X] \rangle = \langle f^*(\alpha \cup \beta), [X] \rangle = \langle \alpha \cup \beta, f_*[X] \rangle = D\langle \alpha \cup \beta, [Y] \rangle = D,$$

a contradiction. The surjectivity of $f_* : H_*(X; \mathbb{K}) \to H_*(Y; \mathbb{K})$ follows immediately by viewing it as the transpose of $f^* : H^*(Y; \mathbb{K}) \to H^*(X; \mathbb{K})$ under the natural duality between homology and cohomology with field coefficients. Finally, if $n = 4$ and $D > 0$, choose a subspace $V \subset H^2(Y; \mathbb{R})$ of dimension $b_2^+(Y)$ on which the analogue of the quadratic form (7.12) with real coefficients is positive definite. Then for each nonzero $\alpha \in V$, we have $\langle \alpha \cup \alpha, [Y] \rangle > 0$, and thus by the calculation above,

$$\langle f^*\alpha \cup f^*\alpha, [X] \rangle = D\langle \alpha \cup \alpha, [Y] \rangle > 0,$$

implying that $f^*(V) \subset H^2(X; \mathbb{R})$ is a subspace of dimension $b_2^+(Y)$ on which the cup product pairing is positive-definite, hence $b_2^+(X) \geq b_2^+(Y)$. The argument for b_2^- is analogous. $\square$

We now prove some results about the topology of blown-up ruled surfaces, which will therefore apply to the moduli space $\overline{\mathcal{U}}_S(J)$.

Proposition 7.62. *Suppose Σ is a closed, connected and oriented surface, $\pi : X \to \Sigma$ is a smooth Lefschetz fibration with regular fiber S^2 such that each singular fiber has only one critical point, $S \subset X$ is a section and $F \subset X$ is a regular fiber. Then:*

(1) The projection $\pi : X \to \Sigma$ and inclusion $\iota : S \hookrightarrow X$ induce isomorphisms

$$\pi_1(X) \xrightarrow{\pi_*} \pi_1(\Sigma), \qquad H^1(\Sigma; \mathbb{K}) \xrightarrow{\pi^*} H^1(X; \mathbb{K}),$$

$$\pi_1(S) \xrightarrow{\iota_*} \pi_1(X), \qquad H^1(X; \mathbb{K}) \xrightarrow{\iota_*} H^1(S; \mathbb{K})$$

for any field $\mathbb{K}$.

(2) $H_2(X)$ is freely generated by $[S]$, $[F]$, and $[E_1], \ldots, [E_N]$, where for each singular fiber labeled by $i = 1, \ldots, N$, E_i denotes the irreducible component that doesn't intersect S.

(3) $b_2^+(X) = 1$.

(4) If $\gamma \in H^2(X)$ satisfies $\langle \gamma, [F] \rangle = 0$ and there are no singular fibers, then $\gamma \cup \gamma = 0$.

(5) If $\gamma \in H^2(X)$ satisfies $\langle \gamma, [F] \rangle \neq 0$, then $\gamma \cup \alpha \in H^3(X)$ is not torsion for every non-torsion element $\alpha \in H^1(X)$.

Proof. We begin by proving the first three statements under the assumption that $\pi : X \to \Sigma$ has no singular fibers, so X is an honest ruled surface. In this case the homotopy exact sequence of the fibration $S^2 \hookrightarrow X \to \Sigma$ implies that $\pi_* : \pi_1(X) \to \pi_1(\Sigma)$ is an isomorphism, and under the obvious identification of Σ with the section $S \subset X$, its inverse is the map $\iota_* : \pi_1(S) \to \pi_1(X)$. The same statements then hold for the abelianization of π_1, which is H_1, und dualizing these gives the corresponding statements about H^1 with field coefficients.

For $H_2(X)$, we observe first that X admits a cell decomposition having exactly two 2-cells, both of which are cycles in cellular homology. Indeed, pick a cell decomposition of Σ that has only one 0-cell, one 2-cell and some 1-cells. Trivializing the sphere bundle over each cell and decomposing the fibers S^2 in the standard way into a 2-cell attached to a 0-cell, we can then decompose the restriction of X over the closure of each k-cell of Σ into a $(k + 2)$-cell attached to a k-cell: the result is a CW-decomposition of X that includes one 2-cell spanning a section over the 2-cell of Σ, and another whose image is the fiber over the 0-cell of Σ. This decomposition implies that $H_2(X)$ is generated by at most two elements. Now observe that $[F] \cdot [S] = 1$ and $[F] \cdot [F] = 0$. The former implies that both $[F]$ and $[S]$ are non-torsion elements of $H_2(X)$, and the two relations together imply that $[F]$ and $[S]$ are linearly independent in $H_2^{\text{free}}(X)$, thus they generate a free abelian group of rank 2. The intersection form can now be represented as a matrix of the form

$$\begin{pmatrix} [F] \cdot [F] & [F] \cdot [S] \\ [S] \cdot [F] & [S] \cdot [S] \end{pmatrix} = \begin{pmatrix} 0 & 1 \\ 1 & k \end{pmatrix}$$

for some $k \in \mathbb{Z}$. The eigenvalues of this matrix are $\frac{1}{2}\left(k \pm \sqrt{k^2 + 4}\right)$, so one is positive and one negative, implying $b_2^+(X) = b_2^-(X) = 1$.

If $\pi : X \to \Sigma$ has $N \geqslant 1$ singular fibers, then for each singular fiber we can blow down the irreducible component that does not intersect S, producing a smooth S^2-bundle $\check{\pi} : \check{X} \to \Sigma$ such that all of the above applies to $\check{X}$, and $X \cong \check{X} \# N\overline{\mathbb{CP}}^2$. The desired statements about π_1 and H_2 then follow from the $N = 0$ case using the Seifert-van Kampen theorem or a Mayer-Vietoris sequence respectively. In particular, if Q denotes the intersection form of $\check{X}$, then the intersection form of X is the direct sum of Q with a copy of -1 for each exceptional sphere coming from a blowup operation. Thus $b_2^-(X) = 1 + N$ while $b_2^+(X)$ is still 1.

We now prove the last two statements. Given $\gamma \in H^2(X)$, we can use the presentation of $H_2(X)$ in statement (2) and write

$$\mathrm{PD}(\gamma) = k[S] + \ell[F] + \sum_{i=1}^{N} a_i[E_i]$$

for unique integers $k, \ell, a_1, \ldots, a_N$, so

$$\langle \gamma, [F] \rangle = \mathrm{PD}(\gamma) \cdot [F] = k.$$

If $k = N = 0$, this means $\mathrm{PD}(\gamma) = \ell[F]$ for some $\ell \in \mathbb{Z}$ and thus $\langle \gamma \cup \gamma, [X] \rangle = \ell^2[F] \cdot [F] = 0$.

If on the other hand $k \neq 0$, suppose $\alpha \neq 0 \in H^1_{\mathrm{free}}(X)$, in which case α also defines a nonzero element of $H^1(X; \mathbb{Q})$. By statement (1), we can write $\alpha = \pi_* \alpha'$ for a unique $\alpha' \in H^1(\Sigma; \mathbb{Q})$, and the nonsingularity of the cup product on Σ then provides an element $\beta' \in H^1(\Sigma; \mathbb{Q})$ such that $\alpha' \cup \beta' \neq 0 \in H^2(\Sigma; \mathbb{Q})$. Setting $\beta := \pi^* \beta' \in H^2(X; \mathbb{Q})$, we then have $\alpha \cup \beta = \pi^*(\alpha' \cup \beta')$ and thus

$$\langle \gamma \cup \alpha \cup \beta, [X] \rangle = \langle \alpha \cup \beta, \mathrm{PD}(\gamma) \rangle = \langle \alpha' \cup \beta', \pi_*(k[S] + \ell[F] + \sum_i a_i[E_i]) \rangle$$

$$= k \langle \alpha' \cup \beta', [\Sigma] \rangle \neq 0.$$

This implies $\gamma \cup \alpha \neq 0 \in H^3(X; \mathbb{Q})$, hence the corresponding integral class is not torsion. □

Corollary 7.63. *Assume X is the total space of a Lefschetz fibration $\pi : X \to \Sigma$, with regular fiber S^2 and only one critical point in every singular fiber, over a closed connected and oriented suface Σ. Suppose additionally that M is a closed, connected and oriented 4-manifold, and $f : X \to M$ is a continuous map of nonzero degree sending the fiber of X to a non-torsion class in $H_2(M)$. Then either $b_1(M) = 0$ or $b_2(M) \geqslant 2$.*

Proof. Let $[F] \in H_2(X)$ denote the fiber class and $A := f_*[F]$, which is non-torsion by assumption, hence $b_2(M) \geqslant 1$. Arguing by contradiction, suppose $b_1(M) > 0$ and $b_2(M) = 1$. Pick $\gamma \in H^2(M)$ such that $\langle \gamma, A \rangle \neq 0$. Then $\langle f^* \gamma, [F] \rangle = \langle \gamma, A \rangle \neq 0$, so by the fifth statement in Proposition 7.62, $f^* \gamma \cup \alpha \neq$

$0 \in H^3(X; \mathbb{Q})$ for all nonzero classes $\alpha \in H^1(X; \mathbb{Q})$. But since $b_1(M) > 0$, we can pick a nonzero element $\alpha \in H^1(M; \mathbb{Q})$ and observe that $\gamma \cup \alpha = 0 \in H^3(M; \mathbb{Q})$ by Proposition 7.60. Using the fact from Proposition 7.61 that $f^* : H^*(M; \mathbb{Q}) \to H^*(X; \mathbb{Q})$ is injective, we then have $f^*\alpha \neq 0 \in H^1(X; \mathbb{Q})$, yet $f^*\gamma \cup f^*\alpha = f^*(\gamma \cup \alpha) = 0 \in H^3(X; \mathbb{Q})$, a contradiction. $\qquad \square$

Example 7.64. The manifold $\mathbb{CP}^2$ is neither a blowup nor a ruled surface, as it has no nontrivial class $A \in H_2(\mathbb{CP}^2)$ satisfying $A \cdot A = 0$. But $\mathbb{CP}^2$ does satisfy the conclusion of Corollary 7.63, specifically $b_1(\mathbb{CP}^2) = 0$, and it is not hard to find an example of a ruled surface X with a map $f : X \to \mathbb{CP}^2$ of nonzero degree. Indeed, blowing up our favorite Lefschetz pencil on $\mathbb{CP}^2$ at the base point gives a smooth S^2-fibration on $X := \mathbb{CP}^2 \# \overline{\mathbb{CP}}^2$, and the blowdown map $\beta : X \to \mathbb{CP}^2$ has degree one and sends the fiber class of X to the generator of $H_2(\mathbb{CP}^2)$.

Remark 7.65. The existence of the section S in Proposition 7.62 is a vacuous condition in light of Exercise 1.14. Note that if $\pi : X \to \Sigma$ has singular fibers, one can regard X as the blowup of an S^2-bundle $\check{X} \to \Sigma$ and construct a section of X as the preimage of a section of $\check{X}$ via the blowdown map.

Exercise 7.66. Suppose the Lefschetz fibration $\pi : X \to \Sigma$ in Proposition 7.62 has exactly one singular fiber, with irreducible components denoted by E_0 and E_1. Show that if $S_0, S_1 \subset X$ are sections that do not contain the critical point, such that S_0 intersects E_0 and S_1 intersects E_1, then $[S_0] \cdot [S_0] - [S_1] \cdot [S_1]$ is odd. *This implies that the two ruled surfaces obtained by blowing down X along E_0 or E_1 are not topologically equivalent, cf. Example 3.46.*

The following is a digression, but we can now tie up a loose end from the introduction and prove the easy direction of Corollary 1.22:

Proposition 7.67. *If (M, ω) is a blown-up symplectic ruled surface, then after a deformation of its symplectic form, it satisfies $\langle c_1(M, \omega) \cup [\omega], [M] \rangle > 0$.*

Proof. Using the basis of $H_2(M)$ in Proposition 7.62, write

$$C := \mathrm{PD}(c_1(M, \omega)) = m[S] + n[F] + \sum_{i=1}^{N} k_i[E_i],$$

so by the relations $[S] \cdot [F] = 1$ and $[F] \cdot [F] = [F] \cdot [E_i] = [S] \cdot [E_i] = 0$, we have $c_1([F]) = m$, $c_1([E_i]) = -k_i$, and $c_1([S]) = m[S] \cdot [S] + n$. But since ω is compatible with the Lefschetz fibration, each fiber F and irreducible component E_i is a symplectic submanifold, and Remark 3.42 implies that we can assume this for S as well after a deformation of ω, which does not change $c_1(M, \omega)$. Thus the evaluation of $c_1(M, \omega)$ on each of these submanifolds can be computed by choosing a tame almost complex structure to make them J-holomorphic and applying the adjunction formula: this gives $0 = [F] \cdot [F] = c_1([F]) - 2$, $-1 = [E_i] \cdot [E_i] = c_1([E_i]) - 2$, and $[S] \cdot [S] = c_1([S]) - \chi(S)$. It follows that $m = 2$, $k_i = -1$, and

$n = c_1([S]) - 2[S] \cdot [S] = \chi(S) - [S] \cdot [S]$, giving

$$C = 2[S] + (\chi(S) - [S] \cdot [S]) [F] - \sum_{i=1}^{N} [E_i],$$

and thus

$$\langle c_1(M, \omega) \cup [\omega], [M] \rangle = \langle [\omega], C \rangle = 2 \int_S \omega + (\chi(S) - [S] \cdot [S]) \langle [\omega], [F] \rangle - \sum_{i=1}^{N} \int_{E_i} \omega.$$
$$(7.13)$$

Now deform ω using the Thurston trick (cf. Lemma 3.41): denoting the genus zero Lefschetz fibration on M by $\pi : M \to \Sigma$, this amounts fo choosing an area form σ on Σ and considering the family of symplectic forms

$$\omega_K = \omega + K \pi^* \sigma$$

for $K \geq 0$. Replacing ω with ω_K in (7.13) makes the term $\int_S \omega_K$ arbitrarily large for $K \gg 0$ while fixing the other terms, but since ω_K is a continuous family, $c_1(M, \omega_K) = c_1(M, \omega)$, thus proving $\langle c_1(M, \omega_K) \cup [\omega_K], [M] \rangle > 0$ for $K > 0$ sufficiently large. $\square$

Exercise 7.68. Show that Proposition 7.67 is not true without allowing ω to be deformed. *Hint: look for counterexamples of the form $S^2 \times \Sigma$ with product symplectic structures.*

Exercise 7.69. Show that $\langle c_1(\mathbb{CP}^2, \omega_{FS}) \cup [\omega_{FS}], [\mathbb{CP}^2] \rangle = 3\pi$.

Exercise 7.70. Show that for any symplectic ruled surface (M, ω) with base Σ, $\langle c_1(M, \omega) \cup c_1(M, \omega) \rangle = 4\chi(\Sigma)$. This proves the easy direction of Corollary 1.23.

Exercise 7.71. Show that if (M, ω) is any closed connected symplectic 4-manifold and $(\tilde{M}, \tilde{\omega})$ is obtained from it by blowing up along a single Darboux ball, then

$$\langle c_1(\tilde{M}, \tilde{\omega}) \cup c_1(\tilde{M}, \tilde{\omega}), [\tilde{M}] \rangle = \langle c_1(M, \omega) \cup c_1(M, \omega), [M] \rangle - 1.$$

This shows that the minimality assumption in Corollary 1.23 is necessary.

The other main topological ingredient needed for proving Theorem 7.34 is the 4-dimensional case of the *Hirzebruch signature theorem*. Recall that for any real vector bundle $E \to X$, the kth **Pontrjagin class** is defined for $k \in \mathbb{N}$ as

$$p_k(E) = (-1)^k c_{2k}(E \otimes \mathbb{C}) \in H^{4k}(X),$$

where $E \otimes \mathbb{C}$ is the complexification of E, i.e. its real tensor product with the trivial complex line bundle, with complex structure on $E \otimes \mathbb{C}$ defined by $v \otimes c \mapsto v \otimes ic$ for $v \in E$ and $c \in \mathbb{C}$. When X is a smooth manifold, we denote by $p_k(X)$ the kth Pontrjagin class of its tangent bundle.

Exercise 7.72. Let V be a complex n-dimensional vector space with complex structure $J : V \to V$, denote by $V^{\mathbb{R}}$ the underlying real $2n$-dimensional vector space, $V^{\mathbb{C}} = V^{\mathbb{R}} \otimes \mathbb{C}$ the complexification of the latter, and $\overline{V}$ the complex vector space whose underlying real space is $V^{\mathbb{R}}$ but with complex structure $-J$.

(a) Show that $V \oplus \overline{V} \to V^{\mathbb{C}} : (v, w) \mapsto (v + w) \otimes 1 - J(v - w) \otimes i$ is a complex-linear isomorphism.

(b) Deduce that for any complex vector bundle E, there is a natural complex bundle isomorphism between $E \oplus \overline{E}$ and $E^{\mathbb{C}}$, where $\overline{E}$ denotes the **conjugate bundle** (i.e. E with its complex structure changed by a sign) and $E^{\mathbb{C}}$ is the complexification of the underlying real vector bundle.

Theorem (Hirzebruch, see [MS74, §19]). *For any closed, connected and oriented 4-manifold X,*

$$\sigma(X) = \frac{1}{3}\langle p_1(X), [X]\rangle.$$

Corollary 7.73. *For any closed and connected almost complex 4-manifold (X, J),*

$$2\chi(X) + 3\sigma(X) = \langle c_1(TX, J) \cup c_1(TX, J), [X]\rangle.$$

Proof. For any complex vector bundle E of rank 2, the underlying real bundle $E^{\mathbb{R}}$ satisfies

$$p_1(E^{\mathbb{R}}) = -c_2(E^{\mathbb{C}}) = -c_1(E) \cup c_1(\overline{E}) - c_2(E) - c_2(\overline{E}) \qquad (7.14)$$

due to Exercise 7.72 and the formula $c(E \oplus F) = c(E) \cup c(F)$ for the total Chern class. Here $\overline{E}$ is the conjugate bundle of E, which is isomorphic to its complex dual bundle E^*; an explicit isomorphism can be written in the form $\overline{E} \to E^* :$ $v \mapsto \langle v, \cdot \rangle$ for any choice of Hermitian bundle metric $\langle\ ,\ \rangle$. By Milnor and Stasheff [MS74, Lemma 14.9], we have $c_k(\overline{E}) = (-1)^k c_k(E)$ for every $k \in \mathbb{N}$, so (7.14) simplifies to

$$p_1(E^{\mathbb{R}}) = c_1(E) \cup c_1(E) - 2c_2(E).$$

The top Chern class $c_2(E)$ is the same as the Euler class $e(E)$, so applying this to the bundle (TX, J) for an almost complex 4-manifold (X, J), the Hirzebruch signature theorem now gives

$$\sigma(X) = \frac{1}{3}\langle c_1(TX, J) \cup c_1(TX, J), [X]\rangle - \frac{2}{3}\langle e(TX), [X]\rangle$$

$$= \frac{1}{3}\langle c_1(TX, J) \cup c_1(TX, J), [X]\rangle - \frac{2}{3}\chi(X),$$

which is equivalent to the stated formula.   □

Whenever (X, J) is an almost complex 4-manifold, let us abbreviate

$$c_1^2(X) := \langle c_1(TX, J) \cup c_1(TX, J), [X] \rangle,$$

so that Corollary 7.73 is written in the succinct form $2\chi(X) + 3\sigma(X) = c_1^2(X)$. The power of this formula is that all three of the numbers $\chi(X)$, $\sigma(X)$ and $c_1^2(X)$ must be integers, and one can typically deduce further restrictions on $c_1^2(X)$ due to the nonsingularity of the cup product pairing. We now prove some topological lemmas about symplectic or almost complex 4-manifolds that will be needed in the proof of Theorem 7.34.

Lemma 7.74. *Suppose (M, J) is a closed and connected almost complex 4-manifold with $b_1(M) = 0$ and $b_2(M) = b_2^+(M) = 1$. Then $H_2^{\mathrm{free}}(M)$ has a generator e satisfying $e \cdot e = 1$ and $c_1(e) = 3$. In particular, every $A \in H_2(M)$ satisfies $A \cdot A = k^2$ and $c_1(A) = 3k$ for some $k \in \mathbb{Z}$.*

Proof. Poincaré duality implies $b_3(M) = b_1(M) = 0$, so $\chi(M) = 3$, and since $b_2^-(M) = b_2(M) - b_2^+(M) = 0$, $\sigma(M) = 1$. Corollary 7.73 thus gives $c_1^2(M) = 9$. Given a generator $e \in H_2^{\mathrm{free}}(M)$, its Poincaré dual $\mathrm{PD}(e)$ generates $H_{\mathrm{free}}^2(M)$, and the nonsingularity of the intersection pairing implies

$$e \cdot e = \pm 1.$$

Writing $c_1(TM, J) = a\,\mathrm{PD}(e) \in H_{\mathrm{free}}^2(M)$ for a uniquely determined $a \in \mathbb{Z}$, we then have

$$9 = \langle a\,\mathrm{PD}(e) \cup a\,\mathrm{PD}(e), [M] \rangle = a^2 \langle \mathrm{PD}(e) \cup \mathrm{PD}(e), [M] \rangle = a^2(e \cdot e) = \pm a^2,$$

implying $a = \pm 3$ and $e \cdot e = 1$. Now

$$c_1(e) = \langle c_1(TM, J), e \rangle = \pm 3 \langle \mathrm{PD}(e), e \rangle = \pm 3(e \cdot e) = \pm 3,$$

so the desired generator is $\pm e$. $\square$

Lemma 7.75. *Suppose (M, J) is a closed and connected almost complex 4-manifold with $b_1(M) = 0$, $b_2(M) = 2$ and $b_2^+(M) = 1$. Then one of the following holds:*

(1) *$A \cdot A$ is even for every $A \in H_2(M)$.*
(2) *For every $A \in H_2(M)$ there exist integers $k, \ell \in \mathbb{Z}$ such that $A \cdot A = k^2 - \ell^2$ and $c_1(A) = 3k - \ell$.*

Proof. Since $H_2^{\mathrm{free}}(M)$ has rank 2 and $b_2^+(M) = 1$, the intersection form of M fits into the classification scheme of Proposition 7.56, namely there exists a pair of elements e_1, e_2 generating $H_2^{\mathrm{free}}(M)$ such that the matrix of the intersection form is one of the two matrices in that proposition. If it is the second, then for any $A = ke_1 + \ell e_2 \in H_2^{\mathrm{free}}(M)$, we have

$$A \cdot A = (ke_1 + \ell e_2) \cdot (ke_1 + \ell e_2) = 2k\ell,$$

which is always even. In the other case,

$$A \cdot A = (ke_1 + \ell e_2) \cdot (ke_1 + \ell e_2) = k^2 - \ell^2.$$

Now we use Hirzebruch: since $b_1(M) = b_3(M) = 0$ and $b_2(M) = 2$, we have $\chi(M) = 4$ and $\sigma(M) = 0$, thus Corollary 7.73 gives $c_1^2(M) = 8$. Using $\mathrm{PD}(e_1)$ and $\mathrm{PD}(e_2)$ as a basis of $H_{\mathrm{free}}^2(M)$, we can write $c_1(TM, J) = a\,\mathrm{PD}(e_1) + b\,\mathrm{PD}(e_2)$ for unique $a, b \in \mathbb{Z}$. Then

$$8 = \langle(a\,\mathrm{PD}(e_1)+b\,\mathrm{PD}(e_2))\cup(a\,\mathrm{PD}(e_1)+b\,\mathrm{PD}(e_2)), [M]\rangle = a^2-b^2 = (a-b)(a+b).$$

After changing e_1 and/or e_2 by a sign, we can assume $a, b \geq 0$ without loss of generality, and then observe that the only pair of nonnegative integers satisfying $(a - b)(a + b) = 8$ is $a = 3$ and $b = 1$. We conclude

$$c_1(TM, J) = 3\,\mathrm{PD}(e_1) + \mathrm{PD}(e_2),$$

implying

$$\begin{aligned}
c_1(A) &= \langle 3\,\mathrm{PD}(e_1) + \mathrm{PD}(e_2), ke_1 + \ell e_2\rangle \\
&= 3k(e_1 \cdot e_1) + 3\ell(e_1 \cdot e_2) + k(e_2 \cdot e_1) + \ell(e_2 \cdot e_2) = 3k - \ell.
\end{aligned}$$

$\square$

Exercise 7.76. Determine which of the two alternatives in the above lemma applies for each of the manifolds $S^2 \times S^2$ and $\mathbb{CP}^2 \# \overline{\mathbb{CP}}^2$.

Lemma 7.77. *In the setting of Corollary 7.63, assume the following additional conditions:*

- *The Lefschetz fibration $X \to \Sigma$ has no singular fibers;*
- *The 4-manifold M has a symplectic form ω;*
- *The image $A \in H_2(M)$ of the fiber class satisfies $c_1(A) := \langle c_1(M, \omega), A\rangle = 2$ for an ω-compatible almost complex structure J.*

Then $A \cdot A$ is not a positive even number.

Proof. Let us abbreviate $c_1 := c_1(M, \omega) \in H^2(M)$ and $c_1^2 := \langle c_1 \cup c_1, [M]\rangle \in \mathbb{Z}$. The assumption of no singular fibers implies $b_2(X) = 2$, so by Proposition 7.61, $b_2(M)$ is either 1 or 2, and $b_2^+(M) = 1$ since M is symplectic and $b_2^+(X) = 1$. If $b_2(M) = 1$, then Corollary 7.63 implies $b_1(M) = 0$, but then Lemma 7.74 is contradicted since $c_1(A) = 2$ is not divisible by 3. This proves $b_2(M) = 2$.

We can now deduce that $\sigma(M) = 0$ and $\chi(M)$ is even, so after choosing an ω-compatible almost complex structure, the Hirzebruch formula implies that $c_1^2 = 2\chi(M)$ is divisible by 4. Arguing by contradiction, assume $A \cdot A = 2d$ for some $d \in \mathbb{N}$. Then since $A = f_*[F]$ for the fiber F in X, we have

$$\langle f^*(\mathrm{PD}(A) - dc_1), [F]\rangle = \langle \mathrm{PD}(A) - dc_1, A\rangle = A \cdot A - 2d = 0,$$

which implies via the fourth statement in Proposition 7.62 that the class $f^*(\mathrm{PD}(A) - dc_1) \in H^2(X)$ has square zero. Since $f^* : H^*(M; \mathbb{Q}) \to H^*(X; \mathbb{Q})$ is injective, it follows that

$$(\mathrm{PD}(A) - dc_1) \cup (\mathrm{PD}(A) - dc_1) = 0 \in H^4(M).$$

Expanding this relation and evaluating on $[M]$ gives

$$0 = A \cdot A - 2dc_1(A) - d^2c_1^2 = -2d - d^2c_1^2,$$

thus $dc_1^2 = -2$ and c_1^2 cannot be divisible by 4. $\qquad \square$

Exercise 7.78. Show that the closed oriented 4-manifolds S^4 and $\mathbb{CP}^2 \# \mathbb{CP}^2$ do not admit almost complex structures.

7.3.7 Conclusion of the Proof

Let us summarize the situation so far. We have constructed a moduli space $\overline{\mathcal{M}}_S(J)$ with the topology of a closed oriented surface, such that the "universal J-holomorphic curve" obtained by adding an extra marked point to each curve in this space is a Lefschetz fibration

$$\pi_m : \overline{\mathcal{U}}_S(J) \to \overline{\mathcal{M}}_S(J)$$

with genus 0 fibers, hence $\overline{\mathcal{U}}_S(J)$ has the topology of a blown-up ruled surface. Moreover, the evaluation map for the extra marked point

$$\mathrm{ev}_{m+1} : \overline{\mathcal{U}}_S(J) \to M$$

is a map of positive degree $N := \deg(\mathrm{ev}_{m+1})$, and if $m \geq 1$, we have a collection of disjoint sections $\Sigma_1, \ldots, \Sigma_m \subset \overline{\mathcal{U}}_S(J)$ such that $[\Sigma_i] \cdot [\Sigma_i] = -N$ and ev_{m+1} maps Σ_i to the constraint point p_i for $i = 1, \ldots, m$. Applying Propositions 7.61 and 7.62, we now have:

Lemma 7.79. $b_2^+(M) = 1$. $\qquad \square$

Lemma 7.80. If $m \geq 1$, then $b_1(M) = 0$.

Proof. Choosing any of the sections $\Sigma_i \subset \overline{\mathcal{U}}_S(J)$, $H_1(\overline{\mathcal{U}}_S(J))$ is generated by loops on Σ_i, all of which are mapped to a point by ev_{m+1}. Since ev_{m+1} acts surjectively on rational homology, it follows that $H_1(M; \mathbb{Q}) = 0$. $\qquad \square$

Lemma 7.81. If $c_1([S]) > 3$, then the hypotheses of Theorem 7.34 are also satisfied for some other sphere S' with $c_1([S']) \in \{2, 3\}$.

Proof. Suppose $c_1([S]) > 3$, so $m > 1$, and suppose first that $\overline{\mathcal{M}}_S(J)$ does not contain any nodal curves. Then $\overline{\mathcal{U}}_S(J)$ has the topology of a ruled surface, implying (via Proposition 7.62) that $H_2(\overline{\mathcal{U}}_S(J))$ is a free group generated by the fiber class $[F]$ and any section, say $[\Sigma_1]$. Since $[\Sigma_1] \cdot [\Sigma_1] = -N$ but Σ_1 and Σ_2 are disjoint, they cannot be homologous, hence $[\Sigma_2] = [\Sigma_1] + k[F]$ for some nonzero integer k. But then

$$[\Sigma_2] \cdot [\Sigma_2] = ([\Sigma_1] + k[F]) \cdot ([\Sigma_1] + k[F]) = [\Sigma_1] \cdot [\Sigma_1] + 2k = -N + 2k \neq -N,$$

contradicting Lemma 7.55.

It follows that $\overline{\mathcal{M}}_S(J)$ contains at least one nodal curve; call its smooth components u^- and u^+. Both of these have nonnegative index and thus satisfy $c_1([u^\pm]) \geq 1$, but since $c_1([u^+]) + c_1([u^-]) = c_1([S]) \geq 4$, at least one of them (say u^+) must also satisfy $2 \leq c_1([u^+]) \leq c_1([S]) - 1$. We can then feed the J-holomorphic immersion $u^+ : S^2 \looparrowright M$ back into Theorem 7.34, and repeat this argument finitely many times until we find an immersed J-holomorphic sphere with c_1 equal to either 2 or 3. $\qquad\square$

The next observation is that in light of what has been proved so far, it will suffice to establish Theorem 7.34 in the case where (M, ω) is minimal. Indeed, if this is done, then for any (M, ω) satisfying the hypotheses of the theorem, we've shown already (Corollary 7.52) that there exists a J-holomorphic sphere with index at least 2 in some fixed homology class for every J in a dense subset of $\mathcal{J}_\tau(M, \omega)$. Lemmas 7.5 and 7.11 then imply that the blowdown $(\check{M}, \check{\omega})$ of (M, ω) along a maximal collection of disjoint exceptional spheres also satisfies the hypotheses of the theorem. But as soon as we find a nonnegative symplectically embedded sphere S in $(\check{M}, \check{\omega})$, we have one in (M, ω) as well, because one can perform the blowup along balls in $(\check{M}, \check{\omega})$ disjoint from S and deform the result symplectically to (M, ω).

With the preceding understood, we shall assume from now on that (M, ω) is minimal. Our goal is then to prove the following.

Lemma 7.82. *If (M, ω) is minimal, then there exists an embedded J-holomorphic sphere $u : S^2 \to M$ with $\mathrm{ind}(u) \geq 2$.*

The embedded sphere we seek will be found among the smooth components of nodal curves in $\overline{\mathcal{M}}_S(J)$, though we may need to change the original immersion S before it can be found (see Lemma 7.84 below). This will finish the proof, because by the adjunction formula, any embedded sphere of index at least 2 satisfies $[u] \cdot [u] \geq 0$. Note that by minimality, the adjunction formula also implies that no J-holomorphic sphere of index 0 is embedded.

We will prove the lemma by contradiction, thus from now on we impose the following:

Assumption 7.83. *(M, ω) admits no embedded J-holomorphic spheres.*

By Lemma 7.81, we can restrict our attention to the cases $m = 0$ and $m = 1$. Let us deal first with $m = 1$.

Lemma 7.84. *If $m = 1$, the almost complex structure $J \in \mathcal{J}_\tau(M, \omega)$ and the positive symplectic immersion $S \hookrightarrow M$ can be modified so that the following condition holds without loss of generality. Denote the finite set of nodal curves in $\overline{\mathcal{M}}_S(J)$ by $u_1, \ldots, u_n$, each u_i having two smooth components $u_i^\pm$ where $c_1([u_i^-]) = 1$ and $c_1([u_i^+]) = 2$, with the marked point situated on u_i^+. Then if $n > 0$, there exists $k \in \{1, \ldots, n\}$ such that:*

(1) $[u_k^+] \cdot [u_k^+] \leqslant [u_i^+] \cdot [u_i^+]$ *for all* $i = 1, \ldots, n$;
(2) $1 \leqslant [u_k^-] \cdot [u_k^+] \leqslant [u_i^-] \cdot [u_i^+]$ *for all* $i = 1, \ldots, n$.

Proof. The statement is vacuous if $n = 0$, so in this case we leave the immersion $S \hookrightarrow M$ as it is. If $n \geqslant 1$, let $p \in M$ denote the constraint point, and consider the following two sets, which are both finite by Lemma 7.42:

- $\mathcal{M}^1$ is the set of all curves in spaces of the form $\mathcal{M}_{0,0}^*(A; J)$ with $\langle [\omega], A \rangle \leqslant \langle [\omega], [S] \rangle$ and vir-dim $\mathcal{M}_{0,0}(A; J) = 0$;
- $\mathcal{M}^2$ is the set of all curves in spaces of the form $\mathcal{M}_{0,1}^*(A; J; p)$ with $\langle [\omega], A \rangle \leqslant \langle [\omega], [S] \rangle$ and vir-dim $\mathcal{M}_{0,1}(A; J; p) = 0$.

Note that both sets are nonempty since each nodal curve in $\overline{\mathcal{M}}_S(J)$ has one smooth component in $\mathcal{M}^1$ and the other in $\mathcal{M}^2$. The notation has been chosen so that any $u \in \mathcal{M}^i$ has $c_1([u]) = i$ for $i = 1, 2$. By assumption none of these curves are embedded, so the adjunction formula implies that they all satisfy

$$[u] \cdot [u] = 2\delta(u) + c_1([u]) - 2 \geqslant 2\delta(u) - 1 > 0.$$

We claim then that for any $u_1 \in \mathcal{M}^1$ and $u_2 \in \mathcal{M}^2$, $[u_1] \cdot [u_2] > 0$. Note that this intersection number must be nonnegative by positivity of intersections, since u_1 and u_2 are simple curves with distinct first Chern numbers and therefore have non-identical images. If $[u_1]$ and $[u_2]$ are linearly dependent in $H_2(M; \mathbb{R})$, then the claim follows from $[u_1] \cdot [u_1] > 0$. If on the other hand they are linearly independent but $[u_1] \cdot [u_2] = 0$, then $[u_1]$ and $[u_2]$ span a 2-dimensional subspace of $H_2(M; \mathbb{R})$ on which the intersection form is positive-definite, implying $b_2^+(M) \geqslant 2$ and thus contradicting Lemma 7.79.

Now pick $u_2 \in \mathcal{M}^2$ to minimize $[u_2] \cdot [u_2]$, and with this choice fixed, pick $u_1 \in \mathcal{M}^1$ to minimize $[u_1] \cdot [u_2]$. The genericity Assumption 7.41 imply that these two curves are transverse to each other, and they have at least one intersection since $[u_1] \cdot [u_2] > 0$. We can then pick one of the intersection points, let $A = [u_1] + [u_2]$, and consider the nodal curve $u \in \overline{\mathcal{M}}_{0,1}(A; J; p)$ whose smooth components are u_1 and u_2, with a single node placed at the chosen intersection point. By the gluing theorem (Corollary 2.39), a neighborhood of u in $\overline{\mathcal{M}}_{0,1}(A; J; p)$ is then a 2-parameter family of smooth J-holomorphic spheres degenerating to u, and the combination of Corollaries 2.26, 2.30 and 2.32 implies that an open and dense subset of these consists of positively symplectically immersed spheres $S' \looparrowright M$ that satisfy

$$c_1([S']) = c_1(A) = c_1([u_1]) + c_1([u_2]) = 3.$$

Fixing any one of these and using it to define the moduli space $\overline{\mathcal{M}}_{S'}(J)$ as before, this space contains the nodal curve u by construction, which satisfies the desired intersection relations.   □

The next lemma is one of the main steps in which Assumption 7.83 is crucial. Note that it is easy to think up examples in which embedded curves are allowed and the statement is false. The reason the lemma works is that the nonembedded curves in this context always have positive self-intersection number, making it harder to satisfy $b_2^+(M) = 1$ unless many of them are in the same homology class.

Lemma 7.85. *Given the conclusion of Lemma 7.84, the curves $u_1^-, \ldots, u_n^-$ are all homologous in $H_2(M; \mathbb{R})$.*

Proof. If not, then there exists $i \neq k$ such that $[S] = [u_i^-] + [u_i^+] = [u_k^-] + [u_k^+]$ but $[u_k^+] \neq [u_i^+] \in H_2(M; \mathbb{R})$. Since $c_1([u_k^+]) = c_1([u_i^+])$, these two classes cannot be linearly dependent. Lemma 7.84 gives

$$[u_k^+] \cdot [u_i^+] = [u_k^+] \cdot ([u_k^+] + [u_k^-] - [u_i^-])$$

$$= [u_k^+] \cdot [u_k^+] - \left([u_k^+] \cdot [u_i^-] - [u_k^+] \cdot [u_k^-] \right)$$

$$\leqslant [u_k^+] \cdot [u_k^+],$$

and thus

$$([u_k^+] \cdot [u_i^+])^2 \leqslant ([u_k^+] \cdot [u_k^+])^2 \leqslant ([u_k^+] \cdot [u_k^+])([u_i^+] \cdot [u_i^+]),$$

so by Exercise 7.57, the intersection form is positive-semidefinite on the 2-dimensional subspace of $H_2(M; \mathbb{R})$ spanned by $[u_k^+]$ and $[u_i^+]$. Since the intersection form is nonsingular, this implies via Exercise 7.58 that $b_2^+(M) \geqslant 2$ and thus contradicts Lemma 7.79.   □

Recall now that $H_2(\overline{\mathcal{U}}_S(J); \mathbb{R})$ is generated by the section $[\Sigma_1]$, the fiber class $[F]$ and the exceptional spheres in singular fibers which form the domains of the curves $u_1^-, \ldots, u_n^-$. In light of the above lemma and the fact that $(\mathrm{ev}_{m+1})_*[\Sigma_1] = 0$, the surjectivity of $(\mathrm{ev}_{m+1})_*$ now implies in the $m = 1$ case that

$$b_2(M) \in \{1, 2\}.$$

Moreover if $b_2(M) = 2$, then since $(\mathrm{ev}_{m+1})_* : H_2(\overline{\mathcal{U}}_S(J); \mathbb{R}) \to H_2(M; \mathbb{R})$ obviously has nontrivial kernel, we deduce that $\dim H_2(\overline{\mathcal{U}}_S(J); \mathbb{R}) \geqslant 3$ and therefore that the moduli space $\overline{\mathcal{M}}_S(J)$ contains at least one nodal curve. The rest of the job will be done by the Hirzebruch signature theorem.

Proof of Lemma 7.82 for $m = 1$. Recall from Lemmas 7.79 and 7.80 that $b_1(M) = 0$ and $b_2^+(M) = 1$. If additionally $b_2(M) = 1$, then Lemma 7.74 provides an integer $k \in \mathbb{Z}$ such that

$$[S] \cdot [S] = k^2 \quad \text{and} \quad c_1([S]) = 3k.$$

We already know $c_1([S]) = 3$, so this fixes $k = 1$ and thus $[S] \cdot [S] = 1$. But since S is not embedded, the adjunction formula gives

$$[S] \cdot [S] = 2\delta(S) + c_1([S]) - 2 = 2\delta(S) + 1 > 1,$$

which is a contradiction.

If instead $b_2(M) = 2$, then we know there exists a nodal curve in $\overline{\mathcal{M}}_S(J)$, one of whose smooth components u^- is an immersed and somewhere injective but non-embedded index 0 sphere with no marked points, so by the index and adjunction formulas it satisfies

$$c_1([u^-]) = 1 \quad \text{and} \quad [u^-] \cdot [u^-] = 2\delta(u^-) - 1 > 0.$$

Notice that $[u^-] \cdot [u^-]$ is odd, so by Lemma 7.75, there must exist integers k, ℓ such that

$$c_1([u^-]) = 3k - \ell \quad \text{and} \quad [u^-] \cdot [u^-] = k^2 - \ell^2.$$

But if $3k - \ell = 1$, then $k^2 - \ell^2 = k^2 - (3k - 1)^2 = -(8k^2 - 6k + 1)$, and the latter is only positive for $1/4 < k < 1/2$, hence not for any $k \in \mathbb{Z}$, so this is again a contradiction.   □

It remains only to deal with the $m = 0$ case. Here it is useful to observe that if $\overline{\mathcal{M}}_S(J)$ has nodal curves, then the result is already implied by our proof for $m \geq 1$:

Lemma 7.86. *If* $m = 0$ *and* $\overline{\mathcal{M}}_S(J)$ *contains a nodal curve, then the hypotheses of Theorem 7.34 are also satisfied with* $c_1([S]) = 3$.

Proof. By assumption, the smooth curves $u \in \mathcal{M}_S(J)$ are not embedded, so by the adjunction formula,

$$[u] \cdot [u] = 2\delta(u) + c_1([u]) - 2 = 2\delta(u) > 0.$$

Fix $u \in \mathcal{M}_S^{\text{good}}(J)$ and a nodal curve in $\overline{\mathcal{M}}_S(J)$ with smooth components u^- and u^+. The latter are both simple index 0 curves with no marked points and thus satisfy $c_1([u^\pm]) = 1$, and since $[u] \cdot [u] > 0$, at least one of them (say u^+) has nontrivial intersection with u. The definition of $\mathcal{M}_S^{\text{good}}(J)$ guarantees that u and

u^+ intersect transversely (see the paragraph preceding Lemma 7.50). Then picking an intersection point and forming a nodal curve of arithmetic genus 0 with smooth components u and u^+, the gluing theorem (Theorem 2.38) yields a 4-dimensional family of smooth J-holomorphic spheres u' that degenerate to this nodal curve and satisfy

$$c_1([u']) = c_1([u]) + c_1([u^+]) = 3.$$

Combining Corollaries 2.26, 2.30 and 2.32, almost all of the u' are positive symplectic immersions. □

Proof of Lemma 7.82 for $m = 0$. We've just shown that the lemma follows from the proof of the $m = 1$ case unless $\overline{\mathcal{M}}_S(J)$ has no nodal curves. This is the case in which $\pi_m : \overline{\mathcal{U}}_S(J) \to \overline{\mathcal{M}}_S(J)$ is a smooth S^2-bundle, and ev_{m+1} is a map of positive degree which sends the fiber to the class $[S] \in H_2(M)$, satisfying

$$c_1([S]) = 2 \quad \text{and} \quad [S] \cdot [S] = 2\delta(u)$$

for $u \in \mathcal{M}_S(J)$. Assuming there are no embedded J-holomorphic curves, $[S] \cdot [S]$ is then a positive even number. But Lemma 7.77 shows that no such class $[S]$ can exist, and this contradiction finishes the proof of Theorem 7.34. □

Chapter 8
Holomorphic Curves in Symplectic Cobordisms

Contact geometry is often called the "odd-dimensional cousin" of symplectic geometry, and there are many interesting problems that lie in the intersection of the two subjects. In this and the next chapter, we will discuss a few such problems that can be attacked with holomorphic curve methods, some of which take some noticeable inspiration from McDuff's results on rational and ruled symplectic 4-manifolds. Unlike previous chapters, our intention here is not to give a complete exposition, but rather to whet the reader's appetite and point to other sources for further reading. The subject of *symplectic field theory*, which emerges from this discussion as one of the natural tools to approach problems in contact topology, remains a subject of active research.

8.1 The Conjectures of Arnol'd and Weinstein

While Gromov's 1985 paper on pseudoholomorphic curves [Gro85] is often credited with launching the modern field of symplectic topology, many of Gromov's ideas were not without precedent. In Chap. 7 we outlined the *Gromov-Witten invariants*, a theory of enumerative invariants that arose from Gromov's work and were later formalized by Ruan [Rua96] and other authors (see [MS12] for a comprehensive survey). In principle, they are defined on a symplectic manifold (M, ω) by choosing a generic tame almost complex structure $J \in \mathcal{J}_\tau(M, \omega)$ and counting (with signs) the elements in 0-dimensional components of the moduli space of J-holomorphic curves $\mathcal{M}(J)$ satisfying suitable constraints. The fact that this leads to an invariant of the symplectic structure ω depends crucially on the *ellipticity* of the Cauchy-Riemann equation, which gives the moduli space $\mathcal{M}(J)$ the following properties:

- For generic J, $\mathcal{M}(J)$ is a smooth finite-dimensional manifold;
- $\mathcal{M}(J)$ admits a natural compactification $\overline{\mathcal{M}}(J)$ such that the "boundary strata" $\overline{\mathcal{M}}(J) \backslash \mathcal{M}(J)$ have virtual codimension at least 2;

© Springer International Publishing AG, part of Springer Nature 2018
C. Wendl, *Holomorphic Curves in Low Dimensions*, Lecture Notes
in Mathematics 2216, https://doi.org/10.1007/978-3-319-91371-1_8

- Any two generic $J_0, J_1 \in \mathcal{J}_\tau(M, \omega)$ are connected by a generic homotopy $\{J_s\}_{s \in [0,1]}$ of ω-tame almost complex structures such that the compactified parametric moduli space $\{(s, u) \mid s \in [0, 1], u \in \overline{\mathcal{M}}(J_s)\}$ defines a cobordism between $\overline{\mathcal{M}}(J_0)$ and $\overline{\mathcal{M}}(J_1)$.

This means that any topological property of the moduli space $\overline{\mathcal{M}}(J)$ that is invariant under cobordisms is independent of the choice of $J \in \mathcal{J}_\tau(M, \omega)$, though it may very well depend on ω, at least up to deformation. Seen in this light, the Gromov-Witten invariants are well defined for much the same reason that the degree of a smooth map between closed manifolds is well defined. Many of the important results in [Gro85] can be interpreted as computations of Gromov-Witten invariants, e.g. the *non-squeezing* theorem is essentially a corollary of such a computation on split symplectic manifolds of the form $S^2 \times M$, cf. [Gro85, 0.2.A'].

This idea for constructing invariants by counting solutions to an elliptic equation is older than the theory of pseudoholomorphic curves: invariants of this type had appeared earlier in the 1980s in the work of Donaldson on smooth 4-manifolds (see [DK90]). In Donaldson's work, both the setting and the PDE were different, but the basic outline sketched above is the same: instead of holomorphic curves in a symplectic manifold, one counts solutions to the *anti-self-dual Yang-Mills equations* on a principle bundle over a smooth 4-manifold, which yields invariants of the underlying smooth structure. The equations behind Donaldson's invariants came originally from quantum field theory, and related gauge-theoretic invariants in differential topology are still actively studied today, e.g. they inspired the later development of Seiberg-Witten theory (see e.g. [KM07]).

A parallel strand in the story of holomorphic curves and symplectic invariants involves the search for periodic orbits of Hamiltonian systems. One of the motivating problems in this field has been the Arnol'd conjecture, which concerns the minimum number of fixed points of a Hamiltonian symplectomorphism. Recall that given a symplectic manifold (M, ω) and a smooth family of smooth functions $\{H_t : M \to \mathbb{R}\}_{t \in \mathbb{R}}$, the (time-dependent) Hamiltonian vector field X_{H_t} is defined by[1]

$$\omega(X_{H_t}, \cdot) = -dH_t.$$

Its flow defines a 1-parameter family of symplectomorphisms, and a general symplectomorphism is called **Hamiltonian** if it is the time-1 flow of such a vector field. Though it is not immediately obvious from the definition, the Hamiltonian symplectomorphisms form a normal subgroup $\mathrm{Ham}(M, \omega)$ in the identity component of the group $\mathrm{Symp}(M, \omega)$ of symplectomorphisms, see e.g. [MS17, Exercise 3.1.14]. It is also not hard to show that any $\varphi \in \mathrm{Ham}(M, \omega)$ can be

---

[1]The literature is far from unanimous about the necessity of the minus sign in the formula $\omega(X_{H_t}, \cdot) = -dH_t$, and there are related interdependent sign issues that arise in defining Hamiltonian action functionals, the standard symplectic forms on $\mathbb{R}^{2n}$ and on cotangent bundles, and various other things that depend on these. For further discussion of these issues, see [Wenb] and [MS17, Remark 3.1.6].

written as the time-1 flow for a family of Hamiltonians H_t that is 1-periodic in t. Indeed, if φ_t is the flow of X_{H_t}, then for any smooth function $f : \mathbb{R} \to \mathbb{R}$ with $f(0) = 0$, $\widetilde{\varphi}_t := \varphi_{f(t)}$ is the Hamiltonian flow for $\widetilde{H}_t := f'(t)H_{f(t)}$, so that choosing $f : [0, 1] \to [0, 1]$ to be surjective, nondecreasing and constant near the end points produces a family $\{\widetilde{H}_t\}_{t \in [0,1]}$ that has the same time-1 flow as H_t but admits a smooth 1-periodic extension to $t \in \mathbb{R}$, cf. [MS17, Exercise 11.1.11]. In this way, the fixed point problem becomes a periodic orbit problem: writing

$$S^1 := \mathbb{R}/\mathbb{Z},$$

we can restrict to time-periodic Hamiltonians $H_t = H(t, \cdot)$ with $H : S^1 \times M \to \mathbb{R}$ and look for 1-periodic maps $z : S^1 \to M$ satisfying $\dot{z}(t) = X_{H_t}(z(t))$. Such an orbit z is called **nondegenerate** if the linearization of the time-1 flow at $z(0)$, the so-called *linearized return map*, does not have 1 as an eigenvalue. In this case it is necessarily isolated; in particular, if M is compact and all 1-periodic orbits are nondegenerate, then there are at most finitely many. Nondegeneracy is a generic condition, i.e. one can show that any time-dependent Hamiltonian admits a perturbation for which every 1-periodic orbit is nondegenerate. One version of the Arnol'd conjecture can then be expressed as follows.

Conjecture 8.1 (Arnol'd). Suppose $\{H_t : M \to \mathbb{R}\}_{t \in S^1}$ defines a time-dependent Hamiltonian system on a symplectic manifold (M, ω) such that all contractible 1-periodic orbits are nondegenerate. Then the number of contractible 1-periodic orbits is bounded below by the sum of the Betti numbers of M.

Exercise 8.2. Using Morse theory, try to show that Conjecture 8.1 is true for any time-independent Hamiltonian $H : M \to \mathbb{R}$ that is Morse and C^2-small. The 1-periodic orbits in this case are all constant, and located at the critical points of H. *For hints, see* [HZ94, P. 200].

One of the milestones in the study of the Arnol'd conjecture occurred slightly before pseudoholomorphic curves became a topic of interest: in 1982, Conley and Zehnder [CZ83b] proved that Conjecture 8.1 holds when (M, ω) is a standard symplectic torus $(\mathbb{T}^{2n}, \omega_{st}) := (\mathbb{R}^{2n}/\mathbb{Z}^{2n}, \omega_{st})$ for any $n \in \mathbb{N}$. Their proof was based on a variational principle for a functional on the space of contractible loops

$$C^\infty_{contr}(S^1, M) = \{z \in C^\infty(S^1, M) \mid z \text{ is homotopic to a constant}\}.$$

Denote the natural coordinates on $\mathbb{R}^{2n} = \mathbb{C}^n$ by $(z_1, \ldots, z_n)$, and let $\langle \, , \, \rangle_\mathbb{R}$ denote the standard (real) inner product. If $H_t : \mathbb{T}^{2n} \to \mathbb{R}$ is 1-periodic in t, then since any contractible loop $z : S^1 \to \mathbb{T}^{2n}$ can be lifted to a loop in the universal cover $\mathbb{R}^{2n}$, one can identify contractible 1-periodic orbits of the equation $\dot{z}(t) = X_{H_t}(z(t))$ with critical points of the functional

$$\mathcal{A}_H : C^\infty(S^1, \mathbb{R}^{2n}) \to \mathbb{R} : z \mapsto - \int_{S^1} \left(\frac{1}{2}\langle \dot{z}(t), iz(t)\rangle_\mathbb{R} - H_t(z(t)) \right) dt,$$

$$(8.1)$$

where two orbits are regarded as equivalent if they differ only by a translation in $\mathbb{Z}^{2n}$. Using Fourier series, the argument of Conley and Zehnder treats (8.1) as a smooth function on a Hilbert space and considers its negative gradient flow equation

$$\frac{du}{ds} + \nabla \mathcal{A}_H(u) = 0, \qquad (8.2)$$

where $u(s)$ is a function of a real variable $s \in \mathbb{R}$ with values in a suitable Hilbert space of loops in $\mathbb{R}^{2n}$. Note that the usual local existence/uniqueness theory for ordinary differential equations works in infinite-dimensional Banach spaces just as in the finite-dimensional case (see [Lan93, Chapter XIV]), so the flow determined by (8.2) is well defined. Now, if we imagine for a moment that gradient flow lines $u(s)$ take values in a compact space, then they will necessarily converge to critical points of $\mathcal{A}_H$ as $s \to \pm\infty$. Of course this is not true in general, as an infinite-dimensional Hilbert space of loops is quite far from being compact, but it turns out that gradient flow lines still converge asymptotically to critical points whenever they satisfy a suitable "energy" bound. The space of finite-energy gradient flow lines can thus be used to deduce existence results for contractible 1-periodic orbits.

Exercise 8.3. Writing $z_j = p_j + iq_j$ for $j = 1, \ldots, n$, the standard Liouville form on $\mathbb{R}^{2n}$ is defined by $\lambda_{\text{st}} = \frac{1}{2}\sum_{j=1}^{n}(p_j\, dq_j - q_j\, dp_j)$. Show that the first term in (8.1) can be rewritten as $-\int_{S^1} z^*\lambda_{\text{st}}$ and thus (by Stokes' theorem) equals $-\int_{\mathbb{D}^2} u^*\omega_{\text{st}}$ for any smooth map $u : \mathbb{D}^2 \to \mathbb{R}^{2n}$ with $u|_{\partial\mathbb{D}^2} = z$.

Using Exercise 8.3, one can generalize the action functional (8.1) to any symplectic manifold (M, ω) which is **symplectically aspherical**, i.e. satisfies $\int u^*\omega = 0$ for all smooth maps $u : S^2 \to M$. Note that this is a homological condition on $[\omega] \in H^2_{\text{dR}}(M)$: it means that

$$\langle [\omega], A \rangle = 0$$

whenever $A \in H_2(M)$ is in the image of the Hurewicz homomorphism $\pi_2(M) \to H_2(M)$. Under this condition, one can choose for any contractible loop $z : S^1 \to M$ an arbitrary smooth extension $u : \mathbb{D}^2 \to M$ with $u|_{\partial\mathbb{D}^2} = z$, and define the expression

$$\mathcal{A}_H : C^\infty_{\text{contr}}(S^1, M) \to \mathbb{R} : z \mapsto -\int_{\mathbb{D}^2} u^*\omega + \int_{S^1} H_t(z(t))\, dt \qquad (8.3)$$

independently of this choice. The critical points of this functional are again solutions to $\dot{z} = X_{H_t}(z)$. Having written down the functional, it is however not at all obvious how the argument of Conley and Zehnder might be extended beyond $(\mathbb{T}^{2n}, \omega_{\text{st}})$, as a more general symplectic manifold (M, ω) does not furnish any reasonable Hilbert space setting in which to make sense of the gradient flow equation (8.2).

A beautiful solution to this problem was found in 1988 by Floer [Flo88a], using ideas about holomorphic curves that Gromov had introduced in the mean time. Floer's idea was to take (8.2) less literally, and instead of treating it as an ODE on an infinite-dimensional space, regard it as a PDE satisfied by a map $u : \mathbb{R} \times S^1 \to M$, where for each $s \in \mathbb{R}$, $u(s, \cdot) : S^1 \to M$ is the loop formerly denoted by $u(s)$, and $\frac{du}{ds}$ is replaced by the partial derivative $\partial_s u$. If $\mathcal{A}_H$ is to be regarded formally as a functional on an infinite-dimensional manifold of loops, then writing down the gradient requires first choosing a (formal) Riemannian metric: one natural way to do this is by choosing a family of ω-compatible almost complex structures $\{J_t\}_{t \in S^1}$, so that $\omega(\cdot, J_t \cdot)$ defines a metric on M for each t. Then if we pick any loop $z \in C^\infty_{\mathrm{contr}}(S^1, M)$ and think of the space of sections $\Gamma(z^* TM)$ as the tangent space to $C^\infty_{\mathrm{contr}}(S^1, M)$ at z, we can define a formal Riemannian metric on $C^\infty_{\mathrm{contr}}(S^1, M)$ by

$$\langle \xi, \eta \rangle = \int_{S^1} \omega \left(\xi(t), J_t(z(t)) \, \eta(t) \right) \, dt$$

for $\xi, \eta \in \Gamma(z^* TM)$. Using this to write down $\nabla \mathcal{A}_H$, the gradient flow equation (8.2) now becomes the first-order elliptic PDE

$$\frac{\partial u}{\partial s} + J_t(u(s, t)) \frac{\partial u}{\partial t} - J_t(u(s, t)) X_{H_t}(u(s, t)) = 0 \tag{8.4}$$

for a map $u : \mathbb{R} \times S^1 \to M$. Except for the time-dependence of J and the extra zeroth-order term, this is essentially a nonlinear Cauchy-Riemann equation on the standard Riemann cylinder

$$(\mathbb{R} \times S^1, i) := \mathbb{C}/i\mathbb{Z},$$

and just as for ordinary J-holomorphic curves, one can analyze the moduli spaces of solutions using elliptic regularity theory, Fredholm theory, and a variation on the "bubbling off analysis" that appears in Gromov's compactness theorem. Since $\mathbb{R} \times S^1$ is not compact, however, one must first gain some control over the asymptotic behavior of solutions $u : \mathbb{R} \times S^1 \to M$, and for this, a bound of the form

$$\int_{\mathbb{R} \times S^1} u^* \omega < \infty \tag{8.5}$$

turns out to be sufficient: any solution to (8.4) with finite energy in this sense has the property that $u(s, \cdot)$ converges to a 1-periodic orbit of X_{H_t} as $s \to \pm\infty$.

The compactness theory for the Floer equation (8.4) resembles Gromov's theory for J-holomorphic curves but has an important additional feature. Aside from the usual "bubbling" phenomena familiar from Gromov's theory, a sequence of Floer cylinders may also converge to a so-called *broken* cylinder (see Fig. 8.1). This is a finite ordered set of finite-energy Floer cylinders $u_1, \ldots, u_N$ with the property that for each $i = 1, \ldots, N - 1$, there are matching periodic orbits

$$\lim_{s \to -\infty} u_i(s, \cdot) = \lim_{s \to +\infty} u_{i+1}(s, \cdot).$$

Fig. 8.1 A sequence of
Floer cylinders can
degenerate to a "broken Floer
cylinder," consisting of
multiple Floer cylinders
connected to each other along
additional periodic orbits

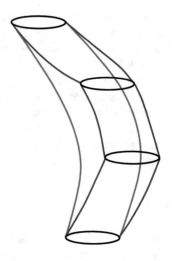

The breaking phenomenon gives rise to a fundamental algebraic difference between
Floer theory and Gromov-Witten theory. In the latter, counts of holomorphic
curves are invariant because the "degenerations" are confined to codimension 2
subsets of the compactified moduli space. We saw various manifestations of this in
Chaps. 6 and 7, where 2-dimensional moduli spaces of holomorphic spheres ended
up filling entire symplectic 4-manifolds in spite of nodal degenerations, because
the latter turned out to be a discrete phenomenon that could happen only finitely
many times. Things are quite different in Floer theory, because the breaking of
Floer cylinders is a *codimension one* phenomenon: this has the consequence that
counts of Floer cylinders do not generally define invariants, as the counts may
be different for distinct choices of auxiliary data. One can however use a count
of Floer cylinders to define a chain complex such that the resulting *homology* is
invariant, leading to the definition of **Floer homology**. Floer's proof of the Arnol'd
conjecture for a wide range of symplectic manifolds rests on the theorem that Floer
homology is well defined and isomorphic to the usual singular homology of M.
For a good survey on Floer homology for Hamiltonian symplectomorphisms, see
[Sal99] or [AD14]. Note that the ideas behind Floer homology are by no means
confined to the realm of holomorphic curves: Floer simultaneously defined a gauge-
theoretic version for smooth 3-manifolds known as *instanton homology* ([Flo88b],
see also [Don02]), and since then, an abundance of "Floer-type" theories have
proliferated in various settings. More recent prominent examples include Heegaard
Floer homology [OS04b] and Seiberg-Witten Floer homology [KM07].

It was soon realized that the connection between holomorphic curves and
periodic orbits arising in Floer homology could also be exploited to study a well-
known dynamical question in contact topology. The Weinstein conjecture had
emerged in the late 1970s from studies of *autonomous* (i.e. time-independent)
Hamiltonian systems restricted to hypersurfaces of constant energy. Recall that if
$H : M \rightarrow \mathbb{R}$ is a time-independent Hamiltonian function on a symplectic manifold
(M, ω), then the flow of X_H preserves every level set of H. Moreover, any two

Hamiltonians with a matching level set $\Sigma \subset M$ induce the same orbits on Σ up to parametrization, as the direction of any Hamiltonian vector field restricted to Σ is determined by the **characteristic line field**

$$\ell_\Sigma := \ker(\omega|_{T\Sigma}) = \{X \in T\Sigma \mid \omega(X, \cdot)|_{T\Sigma} = 0\}.$$

The question of whether a given hypersurface admits periodic orbits therefore depends only on the symplectic geometry of a neighborhood of that hypersurface, not on a choice of Hamiltonian. There are easy examples of hypersurfaces in a symplectic manifold that have no periodic orbits:

Example 8.4. Let M denote the quotient of $S^1 \times \mathbb{R} \times \mathbb{T}^2$ by the $\mathbb{Z}$-action generated by the diffeomorphism

$$\psi : S^1 \times \mathbb{R} \times \mathbb{T}^2 \to S^1 \times \mathbb{R} \times \mathbb{T}^2 : (s, t, \phi, \theta) \mapsto (s, t + 1, \phi + c, \theta)$$

for some $c \in \mathbb{R}$. The symplectic form $\omega := ds \wedge dt + d\phi \wedge d\theta$ satisfies $\psi^*\omega = \omega$ and thus descends to a symplectic form on M. For any fixed $s \in S^1$, let $\Sigma_s \subset M$ denote the projection to $(S^1 \times \mathbb{R} \times \mathbb{T}^2)/\mathbb{Z}$ of the hypersurface $\{s\} \times \mathbb{R} \times \mathbb{T}^2 \subset S^1 \times \mathbb{R} \times \mathbb{T}^2$; this is diffeomorphic to the mapping torus of the diffeomorphism $\mathbb{T}^2 \to \mathbb{T}^2 : (\phi, \theta) \mapsto (\phi + c, \theta)$. The vector field ∂_t on $S^1 \times \mathbb{R} \times \mathbb{T}^2$ also satisfies $\psi_*\partial_t = \partial_t$ and thus descends to a vector field on M whose flow preserves Σ_s, and it generates the characteristic line field on Σ_s. Periodic orbits of this vector field on Σ_s correspond to periodic points of the diffeomorphism $(\phi, \theta) \mapsto (\phi + c, \theta)$ on $\mathbb{T}^2$, so in particular, there are none if $c \notin \mathbb{Q}$.

In contrast to the above example, Weinstein [Wei78] and Rabinowitz [Rab78] discovered in 1978 that certain classes of hypersurfaces *always* admit closed orbits. For example:

Theorem 8.5 (Rabinowitz [Rab78]). *Every star-shaped hypersurface in* $(\mathbb{R}^{2n}, \omega_{st})$ *admits a periodic orbit.*

The key symplectic feature of the "star-shaped" condition is the following. Using the coordinates $z_j = p_j + iq_j$ with $j = 1, \ldots, n$ on $\mathbb{C}^n = \mathbb{R}^{2n}$, consider the radial vector field

$$V = \frac{1}{2} \sum_{j=1}^n \left(p_j \frac{\partial}{\partial p_j} + q_j \frac{\partial}{\partial q_j} \right). \tag{8.6}$$

This satisfies $\mathcal{L}_V \omega_{st} = \omega_{st}$, which means that its flow φ_V^t is a *symplectic dilation*, that is, $(\varphi_V^t)^*\omega_{st} = e^t \omega_{st}$. A vector field with this property on a symplectic manifold is in general called a **Liouville vector field**. Relatedly, its ω_{st}-dual 1-form is the **standard Liouville form**

$$\lambda_{st} := \omega_{st}(V, \cdot) = \frac{1}{2} \sum_{j=1}^n (p_j \, dq_j - q_j \, dp_j),$$

Fig. 8.2 A star-shaped
hypersurface in $\mathbb{R}^{2n}$ generates
a 1-parameter family of
hypersurfaces with the same
Hamiltonian orbits up to
parametrization

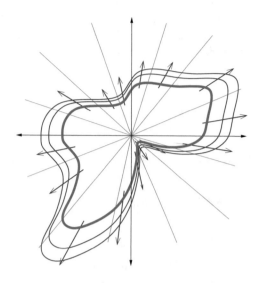

which satisfies $d\lambda_{st} = \omega_{st}$. A closed hypersurface $\Sigma \subset \mathbb{R}^{2n}$ is then star-shaped if
and only if V is everywhere positively transverse to Σ (see Fig. 8.2), and this is true
if and only if the 1-form $\alpha := \lambda_{st}|_{T\Sigma}$ satisfies

$$\alpha \wedge (d\alpha)^{n-1} > 0 \quad \text{on } \Sigma. \tag{8.7}$$

Any 1-form on an oriented $(2n - 1)$-dimensional manifold Σ satisfying (8.7)
is called a (positive) **contact form**, and since it is necessarily nowhere zero,
it determines a smooth co-oriented hyperplane field $\xi := \ker \alpha \subset T\Sigma$. A
co-oriented hyperplane field obtained in this way is called a (positive and co-
oriented) **contact structure**, and this makes the pair (Σ, ξ) a **contact manifold**.
Isomorphisms between contact manifolds are called contactomorphisms: specifically, an orientation-preserving diffeomorphism $\varphi : \Sigma \to \Sigma'$ is called a
contactomorphism from (Σ, ξ) to (Σ', ξ') if it satisfies $\varphi_*\xi = \xi'$ and also
preserves co-orientations. Given contact forms α, α' with $\xi = \ker \alpha$ and $\xi' = \ker \alpha'$,
this is equivalent to the condition

$$\varphi^*\alpha' = f\alpha$$

for an arbitrary smooth function $f : \Sigma \to (0, \infty)$. The contact form α on Σ also
determines a distinguished vector field R_α via the conditions

$$d\alpha(R_\alpha, \cdot) \equiv 0, \qquad \alpha(R_\alpha) \equiv 1.$$

This is called a **Reeb vector field**, and as the exercise below demonstrates, its orbits
(up to parametrization) are precisely the orbits of any Hamiltonian that has Σ as a
regular level set.

Exercise 8.6. Assume $\Sigma \subset \mathbb{R}^{2n}$ is a star-shaped hypersurface, and write $\alpha :=$ $\lambda_{st}|_{T\Sigma}$ with its induced Reeb vector field R_α as described above. Let φ_V^t denote the flow of the radial Liouville vector field V on $\mathbb{R}^{2n}$, and for some $\epsilon > 0$, consider the embedding

$$\Phi : (-\epsilon, \epsilon) \times \Sigma \hookrightarrow \mathbb{R}^{2n} : (t, z) \mapsto \varphi_V^t(z).$$

(a) Show that $\Phi^*\lambda_{st} = e^t\alpha$, hence $\Phi^*\omega_{st} = d(e^t\alpha)$.
(b) Suppose $H : \mathbb{R}^{2n} \to \mathbb{R}$ is a smooth function such that $H \circ \Phi(t, z) = e^t$ for $(t, z) \in (-\epsilon, \epsilon) \times \Sigma$. Show that under the above identification of a neighborhood of Σ in $\mathbb{R}^{2n}$ with $(-\epsilon, \epsilon) \times \Sigma$, the induced Hamiltonian vector field near Σ takes the form $X_H = R_\alpha$.
(c) Show that if $H : \mathbb{R}^{2n} \to \mathbb{R}$ is any smooth function that has Σ as a regular level set with $dH > 0$ in the outward direction, then the restriction of X_H to Σ takes the form $f R_\alpha$ for some smooth function $f : \Sigma \to (0, \infty)$.

Remark: all of the above depends only on the fact that V is transverse to Σ and satisfies $\mathcal{L}_V\omega_{st} = \omega_{st}$.

Exercise 8.7. Show that if (Σ, ξ) is any contact manifold with a contact form α, then $(\mathbb{R} \times \Sigma, d(e^t\alpha))$ is a symplectic manifold in which the unit vector field ∂_t in the $\mathbb{R}$-direction is a Liouville vector field. Show moreover that for any smooth function $f : M \to (0, \infty)$, the contact form $f\alpha$ is realized as the restriction of $\iota_{\partial_t} d(e^t\alpha)$ to the embedding

$$\Sigma \hookrightarrow \mathbb{R} \times \Sigma : p \mapsto (\log f(p), p).$$

Conclude from this that for any two contact forms α and α' defining the same contact structure ξ, the manifolds $(\mathbb{R} \times \Sigma, d(e^t\alpha))$ and $(\mathbb{R} \times \Sigma, d(e^t\alpha'))$ are symplectomorphic. We call these the **symplectization** of (Σ, ξ).

Exercise 8.8. The definition of "symplectization" given in the previous exercise is a bit clumsy, as one must first choose a contact form α to write it down and then show that the result depends only on the contact structure $\xi = \ker\alpha$ up to symplectomorphism. Here is a more natural construction of the same object, without the need for a choice of contact form. Given M with a co-oriented contact structure ξ, let $S\xi \subset T^*M$ denote the subset consisting of nonzero covectors $p \in T^*M$ such that $\ker p = \xi$ and $p(X) > 0$ for every $X \in TM$ positively transverse to ξ. Show that if T^*M is given its canonical symplectic structure, then $S\xi \subset T^*M$ is a symplectic submanifold, and any choice of contact form α induces a natural symplectomorphism of $S\xi$ to $(\mathbb{R} \times M, d(e^t\alpha))$.

The notion of a star-shaped hypersurface in $(\mathbb{R}^{2n}, \omega_{st})$ now admits the following natural generalization to any symplectic manifold (M, ω). We say that an oriented hypersurface $\Sigma \subset (M, \omega)$ is (symplectically) **convex**,[2] or of **contact type**, if a neighborhood of Σ in M admits a Liouville vector field that is positively transverse to Σ. Similarly, if $\partial M \neq \varnothing$, we say that (M, ω) has **convex** (or **contact-type**) **boundary** if ∂M with its natural boundary orientation is a contact-type hypersurface, meaning there is a Liouville vector field near ∂M that points transversely outwards. If instead there is a Liouville vector field pointing transversely inwards, we say that (M, ω) has **concave boundary**. In general, a symplectic manifold with boundary may have some convex and some concave boundary components, and some that are neither.

Exercise 8.9. Assume (M, ω) is a symplectic manifold, $\Sigma \subset M$ is an oriented hypersurface and V is a vector field defined on a neighborhood of Σ in M. Consider the ω-dual 1-form

$$\lambda = \omega(V, \cdot).$$

Show that $\mathcal{L}_V \omega = \omega$ if and only if $d\lambda = \omega$. Moreover, assuming $d\lambda = \omega$, show that V is positively/negatively transverse to Σ if and only if $\lambda \wedge (d\lambda)^{n-1}|_{T\Sigma}$ defines a positive/negative volume form on Σ, i.e. λ restricts to Σ as a positive/negative contact form.

Proposition 8.10. *Suppose (M, ω) is a compact $2n$-dimensional symplectic manifold with boundary.*

> *(1) If $n = 1$, then every boundary component of (M, ω) is both convex and concave.*
> *(2) If $n > 1$, then no boundary component of (M, ω) is both convex and concave.*

Proof. Suppose (M, ω) is 2-dimensional and $\Sigma \subset \partial M$ is a boundary component. Then Σ admits a collar neighborhood $\mathcal{N}(\Sigma) \subset M$ with

$$(\mathcal{N}(\Sigma), \omega) \cong ((-\epsilon, 0] \times S^1, ds \wedge dt)$$

for sufficiently small $\epsilon > 0$, with $s \in (-\epsilon, 0]$ and $t \in S^1 = \mathbb{R}/\mathbb{Z}$. We thus have $\omega = d\lambda_+ = d\lambda_-$ in $\mathcal{N}(\Sigma)$, where

$$\lambda_+ := (s + 1)\, dt, \quad \text{and} \quad \lambda_- := (s - 1)\, dt,$$

so $\lambda_+|_{T(\partial M)} = dt$ is a positive contact form on Σ, implying via Exercise 8.9 that Σ is convex. (Note that a 1-form on S^1 is a positive contact form if and only if it is positive.) Similarly, $\lambda_-|_{T(\partial M)} = -dt$ is a negative contact form, implying via Exercise 8.9 that Σ is concave. In particular, the ω-dual vector fields to λ_+ and λ_- give Liouville vector fields that point transversely outward and inward respectively.

---

[2] Note that a symplectically convex hypersurface in $(\mathbb{R}^{2n}, \omega_{st})$ need not be geometrically convex in the usual sense, e.g. every star-shaped hypersurface is symplectically convex.

The following argument for the case dim $M = 2n \geqslant 4$ was explained to me by Janko Latschev. Arguing by contradiction, suppose ω admits two primitives λ_+ and λ_- near Σ such that if we write $\alpha_\pm := \lambda_\pm|_{TM}$,

$$\alpha_+ \wedge (d\alpha_+)^{n-1} > 0 \quad \text{and} \quad \alpha_- \wedge (d\alpha_-)^{n-1} < 0.$$

By assumption, $d\lambda_+ = d\lambda_- = \omega$, thus $d\alpha_+ = d\alpha_-$ and in particular $\ker d\alpha_+ = \ker d\alpha_-$, so the Reeb vector fields R_{α_+} and R_{α_-} on Σ are colinear. Moreover, they point in opposite directions: this follows from the orientation difference above, since $d\alpha_+ = d\alpha_-$ implies that projection along the Reeb direction defines an orientation-*preserving* bundle isomorphism between the two contact hyperplane fields. Now define a closed 1-form on Σ by

$$\beta = \alpha_+ - \alpha_-.$$

We then have $\beta(R_{\alpha_+}) > 0$, thus

$$\int_\Sigma \beta \wedge (d\alpha_+)^{n-1} > 0. \tag{8.8}$$

But $\beta \wedge (d\alpha_+)^{n-1}$ is an exact form, as

$$\beta \wedge d\alpha_+ \wedge \ldots \wedge d\alpha_+ = -d\left(\beta \wedge \alpha_+ \wedge d\alpha_+ \wedge \ldots \wedge d\alpha_+\right),$$

so (8.8) violates Stokes' theorem.   □

As we saw in Exercise 8.9, a choice of Liouville vector field V near a contact-type hypersurface $\Sigma \subset (M, \omega)$ induces a primitive $\lambda := \omega(V, \cdot)$ of ω whose restriction to Σ is a contact form, and the induced Reeb vector field generates the orbits of any Hamiltonian having Σ as a regular level set. As Exercise 8.6 shows, a contact-type hypersurface $\Sigma \subset (M, \omega)$ always belongs to a whole 1-parameter family of contact-type hypersurfaces whose Hamiltonian orbits are all the same (Fig. 8.2). This provides a good intuitive reason to believe Theorem 8.5: for any contact-type hypersurface $\Sigma \subset (M, \omega)$, if one can find closed Hamiltonian orbits on a *nearby* hypersurface in the 1-parameter family containing Σ, then this implies a closed orbit on Σ itself. Hence it suffices to prove an "almost existence" result, e.g. that in any 1-parameter family of hypersurfaces, almost all of them (or even just a dense subset of them) admit closed orbits. There are indeed results of this type on $(\mathbb{R}^{2n}, \omega_{st})$, and they provide one path to a proof of Theorem 8.5; see [HZ94] for details. More generally, it is now not so hard to believe that *every* contact-type hypersurface in every symplectic manifold might admit closed orbits. Since every contact manifold can be realized as a contact-type hypersurface in its own symplectization (cf. Exercise 8.7), that statement would be equivalent to the following:

Conjecture 8.11 (Weinstein). For every closed manifold Σ with a contact form α, the Reeb vector field R_α admits a periodic orbit.

The Weinstein conjecture was formulated in the late 1970s and was proved in dimension three by Taubes [Tau07] in 2006, using the deep relationship between Seiberg-Witten theory and holomorphic curves. In higher dimensions, there is no Seiberg-Witten theory and the Weinstein conjecture remains open, with sporadic results known in special cases, most of which are higher-dimensional generalizations of holomorphic curve arguments that were already known in dimension three before the work of Taubes. We will discuss a few of these earlier 3-dimensional results in Chap. 9.

8.2 Symplectic Cobordisms and Fillings

Observe that if $\Sigma \subset (M, \omega)$ is a contact-type hypersurface, then the Liouville vector field V transverse to Σ is far from unique: indeed, any C^1-small function $H : M \to \mathbb{R}$ yields another Liouville vector field $V + X_H$ that is also transverse to Σ. As a consequence, the contact form induced on Σ is also not unique, in fact the space of contact forms on Σ that arise in this way is very large, but the following exercise shows that it is topologically quite simple.

Exercise 8.12. Show that for any oriented hypersurface Σ in a symplectic manifold (M, ω), the space of Liouville vector fields positively transverse to Σ is convex.

By a basic result in contact geometry called *Gray's stability theorem* (see e.g. [Gei08] or [Mas14]), any 1-parameter family of contact forms on a closed manifold yields an *isotopy* of their underlying contact structures. Combining this fact with Exercise 8.12 gives a partial answer to the question of why we are often interested in studying contact *structures* rather than contact forms:

Proposition 8.13. *On any closed contact-type hypersurface $\Sigma \subset (M, \omega)$, the contact structure induced on Σ is uniquely determined up to isotopy by the germ of ω near Σ.* $\square$

In light of this, one can speak of closed convex and concave boundary components of a symplectic manifold as being contact manifolds in a natural way. Given two (possibly disconnected) closed contact manifolds $(M_{\pm}, \xi_{\pm})$, we say that a compact symplectic manifold (W, ω) is a (strong) **symplectic cobordism from** (M_-, ξ_-) **to** (M_+, ξ_+)[3] if

$$\partial W = -M_- \amalg M_+,$$

with M_- a concave boundary component carrying the induced contact structure ξ_- and M_+ a convex boundary component carrying the induced contact structure ξ_+

---

[3]This terminology varies among different authors: some would describe what we are defining here as a "symplectic cobordism from (M_+, ξ_+) to (M_-, ξ_-)." This difference of opinion can probably only be resolved on the battlefield.

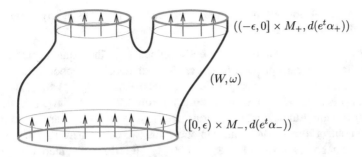

Fig. 8.3 A symplectic cobordism with concave boundary (M_-, ξ_-) and convex boundary (M_+, ξ_+), with symplectic collar neighborhoods defined by flowing along Liouville vector fields near the boundary

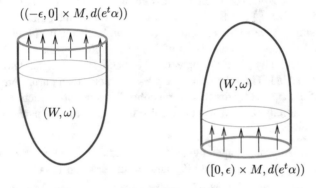

Fig. 8.4 A symplectic filling and symplectic cap respectively of a contact manifold (M, ξ) with contact form α

(see Fig. 8.3). We can also abbreviate this by writing

$$\partial(W, \omega) = (-M_-, \xi_-) \amalg (M_+, \xi_+).$$

If $M_- = \emptyset$, then we call (W, ω) a (strong) **symplectic filling** of (M_+, ξ_+), and if $M_+ = \emptyset$, we call it a (strong) **symplectic cap** of (M_-, ξ_-), see Fig. 8.4. We say also that (W, ω) is an **exact symplectic cobordism** or **Liouville cobordism** (or an **exact/Liouville filling** in the case $M_- = \emptyset$) if the Liouville vector field that determines $\xi_\pm$ on $M_\pm$ can be arranged to extend to a *global* Liouville vector field on (W, ω); in this case the ω-dual of this Liouville field gives a global primitive of ω which restricts to positive and negative contact forms on M_+ and M_- respectively. By Exercise 8.6, one can use flows of Liouville vector fields near $M_\pm$ to construct symplectic collar neighborhoods $\mathcal{N}(M_+), \mathcal{N}(M_-) \subset W$ of M_+ and M_- respectively in the form

$$
\begin{aligned}
(\mathcal{N}(M_+), \omega) &\cong \left((-\epsilon, 0] \times M_+, d(e^t \alpha_+)\right), \\
(\mathcal{N}(M_-), \omega) &\cong \left([0, \epsilon) \times M_-, d(e^t \alpha_-)\right),
\end{aligned}
\tag{8.9}
$$

where $\alpha_\pm$ denote contact forms on $M_\pm$ defined as restrictions of any primitive $\lambda = \omega(V, \cdot)$ determined by a Liouville vector field V.

Remark 8.14. The collar neighborhoods defined near the boundary of a strong symplectic cobordism provide a natural way of attaching cobordisms together: if (W_{01}, ω_{01}) is a strong symplectic cobordism from (M_0, ξ_0) to (M_1, ξ_1) and (W_{12}, ω_{12}) is a strong symplectic cobordism from (M_1, ξ_1) to (M_2, ξ_2), then after possibly rescaling the symplectic forms and deforming them near M_1 so that the induced contact forms match, one can attach the two cobordisms to define a strong symplectic cobordism from (M_0, ξ_0) to (M_2, ξ_2). It follows that there is a transitive relation $\preccurlyeq$ defined on the contactomorphism classes of closed contact manifolds in any given dimension by writing $(M, \xi) \preccurlyeq (M', \xi')$ whenever there exists a strong cobordism from (M, ξ) to (M', ξ'). Note that unlike the situation for topological cobordisms, this is *not* an equivalence relation: the existence of a strong symplectic cobordism from (M, ξ) to (M', ξ') does not imply one in the other direction, since one cannot simply reverse the orientation of a symplectic manifold. For example, it is known that all closed contact 3-manifolds admit symplectic caps [EH02], hence $(M, \xi) \preccurlyeq \varnothing$, but many are not symplectically fillable (see Sect. 9.2) and thus do not satisfy $\varnothing \preccurlyeq (M, \xi)$. The relation $\preccurlyeq$ is a *pre-order*, but not a partial order, as there are plenty of examples of pairs that admit cobordisms in both directions but are not contactomorphic, e.g. [EH02] also proves that this is true for any pair of overtwisted contact 3-manifolds.

Example 8.15. Any star-shaped hypersurface $M \subset (\mathbb{R}^{2n}, \omega_{st})$ is the symplectically convex boundary of a star-shaped domain $W \subset \mathbb{R}^{2n}$ and thus inherits a natural contact structure. Identifying M with the unit sphere $S^{2n-1} = \partial \overline{B}^{2n}$ in the obvious way, the fact that all star-shaped domains can be deformed smoothly to each other implies (via Gray's stability theorem) that the resulting contact structure on S^{2n-1} is uniquely determined up to isotopy: we call this the **standard contact structure** on S^{2n-1}. If we write the standard Liouville form in the usual coordinates $(p_1, q_1, \ldots, p_n, q_n) \in \mathbb{R}^{2n}$ and restrict to the unit sphere, we have

$$\xi_{st} = \ker \alpha_{st}, \qquad \alpha_{st} := \left. \sum_{j=1}^{n} (p_j \, dq_j - q_j \, dp_j) \right|_{TS^{2n-1}}.$$

By this definition, every star-shaped domain in $(\mathbb{R}^{2n}, \omega_{st})$ is an exact symplectic filling of (S^{2n-1}, ξ_{st}) after identifying the latter with the boundary of the domain via a suitable contactomorphism.

Example 8.16. For any closed smooth n-manifold L, the cotangent bundle T^*L has a canonical Liouville form that can be written locally as follows: given local coordinates $(q_1, \ldots, q_n)$ on some neighborhood in L and denoting the induced coordinates on the fibers of T^*L over this neighborhood by $(p_1, \ldots, p_n)$, we have

$$\lambda_{can} = \sum_{j=1}^{n} p_j \, dq_j.$$

We then define the canonical symplectic form on T^*L by $\omega_{\mathrm{can}} = d\lambda_{\mathrm{can}} = \sum_j dp_j \wedge dq_j$. The Liouville vector field ω_{can}-dual to λ_{can} can be written in these same coordinates as

$$V_{\mathrm{can}} = \sum_{j=1}^{n} p_j \frac{\partial}{\partial p_j},$$

so it points radially outward in each fiber and is thus transverse to any closed hypersurface in T^*L whose intersection with every fiber is star-shaped. We shall refer to any domain bounded by a hypersurface of this type as a **star-shaped domain in** T^*L. It follows that such domains are exact symplectic fillings, and since all star-shaped hypersurfaces are isotopic, the induced contact structure on the boundary is uniquely determined up to isotopy. To write this down more concretely, one can choose a Riemannian metric on L and consider the unit disk bundle

$$\mathbb{D}T^*L = \left\{ p \in T^*L \,\middle|\, |p| \leqslant 1 \right\},$$

which is an exact symplectic filling of the **unit cotangent bundle**

$$\mathbb{S}T^*L = \left\{ p \in T^*L \,\middle|\, |p| = 1 \right\}.$$

We define the **canonical contact form** α_{can} on $\mathbb{S}T^*L$ as the restriction of λ_{can}, and set

$$\xi_{\mathrm{can}} = \ker \alpha_{\mathrm{can}}.$$

One can check that the Reeb flow defined by α_{can} on $\mathbb{S}T^*L$ is the natural lift of the geodesic flow determined by the chosen metric.

The following special case will arise in Chap. 9. Let $L = \mathbb{T}^2 = \mathbb{R}^2/\mathbb{Z}^2$, and fix the standard flat metric. Since $T^*\mathbb{T}^2$ is a trivial bundle, we can identify it with $\mathbb{T}^2 \times \mathbb{R}^2$ and define global "coordinates" (q_1, q_2, p_1, p_2) with $q_1, q_2 \in S^1 = \mathbb{R}/\mathbb{Z}$, such that $\lambda_{\mathrm{can}} = p_1 \, dq_1 + p_2 \, dq_2$.[4] We then have a natural diffeomorphism

$$\mathbb{T}^3 = S^1 \times \mathbb{T}^2 \to \mathbb{S}T^*\mathbb{T}^2 \subset \mathbb{T}^2 \times \mathbb{R}^2 : (t, \phi, \theta) \mapsto (\theta, \phi, \cos(2\pi t), \sin(2\pi t)), \tag{8.10}$$

and under this identification of $\mathbb{S}T^*\mathbb{T}^2$ with $\mathbb{T}^3$,

$$\alpha_{\mathrm{can}} = \cos(2\pi t) \, d\theta + \sin(2\pi t) \, d\phi.$$

---

[4]Minor annoyance: the natural orientation of $\mathbb{T}^2 \times \mathbb{R}^2$ is actually the opposite of the one defined by ω_{can} on $T^*\mathbb{T}^2$. This is the reason for reversing the order of θ and ϕ in (8.10).

This is also called the **standard contact form** on $\mathbb{T}^3$ and is used to define its standard contact structure,

$$\xi_{st} := \ker \alpha_{st}, \qquad \alpha_{st} := \alpha_{can} \text{ on } \mathbb{T}^3 = \mathbb{S}T^*\mathbb{T}^2.$$

Example 8.17. Suppose (W, ω) is a strong symplectic filling of (M, ξ) and $L \subset W$ is a Lagrangian submanifold in the interior. By Weinstein's Lagrangian neighborhood theorem (see e.g. [MS17, Theorem 3.4.13]), some neighborhood $\mathcal{U}$ of L in (W, ω) can be identified symplectically with a neighborhood of the zero-section in (T^*L, ω_{can}), and we can therefore arrange $\mathcal{U}$ so that $\partial\overline{\mathcal{U}}$ is a convex hypersurface inheriting a contact structure contactomorphic to $(\mathbb{S}T^*L, \xi_{can})$. This makes $(W \backslash \mathcal{U}, \omega)$ a strong symplectic cobordism from $(\mathbb{S}T^*L, \xi_{can})$ to (M, ξ). If additionally (W, ω) is an exact filling with primitive λ, then one says that $L \subset (W, d\lambda)$ is an **exact Lagrangian** if $\lambda|_{TL}$ is exact. When this holds, a standard argument (see e.g. the proof of Corollary 3.10 in [GZ13]) produces a primitive making $(W \backslash \mathcal{U}, \omega)$ an exact cobordism from $(\mathbb{S}T^*L, \xi_{can})$ to (M, ξ). Conversely, every strong cobordism (W', ω') from from $(\mathbb{S}T^*L, \xi_{can})$ to (M, ξ) comes from this construction, as one can stack (W', ω') on top of $(\mathbb{D}T^*L, \omega_{can})$ as in Remark 8.14 to produce a filling (W, ω) that contains the zero-section $L \subset T^*L$ as a Lagrangian submanifold, and it will be an exact Lagrangian if and only if (W', ω') is an exact cobordism.

As the reader might infer from the appearance of the word "strong" in the above definitions, one can also speak of *weak* symplectic cobordisms, fillings and caps. In dimension four, we say that a compact symplectic manifold (W, ω) with oriented boundary $\partial W = -M_- \amalg M_+$ is a **weak symplectic cobordism** from (M_-, ξ_-) to (M_+, ξ_+) if the $\xi_\pm$ are positive contact structures on $M_\pm$ such that $\omega|_{\xi_\pm} > 0$. The special cases with M_- or M_+ empty are called **weak symplectic fillings** or **caps** respectively. It is easy to check that a strong cobordism is also a weak cobordism, but the converse is false: for example, the symplectic form at the boundary of a weak cobordism need not be exact, and relatedly, weak cobordisms cannot always be stacked in the sense of Remark 8.14, so they do not give rise to a pre-order on contact manifolds. Another significant difference is that the isotopy classes of contact structures on the boundary components of a weak cobordism (W, ω) are not always uniquely determined by ω, e.g. Giroux [Gir94] gave examples of infinitely many contact structures on the 3-torus that are not contactomorphic but are all weakly filled by the same symplectic manifold (see Exercise 9.21). These also served as the first known examples of contact manifolds that admit weak symplectic fillings but not strong ones (see [Eli96]), and many others are now known. A generalization of weak fillings and cobordisms to higher dimensions was introduced in [MNW13] and can be expressed in terms of *pseudoconvexity*; see Remark 9.13.

We will have relatively little to say about weak fillings in this book, but we should note one case in which proving results about strong fillings yields results about weak fillings for free. The following local deformation lemma is due originally to Eliashberg [Eli91, Proposition 3.1]; see also [OO05, Lemma 1.1] and [Eli04, Prop. 4.1], plus [MNW13, Prop. 6] for the higher-dimensional version.

Proposition 8.18. *Suppose* (W, ω) *is a symplectic 4-manifold with oriented boundary* $M = \partial W$ *such that* $\omega|_\xi > 0$ *for some positive contact structure* $\xi \subset TM$ *and* $\omega = d\lambda$ *near* ∂W *for a 1-form* λ. *Then there exists a smooth family of 1-forms* $\{\lambda_s\}_{s\in[0,1]}$ *such that* $\lambda_0 \equiv \lambda$, $\lambda_s = \lambda$ *outside an arbitrarily small neighborhood of* ∂W *for all s,* $\omega_s := d\lambda_s$ *is symplectic and satisfies* $\omega_s|_\xi > 0$ *for all s, and* $\lambda_1|_{TM}$ *is a contact form for* ξ.

Proof. Choose any contact form α for ξ, denote its Reeb vector field by R_α, and define a 2-form on M by $\Omega = \omega|_{TM}$. The condition $\omega|_\xi > 0$ implies $\alpha \wedge \Omega > 0$ on M. Identify a collar neighborhood of ∂W in W smoothly with $(-\epsilon, 0] \times M$, with the coordinate on $(-\epsilon, 0]$ denoted by t, such that ∂_t and R_α span the symplectic complement of ξ at ∂W and satisfy $\omega(\partial_r, R_\alpha) = 1$. Then if $\epsilon > 0$ is sufficiently small, $\Omega + d(t\alpha)$ defines a symplectic form on $(-\epsilon, 0] \times M$ that is cohomologous to ω and matches it precisely at $t = 0$. It follows via a Moser deformation argument that ω and $\Omega + d(t\alpha)$ are isotopic on some neighborhood of M, thus we can now assume after changing the collar neighborhood and shrinking $\epsilon > 0$ that $\omega = \Omega + d(t\alpha)$ on the collar $(-\epsilon, 0] \times M \subset W$ near ∂W. This implies

$$d\alpha \wedge \omega = dt \wedge \alpha \wedge d\alpha > 0$$

on the collar, and after shrinking $\epsilon > 0$ further if necessary, $\alpha \wedge \Omega > 0$ implies

$$dt \wedge \alpha \wedge \omega > 0.$$

Now consider a 1-form of the form

$$\Lambda = f(t)\alpha + g(t)\lambda$$

on $(-\epsilon, 0] \times M \subset W$, where $f, g : (-\epsilon, 0] \to [0, \infty)$ are smooth functions with $f(t) = 0$ and $g(t) = 1$ near $t = -\epsilon$, so that Λ extends smoothly over the rest of W as λ. We have $d\Lambda = dt \wedge (f'(t)\alpha + g'(t)\lambda) + (f(t)d\alpha + g(t)\omega)$, thus

$$\begin{aligned}
d\Lambda \wedge d\Lambda = {}& 2f(t)\left[f'(t)dt \wedge \alpha \wedge d\alpha + g'(t)dt \wedge \lambda \wedge d\alpha\right] \\
& + 2g(t)\left[f'(t)dt \wedge \alpha \wedge \omega + g'(t)dt \wedge \lambda \wedge \omega\right] \\
& + 2f(t)g(t)d\alpha \wedge \omega + [g(t)]^2\omega \wedge \omega.
\end{aligned}$$

The last two terms in this sum are nonnegative (and positive if $g > 0$), and both of the terms in brackets that appear next to $f'(t)$ are strictly positive. Starting from $f \equiv 0$ and $g \equiv 1$ so that $\Lambda \equiv \lambda$, one can now see how to deform the functions f and g smoothly so that $d\Lambda \wedge d\Lambda$ remains positive and thus $d\Lambda$ remains symplectic:

the key is just to make sure that f' is always much larger than $|g'|$. As long as either f or g is always strictly positive, we also have

$$dt \wedge \alpha \wedge d\Lambda = f(t)\,dt \wedge \alpha \wedge d\alpha + g(t)\,dt \wedge \alpha \wedge \omega > 0,$$

which implies $d\Lambda|_\xi > 0$. After a deformation of this type, we can arrange to have $g(t) = 0$ near $t = 0$ at the cost of making f a large and steeply increasing function, and Λ then restricts to ∂W as the contact form $f(0)\alpha$.   □

Corollary 8.19. *If (W, ω) is a weak filling of (M, ξ) and $\omega|_{TM}$ is exact, then ω can be deformed symplectically near the boundary to produce a strong filling of (M, ξ). Moreover, if ω is also globally an exact 2-form, then the deformation can be arranged to produce an exact filling of (M, ξ).*   □

The corollary is often applied in situations where M is a rational homology 3-sphere, so that $H^2_{\mathrm{dR}}(M) = 0$ guarantees the exactness hypothesis for every weak filling; we will apply it this way for fillings of S^3 in Sect. 9.1. Note however that the trick used in the proof of Proposition 8.18, making f a steeply increasing function, would not work on a collar of the form $[0, \epsilon) \times M$, hinting at the fact that the analogue of Corollary 8.19 for symplectic cobordisms is *false*: Example 9.18 exhibits a strong cobordism (W, ω) that cannot be deformed to a Liouville cobordism even though ω is globally exact. The issue is that one must also be able to find a global primitive that is contact at the boundary, and this is not always possible when there are concave boundary components.

Exercise 8.20. Show that there is no such thing as an "exact symplectic cap" of a nonempty contact manifold. *Hint: Stokes' theorem.*

The hierarchy of cobordism notions "exact–strong–weak" can also be extended further in the other direction: a Liouville cobordism $(W, \omega = d\lambda)$ is called a **Weinstein cobordism** if it is equipped with the additional data of a Morse function $\varphi : W \to W$ that is constant at the boundary such that the global Liouville vector field ω-dual to λ is *gradient-like* with respect to φ. The point of this extra data is to produce a Morse-theoretic topological decomposition of (W, ω) into *symplectic handles*, corresponding to the critical points of the Morse function. Up to deformation, Weinstein cobordisms are equivalent to the (a priori much more rigid) notion of **Stein cobordisms**, which come originally from complex geometry, see [CE12]; in particular, a contact manifold is Weinstein fillable if and only if it is Stein fillable. We will not prove any results about Weinstein or Stein fillings in this book, but we will mention them occasionally since they are also an active topic of current research. Except for Stein and Weinstein, it is known that none of the above notions of symplectic fillability are equivalent: we will see some weakly but not strongly fillable examples in Sect. 9.2, while examples that are strongly but not exactly or exactly but not Stein fillable have been found by Ghiggini [Ghi05] and Bowden [Bow12] respectively.

8.3 Background on Punctured Holomorphic Curves

Gromov in his 1985 paper sketched some applications of compact pseudoholomorphic curves to questions of symplectic fillability for certain contact manifolds (cf. Theorem 9.4 below). In order to use Floer's ideas to attack the Weinstein conjecture, it was necessary to develop a more general framework for holomorphic curves on *noncompact* domains that would approach periodic orbits asymptotically. Such a framework was introduced by Hofer in 1993 [Hof93] and produced a proof of the Weinstein conjecture for certain contact 3-manifolds as a corollary of the existence of *finite-energy J-holomorphic planes* in their symplectizations. We shall explain the basic idea of this in Sect. 9.1. The theory of finite-energy punctured holomorphic curves in symplectizations and "completed" symplectic cobordisms was developed further over the course of the 1990s by Hofer et al. [HWZ96a, HWZ95a, HWZ99, HWZ96b], with several striking applications to dynamical questions in 3-dimensional contact geometry (see e.g. Theorem 9.46). At the same time, Eliashberg [Eli98] and other authors (e.g. Chekanov [Che02]) began to develop the potential of this technology for defining Floer-type invariants of contact manifolds. The culmination of this effort was the announcement [EGH00] in 2000 of *symplectic field theory* ("SFT"), a general algebraic framework that combines Gromov-Witten theory and Floer homology to define invariants of contact manifolds and symplectic cobordisms between them by counting punctured holomorphic curves. The analytical details of this theory are formidable due to the problem of achieving transversality for multiply covered holomorphic curves, and this aspect of SFT remains a large project in progress by Fish-Hofer-Wysocki-Zehnder, see [HWZb]. In spite of these complications, the analytically well-established portions of the theory have produced many striking applications, a few of which will be the main topics of Chap. 9.

In preparation for the results surveyed in the next chapter, we now give a quick sketch of the technical apparatus of holomorphic curves in symplectic cobordisms between contact manifolds. More precise statements and proofs of everything that is left to the imagination in this section may be found in [Wend].

8.3.1 Punctures and the Finite Energy Condition

Since contact manifolds are odd-dimensional, they do not admit almost complex structures, and one must first choose a related even-dimensional setting if one wants to make use of holomorphic curves. The object that most obviously lends itself for this purpose is the *symplectization*: recall from Exercise 8.7 that if (M, ξ) is any contact manifold with a contact form α, then $(\mathbb{R} \times M, d(e^t\alpha))$ is a symplectic manifold, and up to symplectomorphism it is independent of the choice of contact form (cf. Exercise 8.8). Now observe that whenever $\gamma \subset M$ is a closed orbit of the Reeb vector field R_α,

$$\mathbb{R} \times \gamma \subset (\mathbb{R} \times M, d(e^t\alpha))$$

is a symplectic submanifold. Moreover, the contact condition implies that $d\alpha|_\xi$ is nondegenerate and thus makes $\xi \to M$ a symplectic vector bundle. It is therefore natural to consider almost complex structures that make the cylinders $\mathbb{R} \times \gamma$ into J-complex curves and restrict to $d\alpha$-compatible complex structures on the bundle $\xi \to M$. We shall write $J \in \mathcal{J}(M, \alpha)$ and say that J is **adapted to** α if it is an almost complex structure on $\mathbb{R} \times M$ such that:

- J is invariant under the flow of the vector field ∂_t in the $\mathbb{R}$-direction (we say that J is "$\mathbb{R}$-invariant");
- $J(\partial_t) = R_\alpha$;
- $J(\xi) = \xi$ and $J|_\xi : \xi \to \xi$ is compatible with $d\alpha|_\xi$.

Observe that an adapted J is automatically compatible with the symplectic form $d(e^t\alpha)$, and for any periodic solution $x : \mathbb{R} \to M$ to $\dot{x} = R_\alpha(x)$ with period $T > 0$, the map

$$u : (\mathbb{R} \times S^1, i) \to (\mathbb{R} \times M, J) : (s, t) \mapsto (Ts, x(Tt)) \qquad (8.11)$$

is a J-holomorphic cylinder, where $(\mathbb{R} \times S^1, i)$ again denotes the standard Riemann cylinder $\mathbb{C}/i\mathbb{Z}$. We refer to curves of this type as **trivial cylinders** (or sometimes *orbit cylinders*).

Remark 8.21. It is important to keep in mind that in the above construction of trivial cylinders over Reeb orbits $x : \mathbb{R} \to M$ with period $T > 0$, there is no need to assume T is the **minimal period** of x, i.e. in general $T = kT_0$ for some integer $k \in \mathbb{N}$ and the smallest number $T_0 > 0$ for which $x(T_0) = x(0)$. This makes the trivial cylinder over the T-periodic orbit x a k-fold cover of the (embedded) trivial cylinder over x as a T_0-periodic orbit. We say in this case that the T_0-periodic orbit is **simply covered**, while the T-periodic orbit has **covering multiplicity** k. In the study of closed Reeb orbits on contact manifolds in general, and in particular in symplectic field theory, simply covered orbits and their multiple covers are regarded as separate objects; note that this distinction does not arise in Hamiltonian Floer homology, since the latter only considers orbits with a fixed period.

Remark 8.22. One can also define a larger space $\mathcal{J}_\tau(M, \alpha)$ with the relaxed condition that $J|_\xi$ should be tamed by (but not necessarily compatible with) $d\alpha|_\xi$. While this definition is clearly natural, the current literature does not clarify whether spaces of J-holomorphic curves with $J \in \mathcal{J}_\tau(M, \alpha)$ satisfy all the analytical properties that are needed. In particular, the standard approach to Fredholm theory for punctured holomorphic curves relies on being able to express the linearized Cauchy-Riemann operator asymptotically in terms of self-adjoint *asymptotic operators*, see [Wend, Chapters 3 and 4], [Sal99, §2.2 and §2.3, especially Lemma 2.4], and [Sch95, Chapter 3]. These operators however are not generally symmetric unless $J \in \mathcal{J}(M, \alpha)$. It is conventional to avoid this issue by always assuming $J \in \mathcal{J}(M, \alpha)$, and we shall do so here as well.

Exercise 8.23. Show that for any given contact form α on M, $\mathcal{J}(M, \alpha)$ is nonempty and contractible.

As in our discussion of Floer homology and the Arnol'd conjecture, it is useful in SFT to impose a dynamical nondegeneracy condition on closed Reeb orbits. Suppose γ denotes a closed orbit of R_α with period $T > 0$ and $\varphi_\alpha^t : M \to M$ denotes the time-t flow of R_α. The conditions defining R_α imply

$$\mathcal{L}_{R_\alpha} \alpha = d\iota_{R_\alpha} \alpha + \iota_{R_\alpha} d\alpha = d(1) + 0 = 0,$$

thus φ_α^t preserves ξ. We then say that γ is **nondegenerate** if for every point p in the image of γ, the linear map

$$d\varphi_\alpha^T(p)|_{\xi_p} : \xi_p \to \xi_p$$

does not have 1 as an eigenvalue. Note that if this condition holds for some particular p, then it holds for every p in the image of γ. Up to the obvious shifts in parametrization, nondegenerate orbits are necesarily isolated, meaning that if we parametrize closed orbits by maps $S^1 \to M$, then no sequence of closed Reeb orbits with distinct images can converge in $C^\infty(S^1, M)$ to one that is nondegenerate. One can then use the Arzelà-Ascoli theorem to show that if M is compact and all closed orbits are nondegenerate, then for every $T > 0$, there are at most finitely many closed Reeb orbits of period less than T. More generally, we say that γ is **Morse-Bott** if its image belongs to a smooth submanifold $N \subset M$ foliated by T-periodic Reeb orbits such that

$$T_p N = \ker \left(d\varphi_\alpha^T(p) - \mathbb{1} \right)$$

for every $p \in N$. This makes nondegeneracy the special case of the Morse-Bott condition in which the submanifold N is 1-dimensional. Conditions of this sort are essential for technical reasons, e.g. one can show that for $J \in \mathcal{J}(M, \alpha)$, the trivial cylinder over a closed Reeb orbit is a *Fredholm regular* J-holomorphic curve if and only if the orbit is Morse-Bott. We say that α is a **nondegenerate** or **Morse-Bott contact form** if all closed orbits of the Reeb vector field R_α are nondegenerate or Morse-Bott respectively. By a standard perturbation result, generic contact forms are nondegenerate (see e.g. the appendix of [ABW10]), though in applications, it is often convenient to work with Morse-Bott contact forms, which are allowed to have more symmetry.

Exercise 8.24. Writing $S^1 = \mathbb{R}/\mathbb{Z}$, let us say that a smooth map $\gamma : S^1 \to M$ is an **even parametrization** of a T-periodic Reeb orbit $x : \mathbb{R} \to M$ if $\gamma(t) = x(Tt)$. Prove:

(a) The period of an evenly parametrized Reeb orbit $\gamma : S^1 \to M$ is $\int_{S^1} \gamma^* \alpha$.

(b) For any smooth 1-parameter family $\{\gamma_\tau : S^1 \to M\}_{\tau \in \mathbb{R}}$ of evenly parametrized Reeb orbits, all γ_τ have the same period.

The most obvious way to adapt Floer's formalism for contact manifolds is now to choose $J \in \mathcal{J}(M, \alpha)$ and consider J-holomorphic cylinders $u : (\mathbb{R} \times S^1, i) \to (\mathbb{R} \times M, J)$ that behave asymptotically like trivial cylinders as $s \to +\infty$ and $s \to -\infty$. This idea is not wrong, but as we'll see when we discuss compactifications below, it is too simplistic: the compactification of the space of J-holomorphic cylinders will generally need to involve noncompact curves that are more general than cylinders. Hofer's paper [Hof93] focused instead on J-holomorphic planes

$$u : (\mathbb{C}, i) \to (\mathbb{R} \times M, J)$$

for which the map $(s, t) \mapsto u(e^{2\pi(s+it)})$ asymptotically approaches a trivial cylinder as $s \to +\infty$. Curves of this type arise naturally in the settings that Hofer was considering, as we will see in Sect. 9.1. Now observe that cylinders and planes can each be regarded as closed Riemann surfaces with finitely many punctures: indeed, $(\mathbb{R} \times S^1, i)$ and $(\mathbb{C}, i)$ are biholomorphically equivalent to $(S^2 \backslash \{0, \infty\}, i)$ and $(S^2 \backslash \{\infty\}, i)$ respectively, where (S^2, i) denotes the standard Riemann sphere $\mathbb{C} \cup \{\infty\}$. It thus becomes natural to consider J-holomorphic curves whose domains are arbitrary closed Riemann surfaces with finitely many punctures.

Before explaining this further, note that one can also generalize the target space by considering two closed contact manifolds $(M_\pm, \xi_\pm)$ with a strong symplectic cobordism (W, ω) from (M_-, ξ_-) to (M_+, ξ_+). Since W is compact, the symplectization is not a special case of this, but it becomes one if we replace (W, ω) with its **completion** $(\widehat{W}, \widehat{\omega})$ defined by

$$(\widehat{W}, \widehat{\omega}) = \left((-\infty, 0] \times M_-, d(e^t \alpha_-) \right) \cup_{M_-} (W, \omega) \cup_{M_+} \left([0, \infty) \times M_+, d(e^t \alpha_+) \right).$$

Here the positive/negative halves of the symplectizations $(\mathbb{R} \times M_\pm, d(e^t \alpha_\pm))$ are attached smoothly to the collar neighborhoods (8.9), see Fig. 8.5. The symplectization of (M, ξ) can now be regarded as the completion of a trivial symplectic cobordism $([0, 1] \times M, d(e^t \alpha))$ from (M, ξ) to itself. We shall write

$$J \in \mathcal{J}_\tau(W, \omega, \alpha_+, \alpha_-) \quad \text{or} \quad J \in \mathcal{J}(W, \omega, \alpha_+, \alpha_-)$$

for any smooth almost complex structure J on $\widehat{W}$ such that $J|_W$ is in $\mathcal{J}_\tau(W, \omega)$ or $\mathcal{J}(W, \omega)$ respectively, while $J|_{(-\infty,0] \times M_-}$ and $J|_{[0,\infty) \times M_+}$ belong to $\mathcal{J}(M_\pm, \alpha_\pm)$. A noncompact almost complex manifold $(\widehat{W}, J)$ constructed in this way is said to have **cylindrical ends**.

Exercise 8.25. Show that $\mathcal{J}_\tau(W, \omega, \alpha_+, \alpha_-) \subset \mathcal{J}_\tau(\widehat{W}, \widehat{\omega})$ and $\mathcal{J}(W, \omega, \alpha_+, \alpha_-) \subset \mathcal{J}(\widehat{W}, \widehat{\omega})$, and both spaces are always nonempty.

Given a closed Riemann surface (Σ, j) and a finite subset $\Gamma \subset \Sigma$, we then define the punctured Riemann surface

$$(\dot{\Sigma}, j) = (\Sigma \backslash \Gamma, j),$$

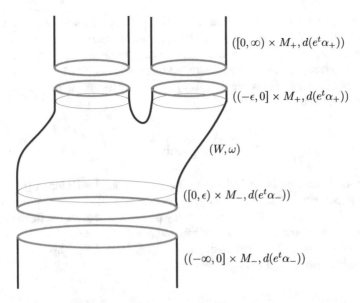

The diagram is labeled with:

$([0, \infty) \times M_+, d(e^t \alpha_+))$

$((-\epsilon, 0] \times M_+, d(e^t \alpha_+))$

(W, ω)

$([0, \epsilon) \times M_-, d(e^t \alpha_-))$

$((-\infty, 0] \times M_-, d(e^t \alpha_-))$

Fig. 8.5 The completion of a symplectic cobordism is constructed by attaching half-symplectizations to form cylindrical ends

choose $J \in \mathcal{J}_\tau(W, \omega, \alpha_+, \alpha_-)$ and consider punctured J-holomorphic curves

$$u : (\dot{\Sigma}, j) \rightarrow (\widehat{W}, J).$$

As with Floer cylinders, an arbitrary noncompact J-holomorphic curve will not have reasonable asymptotic behavior unless it satisfies a suitable energy bound (see Example 8.27 below). The obvious condition to impose, $\int_{\dot{\Sigma}} u^* \widehat{\omega} < \infty$, is however not the right one, as we can see by looking at the trivial cylinder (8.11): the integral of $u^* \widehat{\omega}$ in that example is infinite. We can fix this easily by modifying the symplectic form: let

$$\mathcal{T} = \{\varphi \in C^\infty(\mathbb{R}, (-1, 1)) \mid \varphi' > 0 \text{ and } \varphi(t) = t \text{ for all } t \text{ near } 0\},$$

and for any $\varphi \in \mathcal{T}$, define

$$\omega_\varphi = \begin{cases} d\left(e^{\varphi(t)}\alpha_-\right) & \text{on } (-\infty, 0] \times M_-, \\ \omega & \text{on } W, \\ d\left(e^{\varphi(t)}\alpha_+\right) & \text{on } [0, \infty) \times M_+. \end{cases} \qquad (8.12)$$

Exercise 8.26. Show that for any $\varphi \in \mathcal{T}$, ω_φ is a symplectic form and $\mathcal{J}_\tau(W, \omega, \alpha_+, \alpha_-) \subset \mathcal{J}_\tau(\widehat{W}, \omega_\varphi)$, $\mathcal{J}(W, \omega, \alpha_+, \alpha_-) \subset \mathcal{J}(\widehat{W}, \omega_\varphi)$.

The boundedness of $\varphi \in \mathcal{T}$ means that $\int_{\mathbb{R} \times S^1} u^* \omega_\varphi$ will now satisfy a uniform bound for all $\varphi \in \mathcal{T}$ whenever u is a trivial cylinder, and the same is then true for any punctured J-holomorphic curve that behaves asymptotically like a trivial cylinder near its punctures. Since there is clearly no canonical choice of $\varphi \in \mathcal{T}$, we define the **energy** of a punctured curve $u : (\dot{\Sigma}, j) \to (\widehat{W}, J)$ by taking the supremum over all possible choices:

$$E(u) = \sup_{\varphi \in \mathcal{T}} \int_{\dot{\Sigma}} u^* \omega_\varphi. \tag{8.13}$$

This notion of energy is equivalent[5] to the one defined by Hofer in [Hof93] and is thus sometimes called the *Hofer energy*. One can show that any nonconstant curve satisfying $E(u) < \infty$ has the following property whenever the Reeb orbits are nondegenerate or Morse-Bott[6]: for each puncture $z \in \Gamma$, fix a neighborhood $\mathcal{U}_z \subset \Sigma$ and a biholomorphic identification of $(\mathcal{U}_z, j, z)$ with $(\mathbb{D}^2, i, 0)$. This determines two holomorphic embeddings of half-cylinders,

$$\psi_+ : [0, \infty) \times S^1 \to \mathbb{D}^2 \backslash \{0\} = \mathcal{U}_z \backslash \{z\} \hookrightarrow \dot{\Sigma} : (s, t) \mapsto e^{-2\pi(s+it)},$$

$$\psi_- : (-\infty, 0] \times S^1 \to \mathbb{D}^2 \backslash \{0\} = \mathcal{U}_z \backslash \{z\} \hookrightarrow \dot{\Sigma} : (s, t) \mapsto e^{2\pi(s+it)}.$$

The statement is then that if $E(u) < \infty$, the set of punctures can be partitioned into three subsets $\Gamma = \Gamma^0 \cup \Gamma^+ \cup \Gamma^-$ such that:

- For $z \in \Gamma^0$, u admits a smooth extension over z, i.e. the puncture is **removable** (cf. Theorem 2.36).
- For $z \in \Gamma^+$, $u \circ \psi_+$ maps $[c, \infty) \times S^1$ into $[0, \infty) \times M_+$ for some $c \geqslant 0$, and up to a fixed translation in the $\mathbb{R}$-component, the restriction of $u \circ \psi_+$ to this half-cylinder can be made arbitrarily C^∞-close to a similarly restricted trivial cylinder by taking c sufficiently large. We say that this puncture is **positive**.
- For $z \in \Gamma^-$, $u \circ \psi_-$ maps $(-\infty, -c] \times S^1$ into $(-\infty, 0] \times M_-$ for some $c \geqslant 0$, and up to a fixed translation in the $\mathbb{R}$-component, the restriction of $u \circ \psi_-$ to this half-cylinder can be made arbitrarily C^∞-close to a similarly restricted trivial cylinder by taking c sufficiently large. We say that this puncture is **negative**.

---

[5]The word "equivalent" in this context does not mean that Hofer's definition was the same, but simply that any uniform bound on Hofer's energy implies a uniform bound on the version defined here, and vice versa. Thus for applications to compactness theory and asymptotics, the two notions are interchangeable.

[6]In general, [Hof93] proved that every finite-energy punctured holomorphic curve has positive and negative punctures asymptotic to closed Reeb orbits, but the asymptotic approach to these orbits is much harder to describe if the orbits are not assumed to be at least Morse-Bott. In fact, the asymptotic orbit at each puncture may even fail to be unique up to parametrization, see [Sie17].

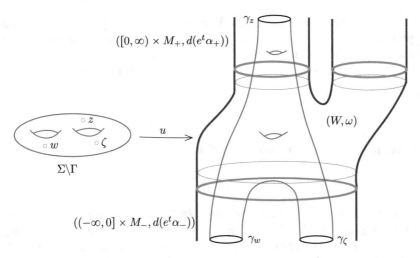

Fig. 8.6   A finite-energy punctured holomorphic curve $u : \Sigma \backslash \Gamma \to \widehat{W}$ in the completion of a strong symplectic cobordism (W, ω), with genus 2, one positive puncture $\Gamma^+ = \{z\}$ and two negative punctures $\Gamma^- = \{w, \zeta\}$

We will assume from now on that all removable punctures are already removed, so $\Gamma^0 = \emptyset$, in which case the conditions on positive and negative punctures imply that $u : \dot{\Sigma} \to \widehat{W}$ is a proper map, and at each puncture $z \in \Gamma^\pm$ it has an **asymptotic Reeb orbit** γ_z in $M_\pm$; see Fig. 8.6. The neighborhoods $\mathcal{U}_z$ of the positive/negative punctures $z \in \Gamma^\pm$ are called the positive/negative **cylindrical ends** of $(\dot{\Sigma}, j)$. Note that the data carried by each asymptotic orbit γ_z includes not only its image in M but also its covering multiplicity, cf. Remark 8.21.

Example 8.27. If $J \in \mathcal{J}(M, \alpha)$ and $x : \mathbb{R} \to M$ is any orbit of the Reeb flow $\dot{x}(t) = R_\alpha(x(t))$, periodic or not, then the map

$$u : (\mathbb{C}, i) \to (\mathbb{R} \times M, J) : s + it \mapsto (Ts, x(Tt))$$

is a J-holomorphic curve with $E(u) = \infty$. This shows that curves with infinite energy always exist and thus give no interesting information. By contrast, finite-energy curves guarantee the existence of a *periodic* Reeb orbit and thus prove the Weinstein conjecture whenever they exist.

While punctured holomorphic curves do not generally represent cycles in $H_2(\widehat{W})$ as in the closed case, a finite-energy curve $u : (\dot{\Sigma} = \Sigma \backslash \Gamma, j) \to (\widehat{W}, J)$ with positive and/or negative punctures $\Gamma = \Gamma^+ \cup \Gamma^-$ asymptotic to Reeb orbits $\{\gamma_z\}_{z \in \Gamma}$ represents a **relative homology class**, meaning the following. Let $\bar{\gamma}^\pm \subset M_\pm$ denote the closed 1-dimensional submanifold defined as the union over $z \in \Gamma^\pm$ of the images of the orbits γ_z. Fixing biholomorphic identifications of suitable neighborhoods $\mathcal{U}_z \subset \dot{\Sigma}$ of each puncture with $[0, \infty) \times S^1$ for $z \in \Gamma^+$ or $(-\infty, 0] \times S^1$ for $z \in \Gamma^-$, let $\overline{\Sigma}$ denote the so-called **circle compactification** of $\dot{\Sigma}$, that is, the compact oriented topological surface with boundary obtained from $\dot{\Sigma}$ by appending $\{\pm\infty\} \times S^1$ to each of the cylindrical ends. Now let $\pi : \widehat{W} \to W$ denote the retraction defined as the identity on W and $\pi(r, x) = x \in M_{\pm} \subset \partial W$ for (r, x) in $[0, \infty) \times M_+$ or $(-\infty, 0] \times M_-$. Then $\pi \circ u : \dot{\Sigma} \to W$ has a natural continuous extension

$$\bar{u} : (\overline{\Sigma}, \partial\overline{\Sigma}) \to (W, \bar{\gamma}^+ \cup \bar{\gamma}^-)$$

and thus represents a relative homology class

$$[u] \in H_2(W, \bar{\gamma}^+ \cup \bar{\gamma}^-).$$

Just as closed curves in a fixed homology class satisfy a uniform energy bound, it is an easy exercise in Stokes' theorem to prove that any set of finite-energy curves that represent a fixed relative homology class and have a uniformly bounded number of positive ends asymptotic to orbits with uniformly bounded period also satisfy a uniform energy bound. If the completed cobordism $\widehat{W}$ is replaced by the symplectization $\mathbb{R} \times M$ of a single contact manifold M, then it is more natural to project everything to M and define

$$[u] \in H_2(M, \bar{\gamma}^+ \cup \bar{\gamma}^-).$$

From this point, the analytical development of the theory of punctured holomorphic curves closely parallels the closed case. The following subsections provide a brief summary of how the technical results in Chap. 2 need to be modified for punctured curves. For the rest of this chapter, assume $n \geqslant 2$, $(M_{\pm}, \xi_{\pm})$ are two closed contact $(2n - 1)$-manifolds with chosen contact forms $\alpha_{\pm}$ for $\xi_{\pm}$, $(\widehat{W}, \widehat{\omega})$ is the completion of a $2n$-dimensional strong symplectic cobordism (W, ω) from (M_-, ξ_-) to (M_+, ξ_+) with collars near the boundary on which $\omega = d(e^t \alpha_{\pm})$, and $J \in \mathcal{J}_\tau(W, \omega, \alpha_+, \alpha_-)$. We will sometimes also refer to "**the $\mathbb{R}$-invariant case,**" meaning that $(\widehat{W}, \widehat{\omega})$ is replaced by the symplectization $(\mathbb{R} \times M, d(e^t \alpha))$ of a closed contact $(2n - 1)$-manifold (M, ξ) with contact form α, and $J \in \mathcal{J}(M, \alpha)$.

Most of the nontrivial results mentioned below require closed Reeb orbits to be either nondegenerate or Morse-Bott, so let us assume that the Morse-Bott condition always holds. We will occasionally go further and specify nondegeneracy, though this condition can usually be relaxed with some effort (see Remark 8.52).

8.3.2 Simple and Multiply Covered Curves

If $\varphi : (\Sigma, j) \to (\Sigma', j')$ is a holomorphic map of degree $k := \deg(\varphi) \geqslant 2$ between two closed Riemann surfaces and $u' : (\dot{\Sigma}' := \Sigma' \backslash \Gamma', j') \to (\widehat{W}, J)$ is a finite-energy punctured J-holomorphic curve, then one can define a k-**fold cover** of u' as a finite-energy curve

$$u = u' \circ \varphi : (\dot{\Sigma}, j) \to (\widehat{W}, J)$$

with domain $\dot{\Sigma} := \Sigma \backslash \Gamma$ for $\Gamma := \varphi^{-1}(\Gamma')$. With this notion understood, Proposition 2.6 continues to hold in the punctured case: all nonconstant curves with Morse-Bott asymptotic orbits are either multiply covered or somewhere injective, i.e. **simple**, and in the latter case they have at most finitely many self-intersections and non-immersed points. This fact has been considered standard for many years, though a complete proof of it has been difficult to find in the literature until relatively recently, and it requires a little bit more than the local results underlying Proposition 2.6, i.e. the fact that self-intersections and non-immersed points of a somewhere injective J-holomorphic curve are always isolated. It also requires asymptotic results to prevent non-immersed points or self-intersections from accumulating near infinity. Results of this kind are proved in [Sie08], and on this basis, complete proofs of the punctured version of Proposition 2.6 can be found in [Nel15, §3.2] or [Wend, §6.4].

Exercise 8.28. Suppose Σ and Σ' are closed, connected and oriented surfaces, $\varphi : \Sigma \to \Sigma'$ is a branched cover of degree $k \in \mathbb{N}$, $\Gamma' \subset \Sigma'$ is a finite subset and $\Gamma = \varphi^{-1}(\Gamma') \subset \Sigma$. Denote

$$\dot{\Sigma} := \Sigma \backslash \Gamma, \quad \dot{\Sigma}' := \Sigma' \backslash \Gamma', \quad \dot{\varphi} := \varphi|_{\dot{\Sigma}} : \dot{\Sigma} \to \dot{\Sigma}'.$$

Use the Riemann-Hurwitz formula (Proposition 4.9) for φ to show that the algebraic count of critical points $Z(d\dot{\varphi}) \geqslant 0$ of $\dot{\varphi}$ satisfies the analogous formula

$$Z(d\dot{\varphi}) = -\chi(\dot{\Sigma}) + k\chi(\dot{\Sigma}'). \tag{8.14}$$

Exercise 8.29. Recall from Remark 8.21 that each of the asymptotic orbits $\{\gamma_z\}_{z\in\Gamma}$ of a finite-energy curve $u : \dot{\Sigma} = \Sigma \backslash \Gamma \to \widehat{W}$ also has a covering multiplicity, i.e. for each puncture z, γ_z is a k_z-fold cover of some simply covered orbit γ'_z for some $k_z \in \mathbb{N}$. Show that if all of the positive or all of the negative asymptotic orbits of u are distinct and simply covered, then u is simple. (The converse, by the way, is false in general.)

Proposition 2.8 is a statement explicitly about curves that represent cycles in homology and is thus not immediately relevant to the punctured case, but the basic fact behind it generalizes easily: a finite-energy curve is constant if and only if its energy vanishes. This is immediate from the fact that every $J \in \mathcal{J}_\tau(W, \omega, \alpha_+, \alpha_-)$ is tamed by the symplectic form ω_φ defined in (8.12) for every $\varphi \in \mathcal{T}$, cf. Exercise 8.26. Here is a related statement that pertains specifically to the $\mathbb{R}$-invariant case, and is an easy application of Stokes' theorem.

Proposition 8.30. *Suppose $J \in \mathcal{J}(M, \alpha)$, and $u : (\dot{\Sigma}, j) \to (\mathbb{R} \times M, J)$ is a finite-energy punctured J-holomorphic curve with asymptotic orbits $\{\gamma_z\}_{z\in\Gamma^\pm}$,* *where each γ_z has period $T_z > 0$. Then*

$$\int_{\dot{\Sigma}} u^* d\alpha = \sum_{z \in \Gamma^+} T_z - \sum_{z \in \Gamma^-} T_z \geq 0,$$

with equality if and only if u is (up to parametrization) either a trivial cylinder or a multiple cover of one.   □

8.3.3 Smoothness and Dimension of the Moduli Space

Fix finite ordered tuples of closed Reeb orbits $\boldsymbol{\gamma}^\pm = (\gamma_1^\pm, \ldots, \gamma_{s_\pm}^\pm)$ in $M_\pm$ for some integers $s_+, s_- \geq 0$; here the data describing a Reeb orbit includes its image and its covering multiplicity (see Remark 8.21), but we do not distinguish between two orbits that differ only by a shift in parametrization. Fix also a relative homology class

$$A \in H_2(W, \bar{\gamma}^+ \cup \bar{\gamma}^-),$$

where $\bar{\gamma}^\pm \subset M^\pm$ denotes the 1-dimensional submanifold obtained from the union of the images of all the orbits $\gamma_1^\pm, \ldots, \gamma_{s_\pm}^\pm$. For integers $g, m \geq 0$, we then define the moduli space

$$\mathcal{M}_{g,m}(A; J; \boldsymbol{\gamma}^+, \boldsymbol{\gamma}^-) = \{(\Sigma, j, \Gamma^+, \Gamma^-, u, \Theta)\}/\sim,$$

where

- (Σ, j) is a closed connected Riemann surface with genus g;
- $\Gamma^+ = (z_1^+, \ldots, z_{s_+}^+)$ and $\Gamma^- = (z_1^-, \ldots, z_{s_-}^-)$ are disjoint finite ordered sets of distinct points;
- $u : (\dot{\Sigma} := \Sigma \backslash (\Gamma^+ \cup \Gamma^-), j) \to (\widehat{W}, J)$ is a finite-energy pseudo-holomorphic curve with positive punctures z_i^+ asymptotic to γ_i^+ for $i = 1, \ldots, s_+$ and negative punctures z_i^- asymptotic to γ_i^- for $i = 1, \ldots, s_-$, and representing the relative homology class A;
- $\Theta = (\zeta_1, \ldots, \zeta_m)$ is an ordered set of m distinct points in $\dot{\Sigma}$;
- $(\Sigma_1, j_1, \Gamma_1^+, \Gamma_1^-, u_1, \Theta_1)$ and $(\Sigma_2, j_2, \Gamma_2^+, \Gamma_2^-, u_2, \Theta_2)$ are defined to be equivalent if and only if there exists a biholomorphic map $\varphi : (\Sigma_1, j_1) \to (\Sigma_2, j_2)$ taking $\Gamma_1^\pm$ to $\Gamma_2^\pm$ and Θ_1 to Θ_2 with all orderings preserved, and satisfying $u_1 = u_2 \circ \varphi|_{\dot{\Sigma}_1}$.

The space $\mathcal{M}_{g,m}(A; J; \boldsymbol{\gamma}^+, \boldsymbol{\gamma}^-)$ has a natural topology such that convergence of a sequence $[(\Sigma_k, j_k, \Gamma_k^+, \Gamma_k^-, u_k, \Theta_k)]$ to an element $[(\Sigma, j, \Gamma^+, \Gamma^-, u, \Theta)]$ means the existence for sufficiently large k of representatives

$$(\Sigma, j_k', \Gamma^+, \Gamma^-, u_k', \Theta) \sim (\Sigma_k, j_k, \Gamma_k^+, \Gamma_k^-, u_k, \Theta_k)$$

such that $j'_k \to j$ in C^∞ on Σ while $u'_k \to u$ in C^∞_{loc} on $\dot{\Sigma}$ and in C^0 up to infinity (with respect to some $\mathbb{R}$-invariant choice of metric on the cylindrical ends). As usual, we shall abbreviate the case without marked points by

$$\mathcal{M}_g(A; J; \boldsymbol{\gamma}^+, \boldsymbol{\gamma}^-) := \mathcal{M}_{g,0}(A; J; \boldsymbol{\gamma}^+, \boldsymbol{\gamma}^-),$$

and we will sometimes use the shorthand notation

$$\mathcal{M}_g(J; \boldsymbol{\gamma}^+, \boldsymbol{\gamma}^-) := \bigcup_{A \in H_2(W, \bar{\gamma}^+ \cup \bar{\gamma}^-)} \mathcal{M}_g(A; J; \boldsymbol{\gamma}^+, \boldsymbol{\gamma}^-).$$

If all asymptotic orbits are nondegenerate, the index formula (2.4) now generalizes to

$$\text{vir-dim}\,\mathcal{M}_{g,m}(A; J; \boldsymbol{\gamma}^+, \boldsymbol{\gamma}^-) = (n-3)(2 - 2g - s_+ - s_-) + 2c_1^\tau(A)$$

$$+ \sum_{i=1}^{s_+} \mu_{\text{CZ}}^\tau(\gamma_i^+) - \sum_{i=1}^{s_-} \mu_{\text{CZ}}^\tau(\gamma_i^-) + 2m,$$

$$(8.15)$$

or equivalently, for a finite-energy curve $u : (\dot{\Sigma} = \Sigma \backslash (\Gamma^+ \cup \Gamma^-), j) \to (\widehat{W}, J)$ with asymptotic orbits $\{\gamma_z\}_{z \in \Gamma^\pm}$,

$$\text{ind}(u) = (n-3)\chi(\dot{\Sigma}) + 2c_1^\tau(u^*T\widehat{W}, J) + \sum_{z \in \Gamma^+} \mu_{\text{CZ}}^\tau(\gamma_z) - \sum_{z \in \Gamma^-} \mu_{\text{CZ}}^\tau(\gamma_z). \quad (8.16)$$

To define the terms on the right hand sides of each of these formulas, one first needs to make an arbitrary choice of trivialization for the bundles $\xi_\pm$ along each of the asymptotic orbits, and this choice is denoted here by τ. In general, if $E \to \dot{\Sigma}$ is a complex line bundle and τ denotes a choice of trivialization for E near the punctures, then the **relative first Chern number** $c_1^\tau(E) \in \mathbb{Z}$ is defined by counting the zeroes of a generic section that is nonzero and constant with respect to τ near the punctures, and if E is a higher-rank complex vector bundle, one defines $c_1^\tau(E)$ by requiring c_1^τ to be additive with respect to direct sums. The term $c_1^\tau(u^*T\widehat{W}, J)$ in (8.16) is then explained by the fact that $T\widehat{W}$ over each cylindrical end of $\widehat{W}$ is naturally a direct sum of a trivial complex line bundle with ξ_+ or ξ_-, hence the trivializations τ of $\xi_\pm$ along each orbit naturally determine trivializations of the complex vector bundle $(u^*T\widehat{W}, J) \to \dot{\Sigma}$ near the punctures. The number $c_1^\tau(u^*T\widehat{W}, J)$ can then be shown to depend only on the complex bundle $T\widehat{W} \to \widehat{W}$, the asymptotic trivializations τ and the relative homology class of u, hence the same term is denoted by $c_1^\tau(A)$ in (8.15).

Much could be said about the **Conley-Zehnder index** $\mu_{\text{CZ}}^\tau(\gamma) \in \mathbb{Z}$, which we do not have space for here, so we will be content to know that $\mu_{\text{CZ}}^\tau(\gamma)$ is a homotopy invariant of nondegenerate Reeb orbits that quantifies (with respect to the trivialization τ) the rotation of nearby (non-periodic) orbits about γ. For details, see [Wend, Chapter 3], or the original sources [CZ83a] or [HWZ95a, §3]. Note that changing the trivialization τ always changes $\mu_{CZ}^{\tau}(\gamma)$ by an even integer, hence the **parity**

$$p(\gamma) := [\mu_{CZ}^{\tau}(\gamma)] \in \mathbb{Z}_2$$

of γ is well defined independently of any choices. One can show moreover that changing τ changes the relative first Chern number by an amount that cancels the total changes to the Conley-Zehnder indices, hence the virtual dimension is also independent of τ.

The index formula can be made to look the same in the Morse-Bott case if one first adjusts one's understanding of the terms $\mu_{CZ}^{\tau}(\gamma_z)$, in slightly different ways depending on the sign of the puncture; see Remark 8.52.

The discussion of Fredholm regularity and genericity in Sect. 2.1.3 generalizes to the punctured case in a completely natural way, the only real difference being the technical details of how to define suitable Banach manifolds of **asymptotically cylindrical** maps $\dot{\Sigma} \to \widehat{W}$. In particular, the analogue of Theorem 2.11 in this context holds with only one minor change, which is that the smooth moduli space

$$\mathcal{M}_{g,m}^{\mathrm{reg}}(A; J; \gamma^+, \gamma^-) \subset \mathcal{M}_{g,m}(A; J; \gamma^+, \gamma^-)$$

might fail to be orientable; we will come back to this delicate topic in Sect. 8.3.6 below (cf. Corollary 8.38). The analogue of Theorem 2.12 holds without any changes at all, as we were careful to state it without requiring the target symplectic manifold to be compact. The caveat here is that while Theorem 2.12 can be applied for finite-energy curves in $\widehat{W}$, it requires a choice of a precompact "perturbation domain" $\mathcal{U} \subset \widehat{W}$ on which J may be perturbed generically, and the result then holds only for somewhere injective curves that intersect $\mathcal{U}$.[7] A natural choice for $\mathcal{U}$ is the interior of the compact cobordism W, but the theorem then stops short of the statement we really want, which is that all somewhere injective curves in $\widehat{W}$ will be regular for generic $J \in \mathcal{J}_{\tau}(W, \omega, \alpha_+, \alpha_-)$. One can repair this gap by supplementing Theorem 2.12 with an analogous result that applies specifically to the $\mathbb{R}$-invariant case:

Theorem 8.31. *Fix an open subset $\mathcal{U} \subset M$ with compact closure and an $\mathbb{R}$-invariant almost complex structure $J_0 \in \mathcal{J}(M, \alpha)$, and let*

$$\mathcal{J}(M, \alpha; \mathcal{U}, J_0) = \{J \in \mathcal{J}(M, \alpha) \mid J = J_0 \text{ outside } \mathbb{R} \times \mathcal{U}\}.$$

---

[7]Requiring $\mathcal{U}$ to have compact closure is useful for technical reasons, as the proof of the theorem requires defining a Banach manifold of perturbed almost complex structures, and there is usually no natural way to put Banach space structures on spaces of maps whose domains are noncompact manifolds.

Then given any integers $g, m, s_+, s_- \geqslant 0$, tuples of Reeb orbits $\gamma^{\pm} = (\gamma_1^{\pm}, \ldots, \gamma_{s_{\pm}}^{\pm})$ in M and a relative homology class $A \in H_2(M, \bar{\gamma}^+ \cup \bar{\gamma}^-)$, there exists a comeager subset

$$\mathcal{J}^{\mathrm{reg}}(M, \alpha; \mathcal{U}, J_0) \subset \mathcal{J}(M, \alpha; \mathcal{U}, J_0)$$

such that for all $J \in \mathcal{J}^{\mathrm{reg}}(M, \alpha; \mathcal{U}, J_0)$, every somewhere injective curve in $\mathcal{M}_{g,m}(A; J; \gamma^+, \gamma^-)$ that intersects $\mathbb{R} \times \mathcal{U}$ is Fredholm regular. $\qquad\square$

Theorem 8.31 is a separate result from Theorem 2.12 because the freedom to perturb J within the class $\mathcal{J}(M, \alpha)$ of $\mathbb{R}$-invariant almost complex structures that differ from each other only on the subbundle ξ is much more restrictive than the freedom to perturb it among all ω-tame almost complex structures on $\widehat{W}$. The result is originally due to Dragnev [Dra04], and two more modern alternative proofs (of slightly more general results) can be found in [Wend, Lecture 8] and [Wena]. Now if Theorem 2.12 is applied in $\widehat{W}$ with $\mathcal{U}$ defined as the interior of W, then any somewhere injective curve failing to intersect $\mathcal{U}$ must be contained in one of the cylindrical ends and is therefore subject to Theorem 8.31, so the outcome is exactly as desired: for generic $J \in \mathcal{J}_\tau(W, \omega, \alpha_+, \alpha_-)$, the open subset

$$\mathcal{M}_{g,m}^*(A; J; \gamma^+, \gamma^-) \subset \mathcal{M}_{g,m}(A; J; \gamma^+, \gamma^-)$$

consisting of somewhere injective curves is a smooth manifold with dimension equal to vir-dim $\mathcal{M}_{g,m}(A; J; \gamma^+, \gamma^-)$.

Remark 8.32. The results in Sect. 2.1 about parametric moduli spaces and moduli spaces with marked point constraints also generalize in straightforward ways to the punctured case. In contrast to Remark 2.20, however, vir-dim $\mathcal{M}_{g,m}(A; J; \gamma^+, \gamma^-)$ is not always an even number, hence the generic condition $\mathrm{ind}(u) \geqslant -1$ for simple curves u in 1-parametric moduli spaces cannot generally be used to deduce the better result $\mathrm{ind}(u) \geqslant 0$. There is still at least one common situation where this trick does work and is quite useful; see the exercise below.

Exercise 8.33. Show that if dim $\widehat{W} = 4$ and u is a finite-energy punctured J-holomorphic curve whose asymptotic orbits all have odd Conley-Zehnder index, then $\mathrm{ind}(u)$ is even.

8.3.4 SFT Compactness

The compactification of the moduli space of punctured holomorphic curves [BEH+03] combines the Gromov compactification with the breaking phenomenon of Floer homology. Assuming all closed Reeb orbits are Morse-Bott (cf. Remark 8.34 below), every sequence of punctured curves $u_k : \dot{\Sigma} \to \widehat{W}$ with uniformly bounded energy has a subsequence converging to a **stable** J-

holomorphic building

$$(v_{N_+}^+, \ldots, v_1^+, v_0, v_1^-, \ldots, v_{N_-}^-),$$

where

- v_i^+ for $i = 1, \ldots, N_+$ are (possibly disconnected) punctured nodal holomorphic curves in the $\mathbb{R}$-invariant symplectization $(\mathbb{R} \times M_+, d(e^t \alpha_+))$, defined up to $\mathbb{R}$-translation;
- v^0 is a (possibly disconnected or empty) punctured nodal J-holomorphic curve in $(\widehat{W}, \widehat{\omega})$;
- v_i^- for $i = 1, \ldots, N_-$ are (possibly disconnected) punctured nodal holomorphic curves in the $\mathbb{R}$-invariant symplectization $(\mathbb{R} \times M_-, d(e^t \alpha_-))$, defined up to $\mathbb{R}$-translation.

These nodal curves are called the **levels** of the building, and they connect in the sense that the negative asymptotic orbits of each level match the positive asymptotic orbits of the one below it, so that gluing levels together along these matching orbits gives a nodal surface with the same arithmetic genus as $\dot{\Sigma}$; see Fig. 8.7. We refer to v^0 as

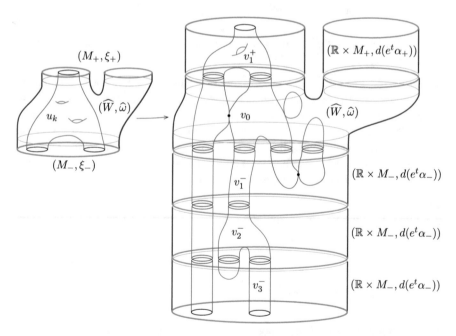

Fig. 8.7 Degeneration of a sequence u_k of finite-energy punctured holomorphic curves with genus 2, one positive end and two negative ends in a completed symplectic cobordism. The limiting holomorphic building $(v_1^+, v_0, v_1^-, v_2^-, v_3^-)$ in this example has one upper level, a main level and three lower levels, each of which is a (possibly disconnected) finite-energy punctured nodal holomorphic curve. The building has arithmetic genus 2 and the same numbers of positive and negative ends as u_k

the **main level**, the levels v_i^+ as **upper levels** and v_i^- as **lower levels**. Note that by convention, N_+ or N_- can be zero, so upper and lower levels may or may not exist, while the main level always exists but is also allowed to be empty (i.e. the domain of v^0 is the empty set), in which case there must be at least one upper or lower level. Supplementing the stability condition of Definition 2.34 by the requirement that no level can consist exclusively of a disjoint union of trivial cylinders without marked points or nodes, the space $\overline{\mathcal{M}}_{g,m}(A; J; \boldsymbol{\gamma}^+, \boldsymbol{\gamma}^-)$ of stable J-holomorphic buildings then has a natural metrizable topology as a compactification of $\mathcal{M}_{g,m}(A; J; \boldsymbol{\gamma}^+, \boldsymbol{\gamma}^-)$, and all its elements have finite automorphism groups. This space is often called the **SFT compactification**.

A slight modification of this construction is appropriate for the $\mathbb{R}$-invariant case, where we prefer to consider the quotient $\mathcal{M}_{g,m}(A; J; \boldsymbol{\gamma}^+, \boldsymbol{\gamma}^-)/\mathbb{R}$ instead of $\mathcal{M}_{g,m}(A; J; \boldsymbol{\gamma}^+, \boldsymbol{\gamma}^-)$ itself. This quotient has a natural compactification consisting of stable holomorphic buildings in which all levels are (possibly disconnected but nonempty) curves in $\mathbb{R} \times M$, but each is now defined only up to $\mathbb{R}$-translation, and while the ordering of the levels still makes sense, there is no longer any meaningful distinction between *main*, *upper* and *lower* levels. We use the same notation $\overline{\mathcal{M}}_{g,m}(A; J; \boldsymbol{\gamma}^+, \boldsymbol{\gamma}^-)$ for this compactification, keeping in mind that it is technically a compactification of $\mathcal{M}_{g,m}(A; J; \boldsymbol{\gamma}^+, \boldsymbol{\gamma}^-)/\mathbb{R}$ rather than $\mathcal{M}_{g,m}(A; J; \boldsymbol{\gamma}^+, \boldsymbol{\gamma}^-)$.

Remark 8.34. As stated in [BEH+03], the theorem that $\overline{\mathcal{M}}_{g,m}(A; J; \boldsymbol{\gamma}^+, \boldsymbol{\gamma}^-)$ is compact requires all contact forms concerned to be nondegenerate or Morse-Bott. The reason is that the breaking orbits appearring between neighboring levels of a building usually cannot be predicted in advance; the only thing we can typically predict about the curves appearing in each level is that they have finite energy (often with a quantitative bound), so it is then essential to know that finite energy implies reasonable asymptotic behavior, which is not true in general without some nondegeneracy condition (see [Sie17]). There are situations in which one can usefully relax this condition a bit, e.g. quantitative bounds on energy imply quantitative bounds on the periods of possible breaking orbits, so sometimes it is enough to know that all orbits up to some fixed period are nondegenerate.

A few subtle issues have been elided in the above sketch, and one of them demonstrates an important difference between the SFT compactification and the simpler space of "broken Floer cylinders" arising in Floer homology. Focusing on the $\mathbb{R}$-invariant case, fix $J \in \mathcal{J}(M, \alpha)$ and three Reeb orbits $\gamma^+, \gamma^-, \gamma^0$ in M, and suppose $(v_+, v_-) \in \overline{\mathcal{M}}_0(J; \gamma^+, \gamma^-)$ is a holomorphic building whose levels are cylinders $v_+ \in \mathcal{M}_0(J; \gamma^+, \gamma^0)/\mathbb{R}$ and $v_- \in \mathcal{M}_0(J; \gamma^0, \gamma^-)/\mathbb{R}$. This building can be the limit of a sequence of smooth holomorphic cylinders $u_k \in \mathcal{M}_0(J; \gamma^+, \gamma^-)$, where convergence means essentially that one can find J-holomorphic parametrizations

$$u_k = (u_k^{\mathbb{R}}, u_k^M) : (\mathbb{R} \times S^1, i) \to (\mathbb{R} \times M, J),$$

$$v_\pm = (v_\pm^{\mathbb{R}}, v_\pm^M) : (\mathbb{R} \times S^1, i) \to (\mathbb{R} \times M, J),$$

satisfying the following:

(1) There exist sequences of constants $c_k^\pm \in \mathbb{R}$ with $c_k^+ - c_k^- \to \infty$ such that $u_k^\pm := (u_k^\mathbb{R} - c_k^\pm, u_k^M)$ converges in $C^\infty_{loc}(\mathbb{R} \times S^1)$ to $v_\pm$.
(2) For any sequence of constants $c_k \in \mathbb{R}$ that does not stay within a uniformly bounded distance of either c_k^+ or c_k^-, the curves $(u_k^\mathbb{R} - c_k, u_k^M)$ converge to a trivial cylinder over either γ^+, γ^0 or γ^-.

Now consider the obvious extensions of the maps $u_k^M : \mathbb{R} \times S^1 \to M$ to continuous maps on the circle compactification

$$\bar{u}_k^M : \overline{Z}_k := [-\infty, \infty] \times S^1 \to M,$$

so that $\bar{u}_k^M$ restricts to $\{\pm\infty\} \times S^1$ as parametrizations of the orbits γ^+ and γ^-. Denote the corresponding extensions of $v_\pm^M : \mathbb{R} \times S^1 \to M$ to $[-\infty, \infty] \times S^1$ by

$$\bar{v}_\pm^M : \overline{Z}_\pm \to M.$$

One consequence of the convergence described above is that the maps $\bar{v}_+^M$ and $\bar{v}_-^M$ glue together to form a continuous map

$$\bar{v}^M : \overline{Z} := \overline{Z}_+ \cup_\Phi \overline{Z}_- \to M,$$

where the attaching map Φ sends $\{-\infty\} \times S^1 \subset \partial \overline{Z}_+$ via the obvious bijection to $\{+\infty\} \times S^1 \subset \partial \overline{Z}_-$, and there exists a sequence of homeomorphisms $\varphi_k : \overline{Z} \to \overline{Z}_k$ such that

$$\bar{u}_k^M \circ \varphi_k \to \bar{v}^M \quad \text{in} \quad C^0(\overline{Z}, M).$$

Observe now that this description of the convergence depends on the choices of parametrizations $u_k : \mathbb{R} \times S^1 \to \mathbb{R} \times M$, so if we change them e.g. by replacing u_k with $u_k(\cdot + a, \cdot + b)$ for some constants $a \in \mathbb{R}$ and $b \in S^1$, then the parametrizations of $v_\pm$ will similarly be replaced by $v_\pm(\cdot + a, \cdot + b)$. The crucial point here is that we are not allowed to change the parametrizations of v_+ and v_- *independently* of one another, as an arbitrary pair of reparametrizations will not necessarily arise as the limit of any sequence of reparametrizations of u_k. In other words, the element $(v_+, v_-) \in \overline{\mathcal{M}}_0(J; \gamma^+, \gamma^-)$ consists of more data than just an element of the Cartesian product $\mathcal{M}_0(J; \gamma^+, \gamma^0) \times \mathcal{M}_0(J; \gamma^0, \gamma^-)$. The extra data is called a **decoration**, and for a given pair of parametrizations $v_\pm : \mathbb{R} \times S^1 \to \mathbb{R} \times M$, one can characterize the decoration as a choice of homeomorphism

$$\partial \overline{Z}_+ \supset \{-\infty\} \times S^1 \xrightarrow{\Phi} \{+\infty\} \times S^1 \subset \partial \overline{Z}_-,$$

subject to the condition that $\bar{v}_+^M$ and $\bar{v}_-^M$ must then glue together to form a continuous map on $\overline{Z}_+ \cup_\Phi \overline{Z}_-$. Two pairs of parametrizations (v_+, v_-) and (v_+', v_-') with decorations Φ and Φ' respectively are then considered to represent the same element of $\overline{\mathcal{M}}_0(J; \gamma^+, \gamma^-)$ if and only if there exist simultaneous reparametrizations identifying v_+ with v'_+ and v_- with v'_- such that the decorations Φ and Φ' are also identified with each other.

The reason this discussion never arises in Floer homology is that in the setting of Floer cylinders asymptotic to 1-periodic Hamiltonian orbits, there is only ever one allowable choice of decoration: orbits of a time-dependent Hamiltonian vector field come with a preferred starting point, and all parametrizations of cylinders are expected to respect this. In SFT this is no longer true, and in particular, the orbit γ^0 could be a k-fold cover of another Reeb orbit for some $k \geqslant 2$, in which case any pair of parametrizations $v_\pm : \mathbb{R} \times S^1 \to \mathbb{R} \times M$ admits k choices of decoration for which the maps can be glued together continuously. Some of these choices may turn out to be equivalent under pairs of biholomorphic reparametrizations for the two curves, but if they are not, then they represent distinct elements of $\overline{\mathcal{M}}_0(J; \gamma^+, \gamma^-)$.

8.3.5 Gluing Along Punctures

As with nodal curves in the Gromov compactification, gluing theorems can be used to describe the neighborhood of a holomorphic building in $\overline{\mathcal{M}}_{g,m}(J; \gamma^+, \gamma^-)$. Historically, the Floer-theoretic version of gluing predates its use in the theory of closed holomorphic curves (cf. Sect. 2.1.7), as e.g. the proof that $\partial^2 = 0$ in Floer homology rests on the fact that not only does the compactification of the space of Floer cylinders contain broken Floer cylinders as in Fig. 8.1, but *every* rigid broken Floer cylinder also arises in this way as the limit of a unique 1-parameter family of smooth Floer cylinders. Gluing will not play a role in the applications to be discussed in Chap. 9, so the reader may prefer to skip this section on first pass, but the basic idea needs to be understood in order to discuss orientations in Sect. 8.3.6 below, and in any case, since the algebra of SFT depends crucially on gluing theorems, a brief discussion is warranted.

In Sect. 8.3.4 above we saw that a holomorphic building cannot always be specified merely in terms of the component smooth curves in its levels, as one must also specify *decorations* for each of its "breaking" orbits, i.e. the orbits along which two neighboring levels connect. One remedy is to enhance the moduli space $\mathcal{M}_{g,m}(A; J; \gamma^+, \gamma^-)$ with slightly more data. Fix an arbitrary choice of point $p_\gamma \in M_\pm$ in the image of each Reeb orbit γ in $M_\pm$, so if γ has covering multiplicity $k \in \mathbb{N}$, then any parametrization of γ passes through p_γ exactly k times. For a J-holomorphic curve $u : (\dot{\Sigma} = \Sigma \backslash (\Gamma^+ \cup \Gamma^-), j) \to (\widehat{W}, J)$ with a puncture $z \in \Gamma^\pm$ asymptotic to γ, an **asymptotic marker** is a choice of a ray $\ell \subset T_z\Sigma$ such that

$$\lim_{t \to 0^+} u(c(t)) = (\pm\infty, p_\gamma)$$

for any smooth path $c(t) \in \Sigma$ with $c(0) = z$ and $0 \neq \dot{c}(0) \in \ell$. Notice that if γ has covering multiplicity $k \in \mathbb{N}$, then there are exactly k choices of asymptotic markers at z, related to each other by the action on $T_z \Sigma$ by the kth roots of unity. We shall denote

$$\mathcal{M}^{\$}_{g,m}(A; J; \boldsymbol{\gamma}^+, \boldsymbol{\gamma}^-) := \{(\Sigma, j, \Gamma^+, \Gamma^-, u, \Theta, \ell)\} / \sim,$$

where $(\Sigma, j, \Gamma^+, \Gamma^-, u, \Theta)$ represents an element of $\mathcal{M}_{g,m}(A; J; \boldsymbol{\gamma}^+, \boldsymbol{\gamma}^-)$, ℓ denotes an assignment of asymptotic markers to every puncture $z \in \Gamma^{\pm}$, and

$$(\Sigma_1, j_1, \Gamma_1^+, \Gamma_1^-, u_1, \Theta_1, \ell_1) \sim (\Sigma_2, j_2, \Gamma_2^+, \Gamma_2^-, u_2, \Theta_2, \ell_2)$$

means the existence of a biholomorphic map $\varphi : (\Sigma_1, j_1) \to (\Sigma_2, j_2)$ which defines an equivalence of $(\Sigma_1, j_1, \Gamma_1^+, \Gamma_1^-, u_1, \Theta_1)$ with $(\Sigma_2, j_2, \Gamma_2^+, \Gamma_2^-, u_2, \Theta_2)$ and satisfies $\varphi_* \ell_1 = \ell_2$. There is a natural surjection

$$\mathcal{M}^{\$}_{g,m}(A; J; \boldsymbol{\gamma}^+, \boldsymbol{\gamma}^-) \to \mathcal{M}_{g,m}(A; J; \boldsymbol{\gamma}^+, \boldsymbol{\gamma}^-)$$

defined by forgetting the markers, and if we restrict to curves in $\mathcal{M}_{g,m}(A; J; \boldsymbol{\gamma}^+, \boldsymbol{\gamma}^-)$ that are somewhere injective and therefore have no nontrivial automorphisms, this surjection is a covering map of finite degree given by the product of the multiplicities of the asymptotic orbits.

Returning now to the broken cylinder example from Sect. 8.3.4, any pair

$$(v_+, v_-) \in \mathcal{M}^{\$}_0(J; \gamma^+, \gamma^0) \times \mathcal{M}^{\$}_0(J; \gamma^0, \gamma^-)$$

naturally specifies an element of $\overline{\mathcal{M}}_0(J; \gamma^+, \gamma^-)$ having v_+ and v_- as its levels; the decoration is uniquely determined by the condition that it map the asymptotic marker of v_+ at $-\infty$ to the marker of v_- at $+\infty$. In this discussion, the additional asymptotic markers at the orbits γ^+ and γ^- are extraneous information which we are free to discard.

With asymptotic markers in the picture, we can now write down a concrete example of what a gluing map in SFT looks like. The following discussion is borrowed from [Wend, Lecture 11].

Figure 8.8 shows the degeneration of a sequence of curves u_k in a moduli space of the form $\mathcal{M}_{3,4}(J; (\gamma_4, \gamma_5), \boldsymbol{\gamma}^-)$ to a building $u_\infty \in \overline{\mathcal{M}}_{3,4}(J; (\gamma_4, \gamma_5), \boldsymbol{\gamma}^-)$ with one main level and one upper level. The main level is a connected curve u_A belonging to $\mathcal{M}_{1,2}(J; (\gamma_1, \gamma_2, \gamma_3), \boldsymbol{\gamma}^-)$, and the upper level consists of two connected curves

$$u_B \in \mathcal{M}_{1,1}(J_+; \gamma_4, (\gamma_1, \gamma_2)), \qquad u_C \in \mathcal{M}_{0,1}(J_+; \gamma_5, \gamma_3),$$

where $J_+ \in \mathcal{J}(M_+, \alpha_+)$ denotes the restriction of J to the positive cylindrical end of $\widehat{W}$. One can endow each of these curves with asymptotic markers compatible with the decoration of u_∞; this is a non-unique choice, but e.g. if one chooses markers for u_A arbitrarily, then the markers at the negative punctures of u_B and u_C are uniquely determined. Let us assume that all three curves are somewhere injective, and that $J|_W \in \mathcal{J}_\tau(W, \omega)$ and $J_+ \in \mathcal{J}(M_+, \alpha_+)$ are generic so that all three curves can be

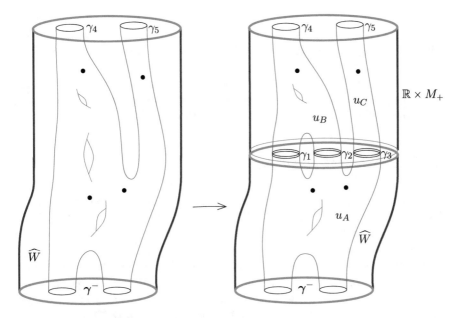

Fig. 8.8 The degeneration scenario behind the gluing map (8.17)

assumed Fredholm regular. Then there are open neighborhoods $\mathcal{U}_A^\$$ and $\mathcal{U}_{BC}^\$$,

$$u_A \in \mathcal{U}_A^\$ \subset \mathcal{M}_{1,2}^\$(J; (\gamma_1, \gamma_2, \gamma_3), \boldsymbol{\gamma}^-),$$

$$[(u_B, u_C)] \in \mathcal{U}_{BC}^\$ \subset \left(\mathcal{M}_{1,1}^\$(J_+; \gamma_4, (\gamma_1, \gamma_2)) \times \mathcal{M}_{0,1}^\$(J_+; \gamma_5, \gamma_3) \right) \Big/ \mathbb{R}$$

which are smooth manifolds of dimensions

$$\dim \mathcal{U}_A^\$ = \text{vir-dim}\, \mathcal{M}_{1,2}(J; (\gamma_1, \gamma_2, \gamma_3), \boldsymbol{\gamma}^-),$$

$$\dim \mathcal{U}_{BC}^\$ = \text{vir-dim}\, \mathcal{M}_{1,1}(J_+; \gamma_4, (\gamma_1, \gamma_2)) + \text{vir-dim}\, \mathcal{M}_{0,1}(J_+; \gamma_5, \gamma_3) - 1.$$

Note here that the $\mathbb{R}$-translation action on $\mathbb{R} \times M_+$ is acting simultaneously on u_B and u_C, i.e. we view them as the connected components of a single disconnected curve on which the translation acts. The **gluing** map is then a smooth embedding

$$\Psi : [R_0, \infty) \times \mathcal{U}_A^\$ \times \mathcal{U}_{BC}^\$ \hookrightarrow \mathcal{M}_{3,4}^\$(J; (\gamma_4, \gamma_5), \boldsymbol{\gamma}^-), \tag{8.17}$$

defined for $R_0 \gg 1$, such that for any $u \in \mathcal{U}_A^\$$ and $v \in \mathcal{U}_{BC}^\$$, $\Psi(R, u, v)$ converges in the SFT topology as $R \to \infty$ to the unique building (with asymptotic markers) having main level u and upper level v. Moreover, every sequence of smooth curves degenerating in this way is eventually in the image of Ψ.

In analogous ways, one can define gluing maps for buildings with a main level and a lower level, or more than two levels, or multiple levels in a symplectization (always dividing symplectization levels by the $\mathbb{R}$-action), and one can combine this with the ideas sketched in Sect. 2.1.7 to include nodal degenerations in the picture. It is important to notice that in all such scenarios, the domain and target of the gluing map have the same dimension, e.g. the dimension of both sides of (8.17) is the sum of the virtual dimensions of the three moduli spaces concerned.

8.3.6 Coherent Orientations

Since vir-dim $\mathcal{M}_{g,m}(A; J; \boldsymbol{\gamma}^+, \boldsymbol{\gamma}^-)$ is not always even, it is immediately clear that the trick sketched in Sect. 2.1.8 for defining orientations on $\mathcal{M}_{g,m}(A; J)$ cannot generally be extended to the punctured case: the linearized Cauchy-Riemann operator $\mathbf{D}_u$ associated to a punctured curve u is not always homotopic through Fredholm operators to its complex-linear part $\mathbf{D}_u^{\mathbb{C}}$. The problem here is that since $\dot{\Sigma}$ is not compact, zeroth-order perturbations of Cauchy-Riemann type operators over $\dot{\Sigma}$ are not compact perturbations, hence there is not even any guarantee that $\mathbf{D}_u^{\mathbb{C}}$ is also Fredholm, or if it is, that its index matches that of $\mathbf{D}_u$. These questions depend in general on the asymptotic behavior of the Cauchy-Riemann operator; this is one of the major reasons why we always need to assume that Reeb orbits are nondegenerate or Morse-Bott.

There is a second problem: even if we can assign orientations to all the moduli spaces $\mathcal{M}_{g,m}(A; J; \boldsymbol{\gamma}^+, \boldsymbol{\gamma}^-)$, the compactification $\overline{\mathcal{M}}_{g,m}(A; J; \boldsymbol{\gamma}^+, \boldsymbol{\gamma}^-)$ contains subsets consisting of "broken curves" built out of other components $\mathcal{M}_{h,k}(B; J; \boldsymbol{\gamma}_1, \boldsymbol{\gamma}_2)$, which are meant to be viewed as "boundary strata" of $\mathcal{M}_{g,m}(A; J; \boldsymbol{\gamma}^+, \boldsymbol{\gamma}^-)$. Floer-theoretic relations such as $\partial^2 = 0$ and the more general algebraic properties of SFT rely on these boundary strata being assigned the *boundary orientation*, meaning the chosen orientations need to be compatible with the gluing maps discussed above in Sect. 8.3.5. Talking about gluing maps means that instead of looking at $\mathcal{M}_{g,m}(A; J; \boldsymbol{\gamma}^+, \boldsymbol{\gamma}^-)$, we need to consider the space $\mathcal{M}_{g,m}^{\$}(A; J; \boldsymbol{\gamma}^+, \boldsymbol{\gamma}^-)$ with the extra data of asymptotic markers at each puncture, and the desired condition is then the following:

Definition 8.35. A system of orientations on the moduli spaces $\mathcal{M}_{g,m}^{\$}(A; J; \boldsymbol{\gamma}^+, \boldsymbol{\gamma}^-)$ is **coherent** if all gluing maps (as in (8.17)) are orientation preserving.

Note that one can make sense of this condition without assuming any transversality or smoothness for $\mathcal{M}_{g,m}^{\$}(A; J; \boldsymbol{\gamma}^+, \boldsymbol{\gamma}^-)$, as an "orientation" of this moduli space can be interpreted to mean a continuously varying choice of orientations for the determinant line bundles of the linearized Cauchy-Riemann operators associated to each curve. We saw in Sect. 2.1.8 that this notion of orientations implies the classical one whenever we restrict our attention to Fredholm regular curves.

It turns out that the solution to the first problem mentioned above is to solve the second one: by an algorithm described by Bourgeois and Mohnke [BM04], one can always find a system of coherent orientations on the moduli spaces $\mathcal{M}^{\$}_{g,m}(A; J; \boldsymbol{\gamma}^{+}, \boldsymbol{\gamma}^{-})$. They are not canonical, except in certain cases, e.g. one can make the necessary choices so that if a curve u happens to have the property that $\mathbf{D}_u$ is homotopic through Fredholm operators to its complex-linear part, then the algorithm assigns to u the natural "complex" orientation, just as in the closed case. The prescription is roughly as follows:

(1) For every choice of nondegenerate asymptotic data for Cauchy-Riemann type operators $\mathbf{D}_{+}$ on the trivial bundle over a plane $\mathbb{C}$ with a positive puncture, choose an orientation arbitrarily for the family of all operators that match the chosen data asymptotically. This is possible because the space of Cauchy-Riemann type operators on a fixed domain with fixed asymptotic behavior is affine, hence contractible. For compatibility with the complex-linear case, we can also arrange this choice so that any complex-linear operator gets the complex orientation.

(2) For any Cauchy-Riemann type operator $\mathbf{D}_{-}$ with nondegenerate asymptotic data on the trivial bundle over a plane $\mathbb{C}$ with a negative puncture, use a "linear gluing" construction to glue it to another Cauchy-Riemann type operator $\mathbf{D}_{+}$ over a plane with positive puncture, producing a Cauchy-Riemann type operator on some bundle over S^2. The latter is homotopic through Fredholm operators to its complex-linear part, so it carries a natural complex orientation, and this together with the orientation chosen for $\mathbf{D}_{+}$ in step (1) uniquely determines an orientation for $\mathbf{D}_{-}$ via the coherence condition.

(3) For any Cauchy-Riemann type operator $\dot{\mathbf{D}}$ with nondegenerate asymptotic data on a bundle over a punctured surface $\dot{\Sigma}$, one can use linear gluing to cap off each of the ends of $\dot{\Sigma}$ with planes having positive or negative ends, producing a Cauchy-Riemann type operator $\mathbf{D}$ on some bundle over a closed surface Σ. Assigning the complex orientation to $\mathbf{D}$, the chosen orientations for operators on planes then determine an orientation for $\dot{\mathbf{D}}$ via the coherence condition.

The loose end in this discussion is that we are actually interested in orienting the space $\mathcal{M}_{g,m}(A; J; \boldsymbol{\gamma}^{+}, \boldsymbol{\gamma}^{-})$, not $\mathcal{M}^{\$}_{g,m}(A; J; \boldsymbol{\gamma}^{+}, \boldsymbol{\gamma}^{-})$, and this is where the story becomes especially interesting. The ideal situation would be if all of the maps

$$\mathcal{M}^{\$}_{g,m}(A; J; \boldsymbol{\gamma}^{+}, \boldsymbol{\gamma}^{-}) \to \mathcal{M}^{\$}_{g,m}(A; J; \boldsymbol{\gamma}^{+}, \boldsymbol{\gamma}^{-}) \tag{8.18}$$

defined by rotating the asymptotic marker at some puncture by a suitable root of unity were orientation preserving, as then the orientations would descend to $\mathcal{M}_{g,m}(A; J; \boldsymbol{\gamma}^{+}, \boldsymbol{\gamma}^{-})$, viewed as a quotient of $\mathcal{M}^{\$}_{g,m}(A; J; \boldsymbol{\gamma}^{+}, \boldsymbol{\gamma}^{-})$. But this condition is not true in general—it depends on the Reeb orbits:

Definition 8.36. A nondegenerate Reeb orbit γ is called a **bad orbit** if it is the k-fold cover of another orbit γ_0 such that $k \in \mathbb{N}$ is even and $\mu^{\tau}_{\text{CZ}}(\gamma) - \mu^{\tau}_{\text{CZ}}(\gamma_0)$ is odd. We say γ is a **good orbit** if it is not bad.

In the 3-dimensional case, one can show that the bad orbits are precisely the even covers of so-called *negative hyperbolic* orbits, i.e. orbits γ_0 for which the linearized return map restricted to the contact bundle has two negative eigenvalues, implying that $\mu_{CZ}^\tau(\gamma_0)$ is odd but $\mu_{CZ}^\tau(\gamma_0^{2k})$ is even for all $k \in \mathbb{N}$.

Proposition 8.37 ([BM04]). *For any choice of coherent orientations, the maps (8.18) defined by adjusting asymptotic markers are all orientation preserving if and only if all of the Reeb orbits in the tuples $\boldsymbol{\gamma}^+$ and $\boldsymbol{\gamma}^-$ are good.* □

Corollary 8.38. *The space $\mathcal{M}_{g,m}(A; J; \boldsymbol{\gamma}^+, \boldsymbol{\gamma}^-)$ is orientable whenever all of the Reeb orbits in the tuples $\boldsymbol{\gamma}^+$ and $\boldsymbol{\gamma}^-$ are good.* □

As a matter of interest, we also mention the following result from [BM04], which is the reason for the "supersymmetric" nature of the algebra in SFT:

Proposition 8.39. *Suppose $\widehat{\boldsymbol{\gamma}}^+$ is an ordered tuple of Reeb orbits obtained from $\boldsymbol{\gamma}^+$ by interchanging two orbits γ_j^+, γ_k^+ in the list, and let*

$$i : \mathcal{M}_{g,m}^\$(A; J; \boldsymbol{\gamma}^+, \boldsymbol{\gamma}^-) \to \mathcal{M}_{g,m}^\$(A; J; \widehat{\boldsymbol{\gamma}}^+, \boldsymbol{\gamma}^-)$$

denote the natural map defined by interchanging the corresponding pair of punctures. Then i is orientation reversing if and only if both of the numbers

$$n - 3 + \mu_{CZ}^\tau(\gamma_i^+)$$

for $i = j, k$ are odd. A similar statement holds for permutations of negative punctures. □

The significance of the number $n - 3 + \mu_{CZ}^\tau(\gamma)$ is that for a suitable choice of trivialization τ, it is the virtual dimension of a moduli space of holomorphic planes positively or negatively asymptotic to γ. The proof of Proposition 8.39 is based on the coherence condition together with the fact that a Cartesian product of two such moduli spaces of planes will change its orientation under change of order if and only if both of them have odd dimension. See [Wend, Lecture 11] for a more detailed sketch of the idea, and [BM04] for the complete proof.

8.3.7 Automatic Transversality

We now specialize to the case $n = 2$, so $\dim \widehat{W} = 4$ or $\dim M = 3$. The generalization of Theorem 2.44 to the punctured case comes from [Wen10a].

Theorem 8.40. *Suppose $n = 2$, and $u : (\dot{\Sigma} = \Sigma \backslash \Gamma, j) \to (\widehat{W}, J)$ is an immersed finite-energy punctured J-holomorphic curve with nondegenerate asymptotic orbits $\{\gamma_z\}_{z \in \Gamma}$ satisfying*

$$\mathrm{ind}(u) > 2g - 2 + \#\Gamma_{\mathrm{even}},$$

where g is the genus of Σ and $\Gamma_{even} \subset \Gamma \subset \Sigma$ is the set of punctures z at which $\mu^\tau_{CZ}(\gamma_z)$ is even. Then u is Fredholm regular. □

Corollary 8.41. *If $n = 2$, every immersed finite-energy punctured J-holomorphic sphere with nonnegative index and only nondegenerate asymptotic orbits with odd Conley-Zehnder index is Fredholm regular.*

The condition on odd Conley-Zehnder indices tends to seem baffling on first glance, so let us sketch a direct proof of Corollary 8.41. The idea closely resembles the proof of Theorem 2.46: since u is immersed, the problem can be reduced to showing that the normal Cauchy-Riemann operator $\mathbf{D}^N_u$ is surjective, where $\mathbf{D}^N_u$ is defined on a suitable Sobolev space of sections of the normal bundle $N_u \to \dot{\Sigma}$. This is a Fredholm operator, and the punctured generalization of the Riemann-Roch formula (see [Sch95] or [Wend, Lecture 5]) gives its Fredholm index

$$\text{ind}\,\mathbf{D}^N_u = \chi(\dot{\Sigma}) + 2c^\tau_1(N_u) + \sum_{z \in \Gamma^+} \mu^\tau_{CZ}(\gamma_z) - \sum_{z \in \Gamma^-} \mu^\tau_{CZ}(\gamma_z).$$

Here τ is the usual arbitrary choice of trivializations for $\xi_\pm$ along each Reeb orbit, and we are using the fact that u resembles a trivial cylinder near infinity to identify N_u in this region with the contact bundle along u and thus define the relative first Chern number $c^\tau_1(N_u) \in \mathbb{Z}$. Using the splitting $u^*T\widehat{W} = T\dot{\Sigma} \oplus N_u$ and writing down the appropriate relative first Chern number of $T\dot{\Sigma} \to \dot{\Sigma}$ in terms of the Euler characteristic, we have

$$c^\tau_1(u^*T\widehat{W}, J) = \chi(\dot{\Sigma}) + c^\tau_1(N_u) \tag{8.19}$$

and thus

$$\text{ind}\,\mathbf{D}^N_u = \text{ind}(u).$$

We will be done if we can show that $\dim \ker \mathbf{D}^N_u \leqslant \text{ind}\,\mathbf{D}^N_u$. The objective is to do this by controlling the algebraic count of zeroes for nontrivial sections $\eta \in \ker \mathbf{D}^N_u \subset \Gamma(N_u)$. Unlike the closed case, such control will not come from topology alone: if the set of punctures is nonempty, then the count of zeroes for generic sections $\eta \in \Gamma(N_u)$ is not homotopy invariant, as there is nothing to prevent zeroes from escaping to infinity under homotopies or emerging from infinity under small perturbations. Put another way, $\dot{\Sigma}$ is an open surface and thus retracts to its 1-skeleton, so $N_u \to \dot{\Sigma}$ is a trivial bundle and there is no apparent control over the count of zeroes for its sections. But the elements of $\ker \mathbf{D}^N_u$ are not just arbitrary sections: they satisfy a linear Cauchy-Riemann type equation, which means (by the similarity principle) that their zeroes are isolated and positive, and crucially, there is also an asymptotic version of this statement.

To express it properly, suppose more generally that $E \to \dot{\Sigma}$ is a complex line bundle with a trivialization τ defined near infinity, and $\eta \in \Gamma(E)$ is a smooth section whose zeroes are contained in a compact subset of $\dot{\Sigma}$. If the zero set $\eta^{-1}(0) \subset \dot{\Sigma}$ is finite, then each individual zero has a well-defined order $o(\eta, z) \in \mathbb{Z}$ and we will denote the algebraic count of zeroes by

$$Z(\eta) = \sum_{z \in \eta^{-1}(0)} o(\eta, z) \in \mathbb{Z}.$$

Whether $\eta^{-1}(0)$ is finite or not, having it confined to a compact subset means also that at each puncture $z \in \Gamma^{\pm}$, η has a well-defined **asymptotic winding number**

$$\mathrm{wind}^{\tau}(\eta, z) \in \mathbb{Z}$$

with respect to τ. It is defined for $z \in \Gamma^{+}$ as the winding number of the loop in $\mathbb{C} \backslash \{0\}$ defined by expressing η with respect to τ along a small circle in $\dot{\Sigma}$ winding once clockwise around z. For $z \in \Gamma^{-}$, the definition is the same, except that the circle around z winds counterclockwise.

Exercise 8.42. Show that if the section $\eta \in \Gamma(E)$ in the above scenario has only finitely many zeroes, then

$$Z(\eta) = c_1^{\tau}(E) + \sum_{z \in \Gamma^{+}} \mathrm{wind}^{\tau}(\eta, z) - \sum_{z \in \Gamma^{-}} \mathrm{wind}^{\tau}(\eta, z).$$

The following asymptotic winding result is a combination of theorems from [HWZ96a] and [HWZ95a].

Proposition 8.43. *If $n = 2$ and $u : (\dot{\Sigma} = \Sigma \backslash \Gamma, j) \to (\widehat{W}, J)$ is an immersed finite-energy punctured J-holomorphic curve with nondegenerate asymptotic orbits $\{\gamma_z\}_{z \in \Gamma}$, then for any nontrivial section $\eta \in \ker \mathbf{D}_u^N$, the zero set $\eta^{-1}(0) \subset \dot{\Sigma}$ is contained in a compact subset. Moreover, its asymptotic winding numbers satisfy*

$$\mathrm{wind}^{\tau}(\eta, z) \leqslant \alpha_{-}^{\tau}(\gamma_z) \quad \text{for} \quad z \in \Gamma^{+},$$

$$\mathrm{wind}^{\tau}(\eta, z) \geqslant \alpha_{+}^{\tau}(\gamma_z) \quad \text{for} \quad z \in \Gamma^{-},$$

where for any nondegenerate Reeb orbit γ with parity $p(\gamma) \in \{0, 1\}$, the integers $\alpha_{\pm}^{\tau}(\gamma) \in \mathbb{Z}$ are uniquely determined by the relations

$$2\alpha_{-}^{\tau}(\gamma) + p(\gamma) = \mu_{\mathrm{CZ}}^{\tau}(\gamma) = 2\alpha_{+}^{\tau}(\gamma) - p(\gamma). \tag{8.20}$$

$\square$

This motivates the definition

$$Z_{\infty}(\eta) = \sum_{z \in \Gamma^{+}} \left[\alpha_{-}^{\tau}(\gamma_z) - \mathrm{wind}^{\tau}(\eta, z) \right] + \sum_{z \in \Gamma^{-}} \left[\mathrm{wind}^{\tau}(\eta, z) - \alpha_{+}^{\tau}(\gamma_z) \right],$$

which is nonnegative for every $\eta \in \ker \mathbf{D}_u^N$ and, by Exercise 8.42, satisfies

$$Z(\eta) + Z_\infty(\eta) = c_1^\tau(N_u) + \sum_{z \in \Gamma^+} \alpha_-^\tau(\gamma_z) - \sum_{z \in \Gamma^-} \alpha_+^\tau(\gamma_z). \tag{8.21}$$

One can now check that the right hand side of this expression is *independent of the trivialization* τ, i.e. while each individual term is manifestly τ-dependent, the dependencies cancel out in the sum. It follows that while $Z(\eta)$ can change as $\eta \in \ker \mathbf{D}_u^N$ is varied, the sum $Z(\eta) + Z_\infty(\eta)$ cannot, which motivates the interpretation of $Z_\infty(\eta)$ as an algebraic count of "hidden" zeroes of η that may emerge from infinity under small perturbations.

One can go further and replace the right hand side of (8.21) with an expression that depends only on the topology of the domain $\dot{\Sigma}$ and the relative homology class of u. Indeed, using (8.19), the right hand side of (8.21) becomes

$$c_N(u) := c_1^\tau(u^*T\widehat{W}) - \chi(\dot{\Sigma}) + \sum_{z \in \Gamma^+} \alpha_-^\tau(\gamma_z) - \sum_{z \in \Gamma^-} \alpha_+^\tau(\gamma_z). \tag{8.22}$$

Note that this expression is well defined for *all* finite-energy curves u with nondegenerate asymptotic orbits; there is no longer a requirement for u to be immersed. We call $c_N(u)$ the **normal Chern number** of u. Its first interpretation comes from (8.21), which now implies that for any immersed curve u' homotopic to u and any nontrivial section η in the kernel of its normal Cauchy-Riemann operator,

$$0 \leqslant Z(\eta) + Z_\infty(\eta) = c_N(u), \tag{8.23}$$

so in particular $c_N(u)$ gives an upper bound on the actual algebraic count of zeroes $Z(\eta)$. We will also see in Sect. 8.3.8 that $c_N(u)$ furnishes the natural replacement for the term $c_1([u]) - \chi(\Sigma)$ in the punctured version of the adjunction formula.

The normal Chern number is typically easiest to compute via its relation to the index:

Proposition 8.44. *For any curve u as in Proposition 8.43,*

$$2c_N(u) = \mathrm{ind}(u) - 2 + 2g + \#\Gamma_{\mathrm{even}},$$

where g is the genus of Σ and $\Gamma_{\mathrm{even}} \subset \Gamma \subset \Sigma$ is the set of punctures with parity 0.

Proof. Denote $\Gamma_{\mathrm{odd}} := \Gamma \backslash \Gamma_{\mathrm{even}}$. We combine the definition of $c_N(u)$ with the index formula (8.16), set $n = 2$, and plug in the relations (8.20):

$$2c_N(u) = 2c_1^\tau(u^*T\widehat{W}) - 2\chi(\dot{\Sigma}) + \sum_{z \in \Gamma^+} 2\alpha_-^\tau(\gamma_z) - \sum_{z \in \Gamma^-} 2\alpha_+^\tau(\gamma_z)$$

$$= 2c_1^\tau(u^*T\widehat{W}) - \chi(\dot{\Sigma}) - (2 - 2g - \#\Gamma) + \sum_{z \in \Gamma^+} [\mu_{\mathrm{CZ}}^\tau(\gamma_z) - p(\gamma_z)]$$

$$- \sum_{z \in \Gamma^-} [\mu_{\mathrm{CZ}}^\tau(\gamma_z) + p(\gamma_z)]$$

$$= \mathrm{ind}(u) - 2 + 2g + \#\Gamma - \#\Gamma_{\mathrm{odd}} = \mathrm{ind}(u) - 2 + 2g + \#\Gamma_{\mathrm{even}}.$$

$\square$

Proof of Corollary 8.41. Assume u is immersed, $g = 0$, $\#\Gamma_{\mathrm{even}} = 0$ and $\mathrm{ind}(u) \geqslant 0$. Proposition 8.44 then gives $2c_N(u) = \mathrm{ind}(u) - 2 = \mathrm{ind}\,\mathbf{D}_u^N - 2$. Note that the condition $\#\Gamma_{\mathrm{even}} = 0$ implies that $\mathrm{ind}(u)$ is even (cf. Exercise 8.33), so we are free to write

$$\mathrm{ind}\,\mathbf{D}_u^N = 2 + 2k$$

with $k = c_N(u)$ an integer greater than or equal to -1. If $k = -1$, then since any nontrivial section $\eta \in \ker\mathbf{D}_u^N$ must satisfy $Z(\eta) \leqslant k$, the similarity principle implies that no such sections exist, hence $\ker\mathbf{D}_u^N$ is trivial and, since $\mathrm{ind}\,\mathbf{D}_u^N = 0$, so is $\mathrm{coker}\,\mathbf{D}_u^N$. Now assume $k \geqslant 0$, pick any $k + 1$ distinct points $z_0, \ldots, z_k \in \dot{\Sigma}$ and consider the map

$$\Phi : \ker\mathbf{D}_u^N \to (N_u)_{z_0} \oplus \ldots \oplus (N_u)_{z_k} : \eta \mapsto (\eta(z_0), \ldots, \eta(z_k)).$$

Since $Z(\eta) \leqslant k$ for every nontrivial $\eta \in \ker\mathbf{D}_u^N$, Φ is an injective map, implying

$$\dim\ker\mathbf{D}_u^N \leqslant 2(k + 1) = \mathrm{ind}\,\mathbf{D}_u^N,$$

hence $\mathrm{coker}\,\mathbf{D}_u^N$ is again trivial. $\square$

There are also generalizations of Theorem 8.40 to allow curves with non-immersed points and/or Morse-Bott asymptotic orbits (see Remark 8.52), as well as a generalization involving pointwise constraints as in Theorem 2.46. The former is contained in the main result of [Wen10a], while the latter is a straightforward exercise combining those arguments with the proof of Theorem 2.46.

8.3.8   Intersection Theory

The above discussion of automatic transversality gives a foretaste of the intersection theory of punctured holomorphic curves, due to Siefring [Sie08, Sie11]. We shall now give a very brief overview of this theory, referring to [Wenf] for a more detailed introduction. The version of the theory discussed here is valid whenever all asymptotic orbits are Morse-Bott, but we should stipulate that "homotopy-invariance" then always refers to homotopies of curves with *fixed asymptotic orbits*, i.e. the orbits are not allowed to move through their respective Morse-Bott families (see the discussion of *constrained* vs. *unconstrained* punctures in Remark 8.52). A more general intersection theory that relaxes this condition is outlined in [Wen10a, §4.1].

Naively counting intersections does not give homotopy-invariant results on noncompact domains, but as in the proof of Corollary 8.41 above, one can compensate by keeping track of extra terms—analogous to $Z_\infty(\eta)$ in (8.23)—which count "hidden" intersections at infinity. Instead of a homological intersection number $[u] \cdot [v]$, any pair of proper smooth maps u and v from punctured Riemann surfaces into $\widehat{W}$ with asymptotic approach to Reeb orbits at their punctures can be assigned an intersection product

$$[u] * [v] \in \mathbb{Z},$$

which depends only on the relative homology classes of u and v and their asymptotic orbits. If u and v have only finitely many intersection points, then this product takes the form

$$[u] * [v] = u \cdot v + \iota_\infty(u, v),$$

where $u \cdot v \in \mathbb{Z}$ denotes the usual signed count of isolated intersections, and the extra term $\iota_\infty(u, v) \in \mathbb{Z}$ can be expressed in terms of the difference between certain asymptotic winding numbers and their theoretical bounds determined by the integers $\alpha_\pm^\tau(\gamma)$ as in Proposition 8.43. If u and v are both finite-energy punctured J-holomorphic curves with Morse-Bott asymptotic orbits and they have non-identical images (i.e. they are not both reparametrizations or covers of the same simple curve), then Siefring's relative asymptotic formula [Sie08] implies that their intersections cannot accumulate near infinity, thus the above decomposition of $[u] * [v]$ is well defined, and positivity of intersections implies that $u \cdot v \geq 0$, while a corresponding asymptotic statement analogous to Proposition 8.43 implies $\iota_\infty(u, v) \geq 0$. One should interpret $\iota_\infty(u, v)$ in this case as the algebraic count of "hidden" intersections that *can emerge from infinity* under small J-holomorphic perturbations of either u or v. The homotopy-invariant condition $[u] * [v] = 0$ now suffices to ensure that u and v are disjoint unless they have identical images, though in contrast to the closed case, it is also possible for u and v to be disjoint (but admit intersecting perturbations) when $[u] * [v] > 0$.

The $*$-pairing admits a natural extension to nodal and/or multi-level objects such as holomorphic buildings, so that homotopy-invariance generalizes to the statement that $*$ is continuous with respect to convergence in the SFT compactification. Relations such as the following formula for nodal curves are then straightforward consequences of the definition:

Proposition 8.45. *Suppose u and v are each J-holomorphic buildings in $\widehat{W}$ with no upper or lower levels, and $u^1, \ldots, u^N$ denote the smooth components of u. Then $[u] * [v] = \sum_{i=1}^N [u^i] * [v]$.* □

While we will not need it for the applications in Chap. 9, we should mention that there exists a similar but more complicated formula for intersection numbers between multi-level buildings; see [Sie11, Theorem 2.1(4)]. It involves extra contributions from what might be called "hidden intersections at intermediate infinity," i.e. intersections that can emerge from breaking orbits when levels are glued together to form smooth curves. This phenomenon has given rise to interesting applications, such as a version of contact homology counting curves in the complement of a fixed Reeb orbit, see [Mom11, HMS15].

In Chap. 9 we will need a criterion for proving $\iota_\infty(u, v) = 0$, and this merits a digression into some further intersection-theoretic quantities that are meaningful only in the $\mathbb{R}$-invariant case.

Recall that if $J : T(\mathbb{R} \times M) \to T(\mathbb{R} \times M)$ belongs to $\mathcal{J}(M, \alpha)$, then it preserves the subbundles ξ and $\mathbb{R} \oplus \mathbb{R}R_\alpha$, so we can consider the fiberwise linear projection to the former along the latter; denote this by

$$\pi_\xi : T(\mathbb{R} \times M) \to \xi.$$

A J-holomorphic curve $u : \dot{\Sigma} \to \mathbb{R} \times M$ then gives rise to a smooth section of the complex line bundle $\mathrm{Hom}_\mathbb{C}(T\dot{\Sigma}, u^*\xi) \to \dot{\Sigma}$, namely

$$\pi_\xi \circ Tu \in \Gamma(\mathrm{Hom}_\mathbb{C}(T\dot{\Sigma}, u^*\xi)).$$

Fix trivializations on $T\dot{\Sigma} \to \dot{\Sigma}$ over the ends, defined via nonzero vector fields that point toward or away from the punctures, so that the relative first Chern number of $T\dot{\Sigma}$ becomes $\chi(\dot{\Sigma})$. This combines with the usual arbitrary choice of trivializations τ for ξ along each Reeb orbit to define asymptotic trivializations for $\mathrm{Hom}_\mathbb{C}(T\dot{\Sigma}, u^*\xi)$, and denoting the latter also by τ, we have

$$c_1^\tau(\mathrm{Hom}_\mathbb{C}(T\dot{\Sigma}, u^*\xi)) = c_1^\tau(u^*\xi) - \chi(\dot{\Sigma}) = c_1^\tau(u^*T(\mathbb{R} \times M)) - \chi(\dot{\Sigma}), \qquad (8.24)$$

where the second equality results from the decomposition of $T(\mathbb{R} \times M)$ into the direct sum of ξ and the trivial complex line bundle $\mathbb{R} \oplus \mathbb{R}R_\alpha$. One interesting consequence of the nonlinear Cauchy-Riemann equation for u is now that in suitable local coordinates near any point, $\pi_\xi \circ Tu$ satisfies a linear Cauchy-Riemann type equation, so by the similarity principle, its zeroes are isolated and positive unless it vanishes identically. Moreover, if all asymptotic orbits of u are Morse-Bott, then the asymptotic formulas of [HWZ96a, HWZ96b] imply controls on the behavior of $\pi_\xi \circ Tu$ near infinity in the same manner as Proposition 8.43, so that in the nontrivial case, it has well-defined asymptotic winding numbers at each puncture and they satisfy the same bounds in terms of the numbers $\alpha_\pm^\tau(\gamma)$. In the notation of Sect. 8.3.7, it is therefore natural to define the (necessarily finite and nonnegative) integer

$$\mathrm{wind}_\pi(u) = Z(\pi_\xi \circ Tu) \geqslant 0,$$

which counts zeroes of $\pi_\xi \circ Tu$, along with the so-called **asymptotic defect**

$$\mathrm{def}_\infty(u) = Z_\infty(\pi_\xi \circ Tu) \geqslant 0,$$

which measures the failure of u to achieve its extremal allowed winding numbers as it approaches each asymptotic orbit. Both of these are well defined if and only if $\pi_\xi \circ Tu$ is not identically zero, which is equivalent to u not being a trivial cylinder or a cover of one. In this situation, we can now combine (8.24) with Exercise 8.42 and the definition of the normal Chern number (8.22) to obtain the useful formula

$$\mathrm{wind}_\pi(u) + \mathrm{def}_\infty(u) = c_N(u). \tag{8.25}$$

Its first obvious application is the following:

Proposition 8.46. *Suppose $J \in \mathcal{J}(M,\alpha)$ and $u = (u^\mathbb{R}, u^M) : \dot\Sigma \to \mathbb{R} \times M$ is a finite-energy punctured J-holomorphic curve with Morse-Bott asymptotic orbits, satisfying $c_N(u) = 0$, such that the image of $u^M : \dot\Sigma \to M$ is not a closed Reeb orbit. Then u^M is immersed and transverse to the Reeb vector field.* □

Indeed, the set of all points at which u^M fails to be immersed and transverse to R_α is precisely the zero-set of $\pi_\xi \circ Tu$. This method for finding surfaces in M transverse to R_α has produced a number of impressive applications to the dynamical study of Reeb flows, see e.g. [HWZ98, HWZ03, HLS15, Bra15]. It is important to understand however that this discussion only makes sense in the $\mathbb{R}$-invariant case: the positivity of the zeroes of $\pi_\xi \circ Tu$ can be understood as an infinitessimal symptom of the fact that since not only u but also its $\mathbb{R}$-translations in $\mathbb{R} \times M$ are all J-holomorphic, they intersect each other positively.

A second application of (8.25) follows from the characterization of $\iota_\infty(u,v)$ in [Sie11] in terms of asymptotic winding numbers, analogously to the definition of $\mathrm{def}_\infty(u)$:

Proposition 8.47. *For $J \in \mathcal{J}(M,\alpha)$, suppose $u : \dot\Sigma \to \mathbb{R} \times M$ is a finite-energy punctured J-holomorphic curve whose asymptotic orbits are all Morse-Bott, γ is an asymptotic orbit of u, $u_\gamma : \mathbb{R} \times S^1 \to \mathbb{R} \times M$ denotes the trivial cylinder over γ, and u and u_γ do not have identical images. If $\mathrm{def}_\infty(u) = 0$, then $\iota_\infty(u, u_\gamma) = 0$.* □

Here is a corollary that will be extremely useful in Chap. 9, see e.g. the proof of Theorem 9.4.

Proposition 8.48. *For $J \in \mathcal{J}(M,\alpha)$, suppose $u = (u^\mathbb{R}, u^M) : \dot\Sigma \to \mathbb{R} \times M$ is a finite-energy punctured J-holomorphic curve with only positive punctures, all asymptotic orbits are Morse-Bott, $c_N(u) = 0$, and the image of $u^M : \dot\Sigma \to M$ is disjoint from that of all of its asymptotic orbits. Assume additionally that $v : \dot\Sigma' \to \mathbb{R} \times M$ is a finite-energy punctured J-holomorphic curve with Morse-Bott asymptotic orbits such that every positive asymptotic orbit of v is also an asymptotic orbit of u. Then $[u] * [v] = 0$.*

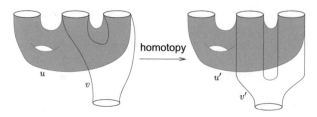

Fig. 8.9 A homotopy through asymptotically cylindrical maps for the proof of Proposition 8.48

Proof. We first observe that since u has no negative punctures, it cannot be a cover of a trivial cylinder, so $\mathrm{wind}_\pi(u)$ and $\mathrm{def}_\infty(u)$ are well defined and $c_N(u) = 0$ thus implies via (8.25) that both vanish. Proposition 8.47 then implies that $\iota_\infty(u, u_\gamma) = 0$ for the trivial cylinder u_γ over every asymptotic orbit γ of u. Since u^M does not intersect γ, u also does not intersect u_γ, hence we deduce

$$[u] * [u_\gamma] = u \cdot u_\gamma + \iota_\infty(u, u_\gamma) = 0.$$

With this understood, we now homotop u and v to new maps u' and v' respectively as depicted in Fig. 8.9: here u' is simply an $\mathbb{R}$-translation of u whose image is contained (since there are no negative punctures) in $[0, \infty) \times M$, while v' is not a J-holomorphic curve, but is instead a smooth asymptotically cylindrical map arranged so that the portion lying in $[0, \infty) \times M$ precisely matches the trivial cylinders over the asymptotic orbits of v. We can now compute $[u] * [v] = [u'] * [v']$ as $u' \cdot v' + \iota_\infty(u', v')$, which gives the same answer as it would if v' were simply a disjoint union of trivial cylinders over asymptotic orbits of u', hence 0. $\square$

Moving back to the general case of a completed cobordism $\widehat{W}$, the adjunction formula generalizes in [Sie11] to somewhere injective finite-energy punctured curves $u : \dot{\Sigma} = \Sigma \backslash \Gamma \to \widehat{W}$ as

$$[u] * [u] = 2\left[\delta(u) + \delta_\infty(u)\right] + c_N(u) + \left[\bar{\sigma}(u) - \#\Gamma\right]. \tag{8.26}$$

Here $\delta(u)$ is defined as in the closed case and is thus nonnegative, with strict inequality unless u is embedded; the fact that it is well defined even though $\dot{\Sigma}$ is not compact depends again on asymptotic results from [Sie08] which prevent self-intersections or non-immersed points from accumulating at infinity. The asymptotic contribution $\delta_\infty(u)$ is also nonnegative, and it is the count of hidden self-intersections that may emerge from infinity under small perturbations. The extra term $\bar{\sigma}(u) \geq \#\Gamma$, called the **spectral covering number** of u, is a measure of covering multiplicities for certain asymptotic eigenfunctions associated with the asymptotic orbits of u. The main thing one needs to know about these quantities for the applications discussed in Chap. 9 is the following:

Proposition 8.49. *If u is a simple curve whose asymptotic orbits are all distinct and simply covered, then $\delta_\infty(u) = \bar{\sigma}(u) - \#\Gamma = 0$.* $\square$

To close this brief survey, we state an analogue for punctured curves of Proposition 2.53, describing curves that locally foliate the target space. Suppose $u : (\dot{\Sigma}, j) \to (\widehat{W}, J)$ satisfies the hypotheses of Corollary 8.41: its genus is zero, $\mathrm{ind}(u) \geqslant 0$ and all its asymptotic orbits have odd Conley-Zehnder index. Suppose additionally that u is embedded and all its asymptotic orbits are distinct and simply covered. The index in this case must be even, so let us write

$$\mathrm{ind}(u) = 2 + 2m,$$

where by Proposition 8.44, $c_N(u) = m$. Now $\delta(u) = 0$ since u is embedded, and Proposition 8.49 implies that $\delta_\infty(u)$ and $\bar{\sigma}(u) - \#\Gamma$ also vanish, so the adjunction formula gives

$$[u] * [u] = m.$$

Using this and the zero-counting trick outlined for the proof of Corollary 8.41 above, the proof of the following statement is entirely analogous to that of Proposition 2.53:

Proposition 8.50. *Suppose* $\dim \widehat{W} = 4$, *and* $u : (\dot{\Sigma} = \Sigma \backslash \Gamma, j) \to (\widehat{W}, J)$ *is an embedded finite-energy punctured J-holomorphic sphere such that*

(i) All asymptotic orbits $\{\gamma_z\}_{z \in \Gamma}$ are nondegenerate, simply covered, and have odd Conley-Zehnder index[8];

(ii) For any two punctures z, ζ of the same sign, the orbits γ_z and γ_ζ are distinct;

(iii) $\mathrm{ind}(u) = 2 + 2m$ for some integer $m \geqslant 0$.

Then for any choice of pairwise distinct points $p_1, \ldots, p_m \in u(\dot{\Sigma})$, u belongs to a smooth 2-parameter family of embedded J-holomorphic curves that all intersect each other transversely at the points $p_1, \ldots, p_m$ and foliate an open neighborhood of $u(\dot{\Sigma}) \backslash \{p_1, \ldots, p_m\}$ in $\widehat{W} \backslash \{p_1, \ldots, p_m\}$. □

In the $\mathbb{R}$-invariant setting, the $m = 0$ case of the above proposition has an especially nice consequence that is worth stating separately. Write

$$u = (u^{\mathbb{R}}, u^M) : (\dot{\Sigma}, j) \to (\mathbb{R} \times M, J),$$

and note that for each $c \in \mathbb{R}$, the curve $u^c := (u^{\mathbb{R}} + c, u^M)$ is also J-holomorphic, thus defining a smooth 1-parameter family of curves. If $\mathrm{ind}(u) = 2$, then u cannot be a trivial cylinder (these have index 0), so u^a and u^b are inequivalent curves for $a \neq b$. Then if $[u] * [u] = 0$, the homotopy invariance of the intersection product

---

[8]The condition in Proposition 8.50 requiring asymptotic orbits to be distinct and simply covered can be relaxed somewhat, e.g. a version of this result for embedded planes asymptotic to a multiply covered Reeb orbit is used to give a dynamical characterization of the standard contact lens spaces in [HLS15].

implies $[u] * [u^c] = 0$ and thus u and u^c are disjoint for all $c \in \mathbb{R}$. It follows that the map $u^M : \dot{\Sigma} \to M$ is injective. We also have $c_N(u) = 0$ in this case, so Proposition 8.46 implies that u^M is also immersed, and thus embedded. One can say slightly more, because Proposition 8.50 gives a smooth 2-parameter family of nearby curves, and the $\mathbb{R}$-translations only account for one parameter. Dividing the 2-parameter family by the $\mathbb{R}$-action, we obtain a smooth 1-parameter family of curves whose projections to M are all embedded and (since none may intersect any $\mathbb{R}$-translation of the others) also disjoint:

Proposition 8.51. *Given $J \in \mathcal{J}(M, \alpha)$ and $u = (u^{\mathbb{R}}, u^M) : (\dot{\Sigma}, j) \to (\mathbb{R} \times M, J)$ satisfying the hypotheses of Proposition 8.50 with $m = 0$, the projections to M of all curves near u in its moduli space from a smooth 1-parameter family of embeddings $\dot{\Sigma} \hookrightarrow M$ that are transverse to the Reeb vector field and foliate a neighborhood of $u^M(\dot{\Sigma})$.*   □

Remark 8.52. Almost every result in this section that is stated under a nondegeneracy assumption can be generalized to allow degenerate but Morse-Bott orbits—in most cases, one can even use all the same formulas (e.g. for the index (8.16) and the normal Chern number (8.22)) as long as one has a suitable interpretation for the Conley-Zehnder index $\mu_{CZ}^{\tau}(\gamma)$ of a Morse-Bott orbit γ and the related winding numbers $\alpha_{\pm}^{\tau}(\gamma)$ introduced in Proposition 8.43. The catch is that this interpretation is context-dependent: the same degenerate orbit can have different **effective Conley-Zehnder indices** depending on whether the puncture asymptotic to it is positive or negative, or for that matter, constrained or unconstrained. The latter distinction means the following: in many applications, it is natural to consider a generalization of the moduli space $\mathcal{M}_{g,m}(A; J; \gamma^+, \gamma^-)$ in which the objects $\gamma_j^{\pm}$ making up the data $\gamma^{\pm}$ are allowed to be whole Morse-Bott submanifolds of Reeb orbits instead of single orbits, so that continuous families of curves in $\mathcal{M}_{g,m}(A; J; \gamma^+, \gamma^-)$ can have continuously varying asymptotic orbits at certain punctures. We call any puncture with this property **unconstrained**, and call it **constrained** if the corresponding data in $\gamma^{\pm}$ specifies a fixed orbit (even if that orbit belongs to a Morse-Bott family). The choice of which punctures to constrain or not obviously has an impact on the virtual dimension of $\mathcal{M}_{g,m}(A; J; \gamma^+, \gamma^-)$, but one can absorb this dependence into the usual index formula (8.16) by defining the effective Conley-Zehnder indices appropriately. All results that depend on the even/odd parity of Conley-Zehnder indices, such as the automatic transversality criterion in Theorem 8.40, must then refer to this effective index. It is defined in practice by adding a small number $\pm\epsilon$ to the (degenerate) asymptotic operator of the orbit; see [Wen10a] for details.

There are also other reasonable approaches to the problem of defining indices for Morse-Bott orbits, e.g. [Bou02] uses the half-integer valued *Robbin-Salamon index* from [RS93], though several of the formulas of this section then require modification for the Morse-Bott case.

Chapter 9
Contact 3-Manifolds and Symplectic Fillings

In this chapter, we survey some fundamental results about contact 3-manifolds that can be proved using the technology sketched in the previous chapter.

9.1 Fillings of S^3 and the Weinstein Conjecture

In Sect. 8.2, we discussed the notion of a compact symplectic manifold (W, ω) with *convex* boundary, which induces a contact structure ξ on $M = \partial W$ canonically up to isotopy, making (W, ω) a *strong symplectic filling* of (M, ξ). The simplest special case is Example 8.15: every star-shaped domain in $(\mathbb{R}^{2n}, \omega_{st})$ is an exact symplectic filling of (S^{2n-1}, ξ_{st}), and in fact, all of them are *deformation equivalent* in the following sense.

Definition 9.1. Two (weak or strong) symplectic fillings (W, ω) and (W', ω') of contact manifolds (M, ξ) and (M', ξ') respectively are said to be (weakly or strongly) **symplectically deformation equivalent** if there exists a diffeomorphism $\varphi : W \to W'$ and smooth 1-parameter families of symplectic structures $\{\omega_s\}_{s \in [0,1]}$ on W and contact structures $\{\xi_s\}_{s \in [0,1]}$ on M such that $\omega_0 = \omega$, $\omega_1 = \varphi^* \omega'$, $\xi_0 = \xi$, $\xi_1 = \varphi^* \xi'$, and (W, ω_s) is a (weak or strong) filling of (M, ξ_s) for every $s \in [0, 1]$. If additionally both fillings are exact, we say they are **Liouville deformation equivalent** if the family can be chosen so that (W, ω_s) is an exact filling of (M, ξ_s) for every $s \in [0, 1]$.

Exercise 9.2. Show that if $\{\omega_s\}_{s \in [0,1]}$ is a smooth 1-parameter family of symplectic forms on W that are each convex at ∂W, then there exists a neighborhood of ∂W admitting a smooth 1-parameter family of vector fields $\{V_s\}_{s \in [0,1]}$ that are all positively transverse to the boundary and satisfy $\mathcal{L}_{V_s} \omega_s = \omega_s$ for all s. In particular, (W, ω_0) and (W, ω_1) are then strongly symplectically deformation equivalent.

© Springer International Publishing AG, part of Springer Nature 2018
C. Wendl, *Holomorphic Curves in Low Dimensions*, Lecture Notes
in Mathematics 2216, https://doi.org/10.1007/978-3-319-91371-1_9

Hint: it may be easier to think in terms of the primitives $\lambda_s = \omega_s(V_s, \cdot)$. *Remember that for any given* ω, *the space of Liouville vector fields positively transverse to the boundary is convex.*

Exercise 9.3. Show that every contact form for the standard contact structure on S^{2n-1} can be realized as the restriction of the standard Liouville form of $(\mathbb{R}^{2n}, \omega_{st})$ to some star-shaped hypersurface.

There is of course no constraint on the topology of a compact manifold with boundary *diffeomorphic* to the sphere; just take any closed manifold and remove a ball. The following theorem reveals that, symplectically, the situation is quite different.

Theorem 9.4 (Gromov [Gro85] and Eliashberg [Eli90]). *Weak symplectic fillings of* (S^3, ξ_{st}) *are unique up to symplectic deformation equivalence and blowup. Moreover, every* exact *symplectic filling of* (S^3, ξ_{st}) *is symplectomorphic to a star-shaped domain in* $(\mathbb{R}^4, \omega_{st})$.

The following dynamical result can be proved by almost the same argument, which is an easy variation on a trick due to Hofer [Hof93]:

Theorem 9.5. *Suppose* (M, ξ) *is a closed contact 3-manifold admitting a strong symplectic cobordism* (W, ω) *to* (S^3, ξ_{st}). *Then every contact form for* (M, ξ) *admits a closed Reeb orbit. Moreover, if the cobordism* (W, ω) *is exact, then the Reeb orbit on* (M, ξ) *may be assumed contractible.*

In particular, this implies that (S^3, ξ_{st}) satisfies the Weinstein conjecture, since it admits a (trivial) exact symplectic cobordism to itself. The theorem can also be used to deduce that every *overtwisted* contact 3-manifold admits contractible Reeb orbits—this was originally proved in [Hof93] by a different argument, but it follows from the result above due to a theorem of Etnyre and Honda [EH02] stating that every overtwisted contact 3-manifold admits Stein cobordisms to every other contact 3-manifold. Note that there are plenty of contact manifolds that always admit a closed Reeb orbit but not necessarily a *contractible* one; this is true for instance of the canonical contact form on the unit cotangent bundle of any Riemannian manifold with no contractible geodesics, e.g. the standard 3-torus (see Example 8.16).

Before sketching proofs of the theorems above, we should discuss a basic fact that lies in the background of Theorem 9.4 and all other results on the classification of fillings: given any (weak or strong) filling (W, ω) of a contact manifold (M, ξ), one can produce another filling of (M, ξ) by blowing up (W, ω) along a ball in the interior. Conversely, just as in the setting of closed symplectic 4-manifolds, we will lose no generality by restricting attention to the minimal case:

Theorem 9.6. *Assume* (W, ω) *is a (weak or strong) symplectic filling of a contact 3-manifold* (M, ξ).

> *(1) If* $E_1, \ldots, E_k \subset W$ *is a maximal collection of pairwise disjoint exceptional spheres in the interior, then the filling* $(W, \check{\omega})$ *obtained by blowing down* (W, ω) *at all of these spheres is minimal.*

(2) Minimality of symplectic fillings is invariant under symplectic deformation equivalence.

This can be proved by essentially the same arguments that we explained for Theorems B and C in Chap. 5: it depends on the fact that for generic choices of suitable tame almost complex structures on (W, ω), each exceptional sphere has a unique J-holomorphic representative, and these representatives deform smoothly with any generic homotopy of suitable almost complex structures. Here we are using the word "suitable" to obscure an important detail: in order to carry out the same argument that worked in the closed case, one must be sure that a family of J-holomorphic spheres in the interior of (W, ω) can never collide with the boundary. It turns out that this can be guaranteed for a natural class of tame almost complex structures on any symplectic filling. In order to state the relevant definition, observe that any contact structure $\xi = \ker \alpha$ on a manifold M carries a canonical conformal class of symplectic vector bundle structures, as $d\alpha|_\xi$ is nondegenerate and independent of the choice of contact form up to scaling.

Definition 9.7.   For any almost complex manifold (W, J) with nonempty boundary, we say that the boundary is J-**convex** (or **pseudoconvex**) if the maximal J-complex subbundle

$$\xi := T(\partial W) \cap JT(\partial W) \subset T(\partial W)$$

is a positive (with respect to the boundary orientation) contact structure on ∂W, and its canonical conformal symplectic structure tames $J|_\xi$.

Exercise 9.8.   Show that if (W, ω) is a (weak or strong) symplectic filling, then it admits an ω-tame almost complex structure J that makes the boundary J-convex, and the space of almost complex structures with this property is contractible.

Choosing $J \in \mathcal{J}_\tau(W, \omega)$ so that the boundary is J-convex has the following consequence. Suppose $f : W \to (0, \infty)$ is any smooth function that has the boundary as a regular level set with $df > 0$ in the outward direction. Then since Definition 9.7 is an open condition, the level sets of f sufficiently close to ∂W form a smooth 1-parameter family of J-convex boundaries of smaller domains in W. It is then not hard to show (see [CE12, Lemma 2.7]) that after replacing f with $\varphi \circ f$ for some smooth function $\varphi : (0, \infty) \to (0, \infty)$ with $\varphi'' \gg 0$, we can arrange f so that the 2-form

$$\omega_f := -d(df \circ J)$$

is a symplectic form taming J near the boundary. Any function with this property is said to be a (strictly) J-**convex** or **plurisubharmonic** function. One of the fundamental properties of J-convex functions, which also follows from a routine computation, is that if $u : \Sigma \to W$ is any J-holomorphic curve with image in the domain of f, then

$$-\Delta(f \circ u) < 0,$$

i.e. $f \circ u$ is *strictly subharmonic*. This implies that $f \circ u$ satisfies a maximum principle (see e.g. [Eva98]), so it can have no local maxima. In particular, u cannot touch ∂W at any point in the interior of its domain:

Proposition 9.9. *If (W, J) is an almost complex manifold with J-convex boundary, then it admits no J-holomorphic curve $u : (\Sigma, j) \to (W, J)$ with $u(z) \in \partial W$ for an interior point $z \in \Sigma$.*   □

With this ingredient added, our proofs of Theorems B and C in Chap. 5 also prove Theorem 9.6, as one can use J-convexity to prevent any J-holomorphic exceptional spheres from colliding with the boundary.

Exercise 9.10. Adapt the proof of Theorem A to show that no 4-dimensional symplectic filling can ever contain a symplectically embedded 2-sphere with nonnegative self-intersection.

Exercise 9.11. Extend the previous exercise in the spirit of Chap. 7 to show that no 4-dimensional symplectic filling can ever contain a positively symplectically immersed sphere $S \looparrowright (W, \omega)$ with $c_1([S]) \geqslant 2$.
Hint: use the aruguments behind Lemmas 7.43 and 7.51, but assuming J-convexity at the boundary.

Exercise 9.12. In Chap. 7, we proved the theorem of McDuff [McD92] that the minimal blowdown of a closed symplectic 4-manifold is unique unless it is rational or ruled. Use Exercise 9.11 to adapt the arguments of Corollary 7.9 and show that the blowdown is similarly unique for 4-dimensional symplectic fillings.

Remark 9.13. The natural generalization of the weak filling condition to higher dimensions turns Exercise 9.8 into a definition: a compact symplectic $2n$-manifold (W, ω) with boundary $\partial W = M$ is called a **weak filling** of (M, ξ) if it admits an ω-tame almost complex structure J for which ∂W is pseudoconvex with $\xi = T(\partial W) \cap JT(\partial W)$. This condition can also be expressed in purely symplectic terms without the auxiliary almost complex structure; see [MNW13, Definition 4]. With this definition, Proposition 8.18 on deforming weak fillings with exact boundary to strong ones extends to all dimensions.

To begin the proof of Theorem 9.4, note that since $H^2_{\mathrm{dR}}(S^3) = 0$, Corollary 8.19 implies that every weak filling of (S^3, ξ_{st}) can be deformed to a strong filling, and in light of the above discussion, we therefore lose no generality by restricting attention to minimal strong fillings. In the following proof, we will show that all of these are diffeomorphic to the ball, hence the symplectic form is globally exact and Corollary 8.19 gives yet another deformation, this time to an exact filling. The main task will therefore be to show that all exact fillings of (S^3, ξ_{st}) are symplectomorphic to star-shaped domains. A slightly weaker version of this result was first proved by Gromov in [Gro85], as a corollary to his version of Theorem E characterizing the symplectic $S^2 \times S^2$. Later, Eliashberg [Eli90] sketched a proof using holomorphic disks with a totally real boundary condition—the full details of this proof are explained in [GZ10] and [CE12, Theorem 16.5]. The proof we shall now outline matches both of these in spirit, but uses slightly different technology.

Sketch of a Proof of Theorem 9.4. We begin with an observation about the standard filling $(\overline{B}^4, \omega_{st})$ of (S^3, ξ_{st}). Using the flow of the standard Liouville vector field (8.6) on $(\mathbb{R}^4, \omega_{st})$, there is an obvious symplectomorphism of $(\mathbb{R}^4, \omega_{st})$ to the completion of $(\overline{B}^4, \omega_{st})$. Identifying $\mathbb{R}^4$ with $\mathbb{C}^2$ as usual, the standard complex structure i then belongs to the space $\mathcal{J}(\overline{B}^4, \omega_{st}, \alpha_{st})$, i.e. it is compatible with ω_{st} on $\overline{B}^4$ and its restriction to the cylindrical end $[0, \infty) \times S^3 \cong \mathbb{R}^4 \backslash B^4$ belongs to $\mathcal{J}(S^3, \alpha_{st})$. Note that the Reeb orbits in (S^3, α_{st}) are precisely the fibers of the Hopf fibration $S^3 \to S^3/S^1 = \mathbb{CP}^1$, so they form a smooth 2-parameter family foliating all of S^3 and thus trivially satisfy the Morse-Bott condition. Let us single out two of these orbits, with even parametrizations (cf. Exercise 8.24) $\gamma_1, \gamma_2 : S^1 \to S^3 \subset \mathbb{C}^2$ given by

$$\gamma_1(t) = (e^{2\pi i t}, 0), \qquad \gamma_2(t) = (0, e^{2\pi i t}).$$

We will consider moduli spaces of holomorphic planes in the completion of $(\overline{B}^4, \omega_{st})$ with positive punctures asymptotic to each of these orbits, meaning the puncture is *constrained* in the sense of Remark 8.52. Fixing a global trivialization τ for $\xi_{st} \to S^3$, which is possible since $c_1(\xi_{st}) = 0 \in H^2(S^3)$, one can compute the effective Conley-Zehnder index

$$\mu_{CZ}^\tau(\gamma_i) = 3 \quad \text{for} \quad i = 1, 2.$$

Now observe that $(\mathbb{C}^2, i)$ has two obvious foliations by embedded holomorphic planes, namely those of the form

$$\begin{aligned} u_w : \mathbb{C} \to \mathbb{C}^2 : z \mapsto (z, w), \qquad w \in \mathbb{C}, \\ v_w : \mathbb{C} \to \mathbb{C}^2 : z \mapsto (w, z), \qquad w \in \mathbb{C}. \end{aligned} \qquad (9.1)$$

All of these planes have finite energy and a positive puncture, with each u_w asymptotic to γ_1 and each v_w asymptotic to γ_2. The index formula (8.16) then gives

$$\text{ind}(u_w) = \text{ind}(v_w) = -\chi(\mathbb{C}) + 3 = 2,$$

and they are Fredholm regular by Corollary 8.41. From Proposition 8.44, we deduce

$$c_N(u_w) = c_N(v_w) = 0,$$

so that Siefring's adjunction formula (8.26) in light of Proposition 8.49 gives

$$[u_w] * [u_w] = [v_w] * [v_w] = 0.$$

Using the homotopy invariance of the intersection product, plus the observation that each u_w has a single transverse intersection with each v_w and approaches a different asymptotic orbit, we obtain the relations

$$[u_w] * [u_{w'}] = [v_w] * [v_{w'}] = 0, \qquad [u_w] * [v_{w'}] = 1 \qquad \text{for all } w, w' \in \mathbb{C}.$$

One should think of these planes as analogous to the two families of holomorphic spheres with zero self-intersection that appear in Gromov's characterization of $S^2 \times S^2$ (Theorem E), and we shall make use of them in much the same way.

If (W, ω) is an arbitrary strong filling of (S^3, ξ_{st}), it is not hard to see that its completion $(\widehat{W}, \widehat{\omega})$ also contains some holomorphic planes with the same intersection properties described above. Indeed, by assumption, a neighborhood of ∂W in (W, ω) admits a Liouville vector field V and hence a primitive $\lambda := \omega(V, \cdot)$ of ω such that $\ker \lambda|_{T(\partial W)} = \xi_{st}$ under some identification of ∂W with S^3. According to Exercise 9.3, this means that $\lambda|_{T(\partial W)}$ can be identified with the restriction of the standard Liouville form λ_{st} on $(\mathbb{R}^4, \omega_{st})$ to some star-shaped hypersurface $\Sigma \subset \mathbb{R}^4$. Using the Liouville flow, one can show that the cylindrical end $(\widehat{W} \backslash W, \widehat{\omega})$ is therefore symplectomorphic to the region Ω outside of Σ in $(\mathbb{R}^4, \omega_{st})$. Choose $R > 0$ large enough so that the ball $\overline{B}_R^4$ of radius R contains Σ: then the cylindrical end of $(\widehat{W}, \widehat{\omega})$ contains $(\mathbb{R}^4 \backslash B_R^4, \omega_{st})$, so we can choose an $\widehat{\omega}$-compatible almost complex structure J whose restriction to $\mathbb{R}^4 \backslash B_R^4$ is the standard complex structure i. Now for any $w \in \mathbb{C}$ sufficiently close to ∞, the planes (9.1) are also contained in $\widehat{W}$ and are J-holomorphic; see Fig. 9.1.

The rest of the proof parallels that of Theorem E: first, perturb J to make it generic on the compact portion of $\widehat{W}$ that is not identified with $\mathbb{R}^4 \backslash B_R^4$. Then let $\mathcal{M}_1$ and $\mathcal{M}_2$ denote the moduli spaces of finite-energy J-holomorphic planes in $\widehat{W}$ that are homotopic to the planes u_w and v_w respectively and approach the same fixed asymptotic orbits. All of these curves are somewhere injective since they have only a single positive asymptotic orbit and it is simply covered (see Exercise 8.29), thus the open subsets of $\mathcal{M}_1$ and $\mathcal{M}_2$ consisting of curves that are not confined to the cylindrical end $\mathbb{R}^4 \backslash B_R^4$ (where J is not generic) are smooth 2-dimensional manifolds. Since the curves u_w satisfy the homotopy-invariant conditions $\delta(u_w) + \delta_\infty(u_w) = [u_w] * [u_w] = 0$, and similarly for v_w, all curves in $\mathcal{M}_1$ and $\mathcal{M}_2$ also satisfy these conditions, hence all of them are embedded, and any two distinct curves u and v that both belong to either $\mathcal{M}_1$ or $\mathcal{M}_2$ satisfy $[u] * [v] = [u] * [u] = 0$, implying they are disjoint. Moreover, every pair $(u, v) \in \mathcal{M}_1 \times \mathcal{M}_2$ satisfies $[u] * [v] = 1$, and since their asymptotic orbits are distinct, $\iota_\infty(u, v) = 0$, implying $u \cdot v = 1$, so that u and v have exactly one intersection point and it is transverse.

To finish the argument, we need to understand the compactness properties of $\mathcal{M}_1$ and $\mathcal{M}_2$. For concreteness, consider $\mathcal{M}_1$, as the required argument for $\mathcal{M}_2$ will be identical. Notice that $\mathcal{M}_1$ is obviously noncompact in a trivial way, because it contains a noncompact family of curves u_w for $w \in \mathbb{C}$ near ∞. One can parametrize this family by a half-cylinder $\{w \in \mathbb{C} \mid |w| \geq r\}$ for $r > 0$ sufficiently large, so it constitutes a cylindrical end in the moduli space $\mathcal{M}_1$. We claim that if (W, ω) is minimal, then the complement of this cylindrical end in $\mathcal{M}_1$ is compact.

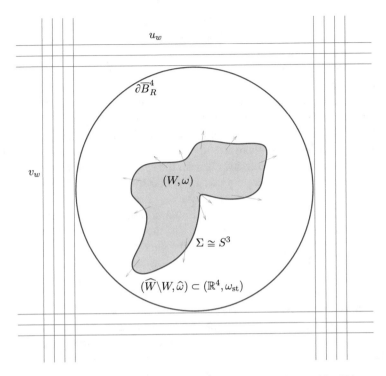

Fig. 9.1 Finding two transverse families of embedded holomorphic planes with self-intersection 0 in the completion of any strong filling (W, ω) of (S^3, ξ_{st})

The claim is based on a uniqueness result for holomorphic curves in the symplectization of (S^3, ξ_{st}). This symplectization can be identified with $\mathbb{C}^2 \backslash \{0\}$ in the same manner that we identified $[0, \infty) \times S^3$ with $\mathbb{R}^4 \backslash B^4$ above, so in particular, the standard complex structure i on $\mathbb{C}^2$ restricts to $\mathbb{C}^2 \backslash \{0\} = \mathbb{R} \times S^3$ as an element of $\mathcal{J}(S^3, \alpha_{st})$. This makes all of the planes u_w for $w \in \mathbb{C} \backslash \{0\}$ into holomorphic curves in the symplectization $\mathbb{R} \times S^3$, and the uniqueness result we need states that these planes and the trivial cylinder over γ_1 are the *only* i-holomorphic curves in $\mathbb{R} \times S^3$ with a single positive end asymptotic to γ_1 and arbitrary negative ends (or none at all). To see this, suppose $v : \dot{\Sigma} \to \mathbb{R} \times S^3$ is a curve with these properties but is not one of the u_w or the trivial cylinder. Then v necessarily has an isolated intersection with one of the curves u_w, as they foliate $\mathbb{R} \times (S^3 \backslash \gamma_1)$, so by positivity of intersections,

$$[v] * [u_w] \geq v \cdot u_w > 0.$$

But this contradicts Proposition 8.48. Note that this also implies a uniqueness result for certain curves in $\widehat{W}$: since J was chosen to match i in the cylindrical region identified with $\mathbb{R}^{2n} \backslash B_R^4$, every J-holomorphic curve that is contained in that region and has a single end asymptotic to γ_1 is of the form u_w for some w. In particular, every element of $\mathcal{M}_1$ is either one of these or has an injective point in the region where J is generic, so they are all Fredholm regular and we now know that $\mathcal{M}_1$ is globally a smooth 2-dimensional manifold.

Now to prove the compactness claim for $\mathcal{M}_1$, suppose $u_k \in \mathcal{M}_1$ is a sequence with no subsequence converging in $\mathcal{M}_1$, and using the SFT compactness theorem, extract a subsequence that converges to a holomorphic building u_∞. Since (W, ω) has a convex boundary component but no concave boundary, u_∞ does not have lower levels, it only has a main level (which could be empty) and a nonnegative number of upper levels. If there are upper levels, then the topmost level of u_∞ contains a nontrivial i-holomorphic curve in the symplectization $\mathbb{R} \times S^3$ with only one positive end, which is asymptotic to γ_1, so by the uniqueness result in the previous paragraph, this curve is one of the u_w. But these have no negative ends, so the only way to produce a connected holomorphic building of this form is if u_∞ has exactly one upper level, consisting of the curve u_w, and its main level is empty. This would imply that as $k \to \infty$, the curves u_k have images contained in $[R_k, \infty) \times S^3 \subset \widehat{W}$ for some sequence $R_k \to \infty$, so the uniqueness result in the previous paragraph implies that they are of the form u_{w_k} for a sequence $w_k \in \mathbb{C}$ diverging to ∞, i.e. they belong to the cylindrical end of $\mathcal{M}_1$.

It remains to understand the case where the limit building u_∞ has no upper levels, meaning it is a nodal curve in $\widehat{W}$. The smooth components of u_∞ must then consist of a single plane u_∞^0 plus a finite collection of spheres $u_\infty^1, \ldots, u_\infty^N$, and the plane is necessarily simple since its asymptotic orbit is simply covered. If this plane is contained in the region where J is not generic, then it must match one of the curves u_w and is therefore regular with index 2; otherwise, the usual generic transversality results imply that it is regular and $\operatorname{ind}(u_\infty^0) \geq 0$. Stokes' theorem implies that the spheres cannot be contained in the cylindrical end since the symplectic form there is exact, thus they also touch the region where J is generic, implying that they are at worst multiple covers of simple regular curves with nonnegative index. We can now use the index relations in Sect. 4.2 again to rule out the possibility that they are multiply covered: by the same index counting arguments that we used in Chap. 4, u_∞ can have at most two smooth components, namely a simple index 0 plane u_∞^0 and a simple index 0 sphere u_∞^1, and Proposition 8.44 gives

$$c_N(u_\infty^0) = c_N(u_\infty^1) = -1.$$

Next we use Proposition 8.45 to break up the relation $[u_\infty] * [u_\infty] = 0$ into a sum over smooth components, and notice that since both are simple, the adjunction formula can be applied to both. The special asymptotic terms on the right hand side of Siefring's formula (8.26) do not appear since u_∞^0 has only one simply covered asymptotic orbit and u_∞^1 has none at all, thus we find

$$\begin{aligned}
0 = [u_\infty] * [u_\infty] &= [u_\infty^0] * [u_\infty^0] + [u_\infty^1] * [u_\infty^1] + 2[u_\infty^0] * [u_\infty^1] \\
&= 2\delta(u_\infty^0) + 2\delta(u_\infty^1) + c_N(u_\infty^0) + c_N(u_\infty^1) + 2[u_\infty^0] * [u_\infty^1] \\
&= 2\delta(u_\infty^0) + 2\delta(u_\infty^1) + 2\left(u_\infty^0 \cdot u_\infty^1 - 1\right),
\end{aligned}$$

where in the last step we've used the fact that u^1_∞ has no punctures and does not have identical image to u^0_∞, implying $[u^0_\infty] * [u^1_\infty] = u^0_\infty \cdot u^1_\infty$. Since the existence of the node implies $u^0_\infty \cdot u^1_\infty \geq 1$, we conclude that $\delta(u^0_\infty) = \delta(u^1_\infty) = 0$ and $u^0_\infty \cdot u^1_\infty = 1$, so in particular, the index 0 J-holomorphic sphere u^1_∞ is embedded. But now applying the adjunction formula again to u^1_∞ reveals that $[u^1_\infty] \cdot [u^1_\infty] = -1$, so it is an exceptional sphere. Choosing $r > 0$ large enough so that u^1_∞ is disjoint from the cylindrical end $[r, \infty) \times S^3 \subset \widehat{W}$, the complement of this cylindrical end is therefore not minimal, but this enlarged domain is symplectically deformation equivalent to the original filling (W, ω), implying via Theorem 9.6 that (W, ω) is not minimal. This finishes the compactness argument: the conclusion is that if (W, ω) is minimal, then each of $\mathcal{M}_1$ and $\mathcal{M}_2$ is a smooth compact 2-dimensional surface with a cylindrical end attached.

By Proposition 8.50 (with Remark 8.52 in mind), both moduli spaces also locally form foliations of $\widehat{W}$, and one can combine this with the compactness result above to argue exactly as in Chap. 6 that they each *globally* foliate $\widehat{W}$ when (W, ω) is minimal. There is therefore a natural map

$$\varphi : \widehat{W} \to \mathcal{M}_1 \times \mathcal{M}_2$$

sending each point $p \in \widehat{W}$ to the unique pair of curves $(u, v) \in \mathcal{M}_1 \times \mathcal{M}_2$ that both pass through p. But since every plane in $\mathcal{M}_1$ intersects every plane in $\mathcal{M}_2$ exactly once transversely, both moduli spaces are diffeomorphic to $\mathbb{C}$, implying that φ is a diffeomorphism to $\mathbb{C}^2$. Since we now know $H^2_{\mathrm{dR}}(W) = 0$, we can apply Corollary 8.19 to assume after a symplectic deformation that the filling (W, ω) is exact. Having done that, a Moser isotopy argument as in the proof of Theorem E can now be used to identify $(\widehat{W}, \widehat{\omega})$ symplectically with $(\mathbb{R}^4, \omega_{\mathrm{st}})$ so that the compact domain $W \subset \widehat{W}$ is identified with a star-shaped domain in $\mathbb{R}^4$.   □

Remark 9.14. The proof above was presented in a way that generalizes well for the classification of strong fillings of $\mathbb{T}^3$ and planar contact manifolds, which we will discuss in the next two sections, but in the special case of (S^3, ξ_{st}), some details can be simplified. For instance, the required uniqueness result for curves in $\mathbb{R} \times S^3$ does not really depend on intersection theory: it suffices rather to observe that any genus zero punctured holomorphic curve $u : \Sigma \to \mathbb{R} \times S^3 = \mathbb{C}^2 \backslash \{0\}$ with a single positive puncture can be extended over its negative punctures to define a holomorphic map $u : \mathbb{C} \to \mathbb{C}^2$ with a pole at infinity, thus a polynomial, and the asymptotic approach to γ_1 or γ_2 dictates that u is affine in one coordinate and constant in the other. This version of the argument does not work for more general contact 3-manifolds, but it does work in higher dimensions, cf. Theorem 9.19 below.

The proof of Theorem 9.5 is a surprisingly easy modification of Theorem 9.4, once one understands the SFT compactness theorem. Arguing by contradiction, suppose one has a strong symplectic cobordism (W, ω) from (M, ξ) to (S^3, ξ_{st}) such that (M, ξ) admits a contact form α with no closed Reeb orbits. The cobordism can then be rescaled and deformed near both boundary components by adding pieces of symplectizations so that, without loss of generality, ω has a primitive λ near ∂W that restricts to S^3 as α_{st} and M as $c\alpha$ for some constant $c > 0$ sufficiently small. Note that if α admits no closed Reeb orbits, then neither does $c\alpha$, so for simplicity of notation, we can rescale α and thus assume $c = 1$. Now choose a compatible almost complex structure $J \in \mathcal{J}(W, \omega, \alpha_{st}, \alpha)$ on the completion $(\widehat{W}, \widehat{\omega})$ whose restriction to the positive end matches the one we used in the proof of Theorem 9.4 above. For this choice, there are again J-holomorphic planes of the form (9.1) in the positive end, and we can define the moduli spaces $\mathcal{M}_1$ and $\mathcal{M}_2$ as before. Now observe that if α has no closed Reeb orbits, then the symplectization $(\mathbb{R} \times M, d(e^t\alpha))$ can have no finite-energy J-holomorphic curves, thus the compactifications of $\mathcal{M}_1$ and $\mathcal{M}_2$ cannot contain any broken holomorphic buildings with levels in the symplectization $(\mathbb{R} \times M, d(e^t\alpha))$. But then the compactifications $\overline{\mathcal{M}}_1$ and $\overline{\mathcal{M}}_2$ are exactly the same as what we found in the filling of (S^3, ξ_{st}), and the argument above implies that (W, ω) is a blowup of a star-shaped domain, which is clearly untrue since it has a concave boundary component. This contradiction implies that α must admit a closed Reeb orbit after all. In fact, the complements of the obvious cylindrical ends in the moduli spaces $\mathcal{M}_1$ and $\mathcal{M}_2$ must fail to be compact, and if α is nondegenerate (see Remark 9.17), then this failure takes the form of a sequence of J-holomorphic planes in $\widehat{W}$ that converge to a broken J-holomorphic building including nontrivial lower levels, which are curves in $(\mathbb{R} \times M, d(e^t\alpha))$ asymptotic to closed orbits of R_α; see Fig. 9.2. If the cobordism (W, ω) is exact, then we can say slightly more: the building must contain at least one J-holomorphic plane in $(\mathbb{R} \times M, d(e^t\alpha))$, whose asymptotic orbit is therefore contractible (Fig. 9.3). This follows from the two exercises below, and thus completes the proof of Theorem 9.5.

Exercise 9.15. Use Stokes' theorem to show that if $(W, d\lambda)$ is an exact symplectic cobordism with $\partial W = (-M_-, \xi_- = \ker\alpha_-) \sqcup (M_+, \xi_+ = \ker\alpha_+)$ and $\lambda|_{TM_\pm} = \alpha_\pm$, then for any $J \in \mathcal{J}_\tau(W, d\lambda, \alpha_+, \alpha_-)$, every nonconstant finite-energy J-holomorphic curve in $\widehat{W}$ has at least one positive end. (In other words, an exact cobordism admits no "holomorphic caps.")

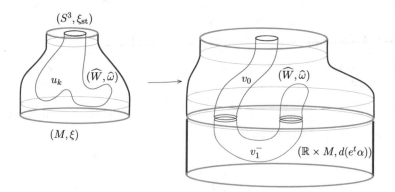

Fig. 9.2 A possible degeneration of holomorphic planes in a strong symplectic cobordism (W, ω) from (M, ξ) to (S^3, ξ_{st}). This particular picture is only possible if the cobordism is not exact, since the main level of the building includes a holomorphic cap (cf. Exercise 9.15)

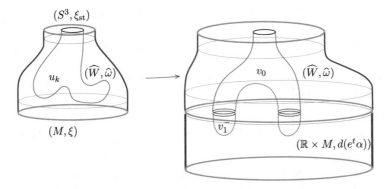

Fig. 9.3 Another possible degeneration as in Fig. 9.2, but this one can also occur if the cobordism is exact, and forces the existence of *contractible* Reeb orbits in (M, ξ)

Exercise 9.16. Assume u is a holomorphic building with arithmetic genus 0 and exactly one positive end, and that each of its connected components has at least one positive end. Show then that each of its levels is a disjoint union of smooth curves without nodes, each having exactly one positive end, and the bottom level is a disjoint union of planes.

Remark 9.17. The proof above for the exact case of Theorem 9.5 added an extra assumption at the last minute: since the SFT compactness theorem requires a nondegeneracy or Morse-Bott hypothesis, we assumed α to be nondegenerate. This assumption can be lifted by a perturbation argument: given an arbitrary contact form α, one can find a sequence of nondegenerate contact forms α_k converging in C^∞ to α, and the above argument provides a contractible orbit for each α_k. Moreover, since (W, ω) is an exact cobordism and the J-holomorphic planes in this proof are all asymptotic to an orbit in $(S^3, \alpha_{\mathrm{st}})$ with fixed period, Stokes' theorem provides a uniform bound for the periods of the contractible orbits we obtain in (M, α_k). By the Arzelà-Ascoli theorem, these then have a subsequence converging to a contractible orbit in (M, α).

Example 9.18. Any Darboux ball in a symplectic 4-manifold contains a Lagrangian torus, e.g. of the form $\gamma \times \gamma \subset \mathbb{C}^2$ for any embedded loop $\gamma \subset \mathbb{C}$ in a neighborhood of the origin. It follows via Example 8.17 that there is a strong symplectic cobordism from $(\mathbb{S}T^*\mathbb{T}^2, \xi_{\mathrm{can}})$ to (S^3, ξ_{st}), but since the flat metric on $\mathbb{T}^2$ has no contractible geodesics, Theorem 9.5 implies that such a cobordism can never be exact. (Note that if the cobordism were exact, then the original Lagrangian torus would be an exact Lagrangian, so the result also follows from Gromov's famous theorem [Gro85] on the nonexistence of exact Lagrangian submanifolds in $(\mathbb{R}^{2n}, \omega_{\mathrm{st}})$.)

In dimensions greater than four, there is no intersection theory of holomorphic curves and thus little hope of proving classification results as strong as Theorem 9.4. The obvious analogue of Theorem 9.5 however is true in higher dimensions, and can be proved by more or less the same argument (see Remark 9.14). The main difference from the 4-dimensional case is that the holomorphic planes in the cobordism need not be embedded and may intersect each other, but one does not actually need such precise control in order to deduce the existence of a contractible Reeb orbit.[1] As far as fillings of (S^{2n-1}, ξ_{st}) are concerned, the classification up to symplectomorphism is unknown, but we do have the following result, often called the *Eliashberg-Floer-McDuff theorem*:

Theorem 9.19 (Eliashberg-Floer-McDuff [McD91]).   *Suppose* (W, ω) *is a strong symplectic filling of* (S^{2n-1}, ξ_{st}) *with* $[\omega]|_{\pi_2(W)} = 0$. *Then* W *is diffeomorphic to a ball.*

One can begin the proof of this theorem the same way as in Theorem 9.4 and carry out the compactness argument as sketched in Remark 9.14, where the condition $[\omega]|_{\pi_2(W)} = 0$ prevents the appearance of nodal curves by precluding the existence in (W, ω) of any nonconstant holomorphic spheres. Without having control over intersections, this argument does not produce a geometric decomposition of W as in the 4-dimensional case, but by making use of the evaluation map and some algebro-topological tools as in Chap. 7, one can still find enough homotopy-theoretic information about W to apply the h-cobordism theorem [Mil65] and thus deduce its diffeomorphism type. This unfortunately tells us nothing about the symplectic structure of the filling, and the only thing known on this subject is a result due to Ivan Smith (see [Sei08a, §6]): if the filling is exact, then it must have vanishing symplectic homology. The h-cobordism approach was recently extended by Barth et al. [BGZ] to give a classification, again up to diffeomorphism, of the symplectically aspherical strong fillings of simply connected contact manifolds that are subcritically Stein fillable, a class in which the standard spheres are the simplest examples. Beyond these outlying cases, there are very few known results constraining the fillings of contact manifolds in higher dimensions, and most of them give much weaker information, e.g. computing the homology of the filling, as in [OV12].

9.2   Fillings of the 3-Torus and Giroux Torsion

Here is another result in the same spirit as Theorem 9.4 but more recent. The torus $\mathbb{T}^3$ has a natural sequence of contact structures defined in coordinates $(t, \phi, \theta) \in S^1 \times S^1 \times S^1 = \mathbb{R}^3/\mathbb{Z}^3$ by

$$\xi_k = \ker \alpha_k, \qquad \alpha_k = \cos(2\pi k t)\, d\theta + \sin(2\pi k t)\, d\phi$$

---

[1] With a bit more care, one can also use the extra intersection-theoretic information in dimension four to prove a stronger result: in the exact cobordism case of Theorem 9.5, (M, ξ) admits a closed Reeb orbit that is *unknotted*, i.e. it is embedded and bounds an embedded disk, see [CW]. There is no obvious analogue of this in higher dimensions.

for $k \in \mathbb{N}$. Recall from Example 8.16 that ξ_1 is also known as the *standard* contact structure ξ_{st} on $\mathbb{T}^3$ and can be identified with the canonical contact structure of the unit cotangent bundle $ST^*\mathbb{T}^2$, thus it has an exact symplectic filling in the form of a unit disk bundle,

$$\partial(\mathbb{D}T^*\mathbb{T}^2, \omega_{can}) = (\mathbb{T}^3, \xi_1).$$

More generally, the star-shaped domains in $(T^*\mathbb{T}^2, \omega_{can})$ form a family of exact fillings of $(\mathbb{T}^3, \xi_1)$, all of which are Liouville deformation equivalent.

The first part of the following result was originally proved by Eliashberg in 1996, who constructed a symplectic cobordism that essentially reduced it to Theorem 9.4. The second part is more recent and is due to the author.

Theorem 9.20 (Eliashberg [Eli96] and Wendl [Wen10b]).

(1) $(\mathbb{T}^3, \xi_k)$ *is not strongly symplectically fillable for* $k \geq 2$.
(2) *Every minimal strong symplectic filling of* $(\mathbb{T}^3, \xi_1)$ *is symplectically deformation equivalent to* $(\mathbb{D}T^*\mathbb{T}^2, \omega_{can})$, *and every exact filling of* $(\mathbb{T}^3, \xi_1)$ *is symplectomorphic to a star-shaped domain in* $(T^*\mathbb{T}^2, \omega_{can})$.

Since $H^2_{dR}(\mathbb{T}^3)$ is nontrivial, weak fillings of $(\mathbb{T}^3, \xi_k)$ cannot always be deformed to strong fillings (cf. Proposition 8.18), so Theorem 9.20 consequently has nothing to say about weak fillings. The following exercise establishes Giroux's observation from [Gir94] that $(\mathbb{T}^3, \xi_k)$ is in fact weakly fillable for every $k \in \mathbb{N}$. It is clear that this construction can be generalized to produce weak fillings with an enormous variety of topologies, so it seems reasonable to expect that in contrast to strong fillings, the problem of classifying weak fillings of $\mathbb{T}^3$ is essentially hopeless.

Exercise 9.21. Let α_k denote the contact forms on $\mathbb{T}^3$ defined above for $k \in \mathbb{N}$.

(a) Show that for every $k \in \mathbb{N}$, $\{\alpha_k + c\,dt\}_{c \in \mathbb{R}}$ is a 1-parameter family of contact forms on $\mathbb{T}^3$, hence the contact structure ξ_k admits a smooth deformation through oriented 2-plane fields $\{\xi_k^s \subset T\mathbb{T}^3\}_{s \in [0,1]}$ such that $\xi_k^0 = \xi_k$, ξ_k^s is contact for every $s \in [0,1)$, and ξ_k^1 is the integrable distribution $\ker dt$.
(b) Show that $\mathbb{D}^2 \times \mathbb{T}^2$ admits a symplectic structure ω such that for every $k \in \mathbb{N}$, $(\mathbb{D}^2 \times \mathbb{T}^2, \omega)$ is a weak filling of $(\mathbb{T}^3, \xi_k')$ for some contact structure ξ_k' isotopic to ξ_k.
Hint: choose ω so that $\omega|_{\ker dt} > 0$ at the boundary.

The uniqueness of minimal strong fillings of $(\mathbb{T}^3, \xi_1)$ follows by the same strategy we used in Theorem 9.4 for (S^3, ξ_{st}), but using holomorphic cylinders instead of planes. Consider first the standard filling $(\mathbb{D}T^*\mathbb{T}^2, \omega_{can})$, whose completion is $(T^*\mathbb{T}^2, \omega_{can})$. The latter can be identified naturally with

$$\mathbb{T}^2 \times \mathbb{R}^2 = S^1 \times S^1 \times \mathbb{R} \times \mathbb{R} \cong (\mathbb{R} \times S^1) \times (\mathbb{R} \times S^1),$$

and we can then define an ω_{can}-compatible almost complex structure[2] on $T^*\mathbb{T}^2$ by $J = i \oplus i$, where i is the standard complex structure on $\mathbb{R} \times S^1 = \mathbb{C}/i\mathbb{Z}$. We then have two transverse families of holomorphic cylinders, each with two positive ends, foliating $T^*\mathbb{T}^2$. One technical difference from the proof of Theorem 9.4 is that the cylinders in each of these families are not all asymptotic to the same Reeb orbits: as can easily be checked, the closed Reeb orbits on $(\mathbb{T}^3, \alpha_1)$ all come in S^1-parametrized Morse-Bott families that foliate 2-tori, and e.g. each of the cylinders $(\mathbb{R} \times S^1) \times \{(s_0, t_0)\}$ is asymptotic to two orbits in distinct Morse-Bott families, with the particular orbits in these families specified by the parameter $t_0 \in S^1$. It is therefore natural to regard these cylinders as elements in a moduli space of curves with *unconstrained* asymptotic orbits in the sense of Remark 8.52, and with this understood, each curve u satisfies

$$\mathrm{ind}(u) = 2, \qquad c_N(u) = 0, \qquad [u] * [u] = 0,$$

where $[u] * [u]$ here denotes the unconstrained Morse-Bott version of Siefring's intersection number, as outlined in [Wen10a, §4.1]. Now given an arbitrary strong filling (W, ω) of $(\mathbb{T}^3, \xi_1)$, the completion $(\widehat{W}, \widehat{\omega})$ matches $(T^*\mathbb{T}^2, \omega_{\mathrm{can}})$ outside a compact subset and thus also contains embedded holomorphic cylinders with self-intersection 0 near infinity. Just as in the case of (S^3, ξ_{st}), the moduli spaces of holomorphic cylinders in $(\widehat{W}, \widehat{\omega})$ that emerge from these turn out to foliate $\widehat{W}$ with two transverse families of smooth holomorphic cylinders if there are no exceptional spheres, providing a diffeomorphism of $\widehat{W}$ to $(\mathbb{R} \times S^1) \times (\mathbb{R} \times S^1) = T^*\mathbb{T}^2$, and in the exact case it can be turned into a symplectomorphism using a Moser isotopy argument as in the proof of Theorem E.

One major detail that differs from the argument for (S^3, ξ_{st}) deserves further comment. In the compactness argument for the two spaces of holomorphic cylinders in $(\widehat{W}, \widehat{\omega})$, one can use the same intersection-theoretic argument as in Theorem 9.4 to rule out nontrivial buildings with upper levels, but for buildings with a main level only, there is now a wider range of nodal curves that can occur. Figure 9.4 shows the two possibilities: both of them have two smooth components which are each simple index 0 curves, but in one case these consist of a cylinder and a sphere, whereas the other case involves two planes asymptotic to two distinct orbits. One can use the Siefring adjunction formula as in the proof of Theorem 9.4 to rule out the first scenario if (W, ω) is minimal, because the spherical component must be an exceptional sphere. The second scenario can also be ruled out, but the general argument for this requires Lefschetz fibrations, which will appear again in the next section and we shall therefore take this as an excuse to postpone the argument (see Remark 9.49). If (W, ω) happens to be a Stein filling, then there is also a cheap trick one can apply to exclude this degeneration: by an earlier result of Stipsicz based

---

[2]One minor headache is that this choice of almost complex structure on the completion of $(\mathbb{D}T^*\mathbb{T}^2, \omega_{\mathrm{can}})$ is not $\mathbb{R}$-invariant on the cylindrical end. This problem can be dealt with easily by a minor modification of J near infinity; see [Wen10b] for details.

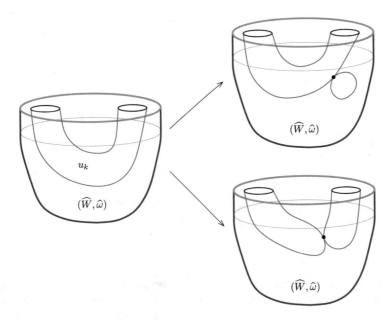

Fig. 9.4 Two types of nodal degenerations can occur for the holomorphic cylinders in a filling of $\mathbb{T}^3$. The first is possible only if the filling is not minimal, and the second only if certain Reeb orbits are contractible in the filling

on Seiberg-Witten theory [Sti02], every Stein filling of $(\mathbb{T}^3, \xi_1)$ is *homeomorphic* to the unit disk bundle $\mathbb{D}T^*\mathbb{T}^2 \cong \mathbb{T}^2 \times \mathbb{D}^2$, via a homeomorphism that identifies the closed Reeb orbits in $(\mathbb{T}^3, \alpha_1)$ with lifts of geodesics of the flat $\mathbb{T}^2$ to its unit cotangent bundle. Since none of these geodesics are contractible, $(\mathbb{T}^3, \alpha_1)$ does not have any closed Reeb orbits that are contractible in any Stein filling, hence in the Stein case, the planes appearing in Fig. 9.4 are topologically impossible. The argument for general fillings gives the same result without appealing to Seiberg-Witten theory—if the reader is willing to accept this statement on faith for now and wait until Remark 9.49 for a justification, then the classification of minimal strong fillings of $(\mathbb{T}^3, \xi_1)$ is now complete.

The same holomorphic cylinders imply that $(\mathbb{T}^3, \xi_k)$ cannot be fillable if $k \geqslant 2$. To see why, consider first the family of embedded holomorphic cylinders in $T^*\mathbb{T}^2 = (\mathbb{R} \times S^1) \times (\mathbb{R} \times S^1)$ defined by

$$(\mathbb{R} \times S^1) \times \{(s_0, t_0)\} \subset T^*\mathbb{T}^2.$$

When $s_0 > 1$ or $s_0 < -1$, these cylinders lie in the cylindrical end $T^*\mathbb{T}^2 \backslash \mathbb{D}T^*\mathbb{T}^2$ and thus define two families of holomorphic cylinders in the symplectization of $(\mathbb{T}^3, \xi_1)$, each with positive ends asymptotic to Reeb orbits that foliate a pair of 2-tori $T_1, T_2 \subset \mathbb{T}^3$. Composing these holomorphic cylinders with the projection $\mathbb{R} \times \mathbb{T}^3 \to \mathbb{T}^3$, they foliate $\mathbb{T}^3 \backslash (T_1 \cup T_2)$ by two S^1-families of embedded cylinders

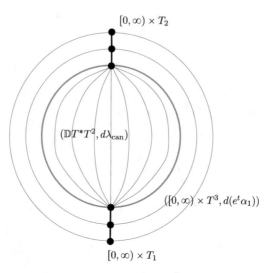

Fig. 9.5 The obvious foliation of $T^*T^2 \cong (\mathbb{R} \times S^1) \times (\mathbb{R} \times S^1)$ by holomorphic cylinders yields a foliation of the symplectization of $(\mathbb{T}^3, \xi_1)$ by two disjoint S^1-families of cylinders with positive ends

whose closures each have one boundary component in T_1 and the other in T_2 (Fig. 9.5).

The above picture of $(\mathbb{T}^3, \xi_1)$ foliated by holomorphic cylinders lifts to $(\mathbb{T}^3, \xi_k)$ under the obvious k-fold covering map

$$(\mathbb{T}^3, \xi_k) \to (\mathbb{T}^3, \xi_1) : (t, \phi, \theta) \mapsto (kt, \phi, \theta),$$

thus there is a set of $2k$ pairwise disjoint 2-tori $T_1, \ldots, T_{2k} \subset \mathbb{T}^3$ and a foliation of $\mathbb{T}^3 \backslash (T_1 \cup \ldots \cup T_{2k})$ by cylinders that each lift to the symplectization $(\mathbb{R} \times \mathbb{T}^3, d(e^t \alpha_k))$ as embedded holomorphic cylinders with positive ends. The case $k = 2$ is shown in Fig. 9.6. To prove the first part of Theorem 9.20, one can now argue as follows: if (W, ω) is a filling of $(\mathbb{T}^3, \xi_k)$ for $k \geqslant 2$, then its completion $(\widehat{W}, \widehat{\omega})$ contains $2k$ families of holomorphic cylinders that have self-intersection 0. Adapting again the arguments we used in Chap. 6, each of these families generates a moduli space of embedded J-holomorphic cylinders that foliate $(\widehat{W}, \widehat{\omega})$ except for finitely many nodal curves, and intersection considerations (cf. Exercise 9.22 below) dictate that each of the $2k$ families in the cylindrical end must be part of the same foliation and thus belong to the same connected component of the moduli space of holomorphic cylinders in $(\widehat{W}, \widehat{\omega})$. But unlike the situation for $(\mathbb{T}^3, \xi_1)$, the distinct families in the symplectization of $(\mathbb{T}^3, \xi_k)$ each have asymptotic orbits in distinct neighboring pairs of the tori $T_1, \ldots, T_{2k}$, so they are clearly not homotopic through asymptotically cylindrical maps in $\widehat{W}$. This is a contradiction, and proves that a strong filling of $(\mathbb{T}^3, \xi_k)$ cannot exist when $k \geqslant 2$.

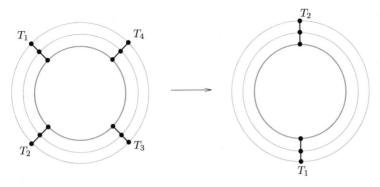

Fig. 9.6 The foliation of (T^3, ξ_1) by two families of holomorphic cylinders lifts to four families under the natural double cover $(T^3, \xi_2) \to (T^3, \xi_1)$

The following exercise involving symplectic ruled surfaces can be solved by deriving a contradiction analogous to the one we found above:

Exercise 9.22. Suppose (M, ω) is a closed symplectic 4-manifold containing two disjoint symplectically embedded spheres $S_1, S_2 \subset (M, \omega)$ with $[S_1] \cdot [S_1] = [S_2] \cdot [S_2] = 0$. Adapt the proof of Theorem A to show that S_1 and S_2 must then be symplectically isotopic.

The above argument proving nonfillability of $(\mathbb{T}^3, \xi_k)$ for $k \geq 2$ can easily be adapted to prove that a contact 3-manifold (M, ξ) is not strongly fillable whenever it contains a contact domain of the form

$$\left([0, 1] \times \mathbb{T}^2, \xi_{\text{GT}}\right) := \left([0, 1] \times S^1 \times S^1, \ker\left[\cos(2\pi t)\, d\theta + \sin(2\pi t)\, d\phi\right]\right),$$

see [Wen10b] for details. We call this a **Giroux torsion domain**, and any contact 3-manifold containing one is said to have **positive Giroux torsion**. Notice for instance that $(\mathbb{T}^3, \xi_k)$ has positive Giroux torsion for every $k \geq 2$, but the obvious contact immersion of $([0, 1] \times \mathbb{T}^2, \xi_{\text{GT}})$ into $(\mathbb{T}^3, \xi_1)$ fails to be an embedding. The fact that Giroux torsion is an obstruction to strong fillability was apparently first conjectured by Eliashberg in the early 1990s (before the term "Giroux torsion" existed), but it was not proved until 2006, by Gay [Gay06]. Gay's proof works by attaching a symplectic cobordism on top of a hypothetical filling of (M, ξ) and then finding symplectically embedded spheres that cause a contradiction similar to Exercise 9.22.[3] Though it was not stated explicitly in his paper, Gay's argument also proves a result about weak fillings: if (M, ξ) has **separating Giroux torsion**, meaning the Giroux torsion domain $[0, 1] \times \mathbb{T}^2 \subset M$ splits M into disjoint

---

[3]This summary of the proof in [Gay06] is a slight rewriting of history: Gay appealed to Seiberg-Witten theory in the final step of his proof, but could just as well have used holomorphic curves as in Exercise 9.22 instead.

components, then (M, ξ) is also not weakly fillable. The reason for this is that if $[0, 1] \times \mathbb{T}^2 \subset M$ separates M and (W, ω) is a weak filling of (M, ξ), then ω is necessarily exact on a neighborhood of $[0, 1] \times \mathbb{T}^2$, in which case one can use a variant of Proposition 8.18 to deform the weak filling to something that looks like a strong filling near the torsion domain; the cobordism of [Gay06] can then be attached and the argument goes through as in the strong case. A different proof of this result using Heegaard Floer homology was given by Ghiggini and Honda [GH], and a short time later two alternative proofs using holomorphic curves appeared in [NW11]; in particular, §2 in that paper explains how to use the notion of *stable Hamiltonian structures* to define symplectic completions of weak fillings that are not necessarily exact at the boundary, so that weak filling obstructions can also be detected via punctured holomorphic curves.

Note that by Eliashberg's flexibility theorem for overtwisted contact structures [Eli89], every overtwisted contact 3-manifold automatically has separating Giroux torsion, as the torsion domain is contained in the standard model of a *full Lutz-twist* on a solid torus, see e.g. [Gei08, §4.3]. One therefore obtains from this obstruction an alternative proof of the very first result on symplectic nonfillability:

Theorem (Gromov [Gro85] and Eliashberg [Eli90]). *If (M, ξ) is a closed overtwisted contact 3-manifold, then it admits no weak symplectic fillings.* □

The original proofs of this result used holomorphic disks and were closely related to Eliashberg's proof of Theorem 9.4; a complete account of this may be found in [Zeh03]. The notion of overtwistedness was recently extended to higher dimensions by Borman-Eliashberg-Murphy [BEM15], and the corresponding generalization of the above nonfillability result is due mainly to Niederkrüger [Nie06]; see also [MNW13, §3].

There is nothing analogous to the uniqueness of fillings of $(\mathbb{T}^3, \xi_1)$ known in higher dimensions, but a higher-dimensional analogue of Giroux torsion and the nonfillability of $(\mathbb{T}^3, \xi_k)$ for $k \geqslant 2$ is explained in [MNW13]. Generalizing in a different direction, one can ask whether the unit cotangent bundle $\mathbb{S}T^*\Sigma_g$ of a closed oriented surface Σ_g of genus g ever admits fillings other than the disk bundle $\mathbb{D}T^*\Sigma_g$. For minimal strong fillings up to symplectic deformation, the answer is "no" when $g \leqslant 1$; for $g = 1$ this is Theorem 9.20, and for $g = 0$ it is a result of McDuff [McD90] and Hind [Hin00], and also a special case of Theorem 9.41 below (see Exercise 9.38). For $g \geqslant 2$ the situation is very different, as McDuff showed in [McD91] that there exist exact fillings diffeomorphic to $[-1, 1] \times \mathbb{S}T^*\Sigma_g$ such that one boundary component is the unit cotangent bundle with its canonical contact structure. This has two somewhat troubling implications: first, one can now produce many non-standard (and generally non-exact) strong fillings of $(\mathbb{S}T^*\Sigma_g, \xi_{can})$ by attaching any of the symplectic caps provided by [EH02] to the other boundary component, and these caps come in an infinite variety of topological types, implying that it is probably hopeless to classify the strong fillings of $(\mathbb{S}T^*\Sigma_g, \xi_{can})$. A second piece of bad news comes from the observation that if the symplectization of $(\mathbb{S}T^*\Sigma_g, \xi_{can})$ contained any nice index 2 holomorphic curves with no negative ends, i.e. the type of curves that are used in the proofs of

Theorems 9.4 and 9.20, then we could use them to contradict the existence of the semi-fillings $[-1, 1] \times \mathbb{S}T^*\Sigma_g$ constructed by McDuff. Indeed, such curves would give rise to a 2-dimensional moduli space that tries to foliate the completion of $[-1, 1] \times \mathbb{S}T^*\Sigma_g$ with curves whose ends are all in the cylindrical end over a single boundary component, eventually producing a curve that touches the other J-convex boundary component tangentially and thus violates Proposition 9.9. It therefore seems unlikely that a better understanding of fillings of $(\mathbb{S}T^*\Sigma_g, \xi_{\mathrm{can}})$ for $g \geqslant 2$ will come from punctured holomorphic curve techniques.

That is the bad news. The good news is that some very recent progress on this question has come from the direction of Seiberg-Witten theory, in papers by Li et al. [LMY17] and Sivek and Van-Horn Morris [SV17] showing that all *exact* fillings of $(\mathbb{S}T^*\Sigma_g, \xi_{\mathrm{can}})$ must, at least topologically, bear a strong resemblance to $\mathbb{D}T^*\Sigma_g$. Even more recently, Li and Özbağcı [LÖ] have shown that for non-orientable surfaces Σ, $(\mathbb{S}T^*\Sigma, \xi_{\mathrm{can}})$ is planar and thus amenable to the holomorphic curve methods discussed in the next section. These developments lend some plausibility to the following conjecture:

Conjecture 9.23. For any closed surface Σ, all exact fillings of the unit cotangent bundle $(\mathbb{S}T^*\Sigma, \xi_{\mathrm{can}})$ are Liouville deformation equivalent to the unit disk bundle $(\mathbb{D}T^*\Sigma, \omega_{\mathrm{can}})$.

The conjecture makes sense of course in all dimensions, but any serious progress on it in dimensions greater than four would seem to require entirely new ideas.

9.3 Planar Contact Manifolds

We conclude this survey with a result on symplectic fillings that is somewhat analogous to the construction of Lefschetz pencils in Theorem F. We must first discuss the relationship between contact structures and open book decompositions. For a more detailed introduction to this subject, especially in dimension three, see [Etn06].

Given a closed, connected and oriented manifold M, an **open book decomposition** of M is a fiber bundle

$$\pi : M\backslash K \to S^1,$$

where $K \subset M$ is a closed oriented submanifold with codimension 2 (the **binding**), which has trivial normal bundle and admits a neighborhood $\mathcal{N}(K) \cong K \times \mathbb{D}^2 \subset M$ in which π takes the form

$$\pi|_{K \times (\mathbb{D}^2\backslash\{0\})} : K \times (\mathbb{D}^2\backslash\{0\}) \to S^1 : (x, re^{2\pi i\phi}) \mapsto \phi.$$

The fibers of π are called **pages** of the open book; they are each embedded hypersurfaces that inherit from M a natural orientation, and their closures are also embedded and have matching (oriented) boundaries

$$\overline{\partial \pi^{-1}(\phi)} = K.$$

We will be most interested in the case dim $M = 3$: then the binding is an oriented link and the pages form an S^1-family of Seifert surfaces spanning K.

The topology of an open book $\pi : M \backslash K \rightarrow S^1$ is fully determined by the topology of its pages and its monodromy, which means the following. Fix a fiber $\pi^{-1}(0) \subset M$ and denote its closure by $P = \overline{\pi^{-1}(0)}$. Choose any vector field transverse to the pages whose flow ψ^t for time t maps $\pi^{-1}(0)$ to $\pi^{-1}(t)$, and which takes the form $\psi^t(x, r) = (x, re^{2\pi i t})$ on the neighborhood $\mathcal{N}(K) = K \times \mathbb{D}^2$. The time-1 flow is then a diffeomorphism

$$\psi : P \rightarrow P$$

that equals the identity near the boundary, and the isotopy class of this diffeomorphism rel boundary is independent of the choice of vector field. We call the resulting mapping class on P the **monodromy** of the open book, and the pair (P, ψ) are known as an **abstract open book**. Any two open books that determine the same abstract open book up to diffeomorphism are themselves diffeomorphic.

The following important definition is due to Giroux [Gir02].

Definition 9.24. Suppose $\pi : M \backslash K \rightarrow S^1$ is an open book decomposition and α is a contact form on M. We call α a **Giroux form** for $\pi : M \backslash K \rightarrow S^1$ if $\alpha|_{TK}$ is a positive contact form on K and the restriction of $d\alpha$ to every page is a symplectic form compatible with the page's natural orientation. A (positive, co-oriented) contact structure ξ on M is said to be **supported by** the open book $\pi : M \backslash K \rightarrow S^1$ if it admits a contact form α that is a Giroux form.

Exercise 9.25. Show that if dim $M = 3$, $\pi : M \backslash K \rightarrow S^1$ is an open book and α is a contact form on M, the following conditions are equivalent:

(1) α is a Giroux form;
(2) $\alpha|_{TK} > 0$ and $d\alpha$ defines a positive area form on every page;
(3) The Reeb vector field R_α is (positively) tangent to K and (positively) transverse to all pages.

In order to avoid unnecessary complications, we shall restrict our attention from now on to dimension three. In 1975, Thurston and Winkelnkemper proved [TW75] that every open book on a 3-manifold gives rise to a contact structure—in modern terminology, we would say that every open book *supports* a contact structure, and moreover, one can check that the resulting contact structure is uniquely determined up to isotopy by the open book. A complete proof of this may be found in [Etn06], and it is quite analogous to Theorem 3.33 on the relationship between Lefschetz fibrations and symplectic structures. Also analogous to the symplectic case, there is a much harder converse of this result: Giroux [Gir] explained in

2001 how to find a supporting open book for any given contact 3-manifold, and a corresponding result in higher dimensions has been announced by Giroux and Mohsen (see [Gir02, Col08]), using similar techniques to Donaldson's existence result for Lefschetz pencils on symplectic manifolds.

Example 9.26. Regarding (S^3, ξ_{st}) as the contact-type boundary of the unit ball in $(\mathbb{C}^2, \omega_{\mathrm{st}})$, $\xi_{\mathrm{st}} \subset TS^3$ is also the maximal complex subbundle in TS^3 with respect to the standard complex structure i. Let $K = \partial \mathbb{D}^2 \times \{0\} \subset S^3 \subset \mathbb{C}^2$ and define the map

$$\pi : S^3 \backslash K \to S^1 = \partial \mathbb{D}^2 : (z_1, z_2) \mapsto \frac{z_2}{|z_2|}.$$

This is an open book decomposition with pages diffeomorphic the disk, and the standard contact form α_{st} is a Giroux form. It follows that any contact 3-manifold supported by an open book with disk-like pages is contactomorphic to (S^3, ξ_{st}). Note that in this case the monodromy plays no role, as all compactly supported diffeomorphisms of the open disk are isotopic.

Example 9.27. Here is a second open book supporting the standard contact structure on S^3. Let $K \subset S^3$ denote the Hopf link $(\partial \mathbb{D}^2 \times \{0\}) \cup (\{0\} \times \partial \mathbb{D}^2)$, and define

$$\pi : S^3 \backslash K \to S^1 = \partial \mathbb{D}^2 : (z_1, z_2) \mapsto \frac{z_1 z_2}{|z_1 z_2|}.$$

This defines an open book with pages diffeomorphic to the cylinder and monodromy isotopic to a right-handed Dehn twist (see Proposition 9.34 and Exercise 9.37), and it again has the standard contact form α_{st} as a Giroux form.

Example 9.28. Consider $S^1 \times S^2$ with coordinates (t, θ, ϕ) where $t \in S^1 = \mathbb{R}/\mathbb{Z}$ and (θ, ϕ) denote spherical polar coordinates on S^2. The standard contact structure ξ_{st} on $S^1 \times S^2$ can then be written as the kernel of

$$f(\theta)\, dt + g(\theta)\, d\phi$$

where $f, g : [0, \pi] \to \mathbb{R}$ are suitable smooth functions such that the path $\theta \mapsto (f(\theta), g(\theta)) \in \mathbb{R}^2$ traces a semicircle counterclockwise around the origin moving from $(1, 0)$ to $(-1, 0)$. Defining $K = \{\theta = 0\} \cup \{\theta = \pi\}$, there is an open book

$$\pi : (S^1 \times S^2) \backslash K \mapsto S^1 : (t, \theta, \phi) \mapsto \frac{\phi}{2\pi}$$

which has cylindrical pages and trivial monodromy. The contact form written above is a Giroux form for this open book, hence every contact manifold supported by an open book with cylindrical pages and trivial monodromy is contactomorphic to $(S^1 \times S^2, \xi_{\mathrm{st}})$.

Example 9.29. For any pair of relatively prime integers $p > q \geq 1$, the **lens space** $L(p, q)$ with its standard contact structure ξ_{st} is defined as the quotient of (S^3, ξ_{st}) by the finite group of contactomorphisms

$$\varphi_k : S^3 \to S^3 : (z_1, z_2) \mapsto \left(e^{2\pi i k/p}z_1, e^{2\pi i k q/p}z_2\right)$$

for $k \in \mathbb{Z}_p$. One can verify that for any $k \geq 1$, the contact manifold supported by an open book with cylindrical pages and monodromy isotopic to the kth power of a right-handed Dehn twist is contactomorphic to $(L(k, k-1), \xi_{st})$.

As Examples 9.26 and 9.27 demonstrate, it is possible for two topologically different open books to support the same contact structure: in fact, any open book can be modified by an operation known as *positive stabilization* to a new one without changing the supported contact structure (see e.g. [vK]). By a deep result known as the *Giroux correspondence*, the contactomorphism classes of contact structures on any closed oriented 3-manifold are in one-to-one correspondence with equivalence classes of open books up to positive stabilization. Stabilization always changes the topology of the pages: it is defined by attaching a topological 1-handle to the page, so in general it increases either the genus or the number of boundary components. This implies that for any given contact structure, there is nothing special about having an open book with pages of high genus, as the genus can always be increased via positive stabilization. On the other hand, the question of whether a supporting open book with *low* genus can be found is quite subtle and has serious implications, as we will see below.

Definition 9.30. An open book on a 3-manifold is called **planar** if its pages have genus zero, i.e. they are diffeomorphic to spheres with finitely many punctures. A contact structure is **planar** if it admits a supporting planar open book.

We've seen in the examples above that the standard contact structures on S^3, $S^1 \times S^2$ and the lens spaces $L(k, k-1)$ are all planar. Etnyre [Etn04] has shown that all overtwisted contact 3-manifolds are planar, but there are also non-planar contact 3-manifolds: the simplest examples are the tori $(\mathbb{T}^3, \xi_k)$ that we considered in Sect. 9.2. As we will see below, the planar contact structures are in some sense the "rational or ruled" objects of the 3-dimensional contact world. The evidence for this begins with the following result regarding the Weinstein conjecture, proved in 2005 by Abbas, Cieliebak and Hofer:

Theorem 9.31 (Abbas-Cieliebak-Hofer [ACH05]). *Suppose (M, ξ) is supported by a planar open book whose pages have $k \geq 1$ boundary components. Then for every contact form for ξ, there exists a compact oriented surface Σ, with genus zero and at least 1 but at most k boundary components, admitting a continuous map $\Sigma \to M$ whose restriction to each component of $\partial \Sigma$ is a closed Reeb orbit.*

Remark 9.32. The $k = 1$ case of Theorem 9.31 reproves the existence of a contractible Reeb orbit for every contact form on (S^3, ξ_{st}), and in fact the slightly more general Theorem 9.50 below gives a second proof of Theorem 9.5.

As in our discussion of (S^3, ξ_{st}) in Sect. 9.1, this dynamical result is closely related to a classification result for symplectic fillings. To understand this, we must examine the intimate relationship between open book decompositions and Lefschetz fibrations over the disk. We will consider Lefschetz fibrations for which both the fiber and the base are compact oriented surfaces with *nonempty boundary*, thus the total space is necessarily a manifold with boundary and corners. To be precise, in the following we assume W to be a smooth, compact, oriented and connected 4-manifold with boundary and corners such that ∂W is the union of two smooth faces

$$\partial W = \partial_h W \cup \partial_v W$$

which intersect at a corner of codimension two. The boundary of W is well defined as a topological 3-manifold, but the corner can also be smoothed so that the boundary inherits a smooth structure, and none of what we will say in the following depends on the choices involved in this smoothing procedure.

Definition 9.33. A **bordered Lefschetz fibration** of W over the disk $\mathbb{D}^2$ is a smooth map $\Pi : W \to \mathbb{D}^2$ with finitely many interior critical points $W_{\text{crit}} \subset \mathring{W}$ and critical values $\mathbb{D}^2_{\text{crit}} \subset \mathring{\mathbb{D}}^2$ such that the following conditions hold:

(1) $\Pi^{-1}(\partial \mathbb{D}^2) = \partial_v W$ and $\Pi|_{\partial_v W} : \partial_v W \to \partial \mathbb{D}^2$ is a smooth fiber bundle;
(2) $\Pi|_{\partial_h W} : \partial_h W \to \mathbb{D}^2$ is also a smooth fiber bundle;
(3) Near each point $p \in W_{\text{crit}}$, there exists a complex coordinate chart (z_1, z_2) and a corresponding complex coordinate z on a neighborhood of $\pi(p) \in \mathbb{D}^2$ in which Π locally takes the form

$$\pi(z_1, z_2) = z_1^2 + z_2^2;$$

(4) All fibers $W_z := \Pi^{-1}(z)$ for $z \in \mathbb{D}^2$ are connected and have nonempty boundary in $\partial_h W$.

As in Chap. 3, we call W_z a **regular fiber** if $z \in \mathbb{D}^2 \backslash \mathbb{D}^2_{\text{crit}}$ and otherwise a **singular fiber**. We say that Π is **allowable** if all the irreducible components of its fibers have nonempty boundary (see Fig. 9.7).

If $\Pi : W \to \mathbb{D}^2$ is a bordered Lefschetz fibration, then ∂W inherits an open book decomposition in a natural way whose pages have the same topology as the fibers of Π. Indeed, observe that $\Pi|_{\partial_h W} : \partial_h W \to \mathbb{D}^2$ is a disjoint union of circle bundles over the disk and is thus necessarily trivial, so it can be identified with the trivial bundle

$$\Pi : K \times \mathbb{D}^2 \to \mathbb{D}^2 : (x, z) \mapsto z$$

where K is the boundary of a fiber $\Pi^{-1}(*) \cong S^1 \amalg \ldots \amalg S^1$. We can thus define an open book $\pi : M \backslash K \to S^1$ by identifying $\partial \mathbb{D}^2$ in the natural way with S^1 and setting

$$\pi|_{\partial_v W} = \Pi|_{\partial_v W} : \partial_v W \to \partial \mathbb{D}^2 = S^1$$

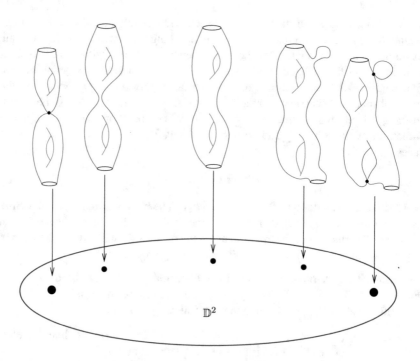

Fig. 9.7 A bordered Lefschetz fibration over the disk, with fibers having genus 2 and two boundary components. The fibration is not allowable due to the singular fiber at the right, which has a closed irreducible component

and

$$\pi|_{\partial_h W \backslash K} : K \times (\mathbb{D}^2 \backslash \{0\}) \to S^1 : (x, re^{2\pi i \phi}) \mapsto \phi.$$

The most important topological fact about this relationship between Lefschetz fibrations and open books is that the monodromy of the open book depends on the Lefschetz critical points. Observe that in the special case where there are no critical points, the fibration $\Pi : W \to \mathbb{D}^2$ is necessarily trivial, hence so is the monodromy of the open book. When there are critical points, we obtain the following statement; see [ÖS04a] for a more precise version and further discussion.

Proposition 9.34. *Suppose* $\Pi : W \to \mathbb{D}^2$ *is an allowable bordered Lefschetz fibration and* $\pi : \partial W \backslash K \to S^1$ *is the induced open book on the boundary. Then the monodromy of* $\pi : \partial W \backslash K \to S^1$ *is a composition of right-handed Dehn twists, one for each Lefschetz critical point in* $\Pi : W \to \mathbb{D}^2$. $\qquad \square$

The following notion is now the symplectic analogue of a supported contact structure for an open book. The conditions are mostly the same as in the case of a closed Lefschetz fibration or pencil (see Definition 3.32), but we need to specify more about its behavior near the boundary.

Definition 9.35. Given a bordered Lefschetz fibration $\Pi : W \to \mathbb{D}^2$, we will say that a symplectic structure ω on W is **compatible** with Π if:

(1) The smooth part of every fiber $W_z \backslash W_{\mathrm{crit}}$ for $z \in \mathbb{D}^2$ is a symplectic submanifold;

(2) For any almost complex structure J defined near W_{crit} that restricts to a smooth positively oriented complex structure on the smooth parts of all fibers $W_z \backslash W_{\mathrm{crit}}$, J is tamed by ω at W_{crit};

(3) Near ∂W, ω can be written as $d\lambda$ where λ is a smooth 1-form whose restrictions to $\partial_v W$ and $\partial_h W$ are each contact forms, and the resulting Reeb vector field on $\partial_h W$ is positively tangent to the fibers.

It is not hard to show that whenever $\Pi : W \to \mathbb{D}^2$ is a bordered Lefschetz fibration with a compatible symplectic structure ω and λ is the primitive defined near ∂W as in the above definition, the corner ∂W can be smoothed and λ modified slightly so that its restriction to the smoothed boundary defines a Giroux form for the induced open book. Moreover, the fact that λ is contact on both smooth faces of ∂W means that its ω-dual Liouville vector field is outwardly transverse to both faces, and thus remains transverse to any smoothing of the boundary. It follows that after smoothing the boundary, (W, ω) becomes a strong symplectic filling of the contact manifold (M, ξ) supported by the induced open book at the boundary. We call $\Pi : W \to \mathbb{D}^2$ in this case a **symplectic bordered Lefschetz fibration** of (W, ω) over $\mathbb{D}^2$.

To find concrete examples of bordered Lefschetz fibrations in nature, it is often convenient to start with Lefschetz fibrations whose fibers and base are each noncompact, and then restrict to suitable compact subdomains with boundary and corners.

Example 9.36. The map $\Pi : \mathbb{C}^2 \to \mathbb{C} : (z_1, z_2) \mapsto z_2$ is trivially a Lefschetz fibration, with no singular fibers, and its restriction to the polydisk $W := \mathbb{D}^2 \times \mathbb{D}^2 \subset \mathbb{C}^2$ is then a bordered Lefschetz fibration $W \to \mathbb{D}^2$ compatible with the standard symplectic structure ω_{st}. It has vertical boundary $\mathbb{D}^2 \times \partial \mathbb{D}^2$ and horizontal boundary $\partial \mathbb{D}^2 \times \mathbb{D}^2$, and the induced open book on $\partial W \cong S^3$ is equivalent to the one in Example 9.26, thus the strong symplectic filling determined by this construction is deformation equivalent to the standard filling of (S^3, ξ_{st}) by a ball.

Exercise 9.37. Find a compact subdomain $W \subset \mathbb{C}^2$ with boundary and corners such that the Lefschetz fibration $\mathbb{C}^2 \to \mathbb{C} : (z_1, z_2) \mapsto z_1 z_2$ (cf. Exercise 3.21) restricts to W as a bordered Lefschetz fibration inducing an open book equivalent to Example 9.27 at ∂W.

Exercise 9.38. Fixing the standard Euclidean metric $\langle \, , \, \rangle$ on $\mathbb{R}^{n+1}$, one can identify T^*S^n with the submanifold

$$TS^n = \left\{ (\mathbf{q}, \mathbf{p}) \in \mathbb{R}^{n+1} \times \mathbb{R}^{n+1} \mid |\mathbf{q}| = 1 \text{ and } \langle \mathbf{p}, \mathbf{q} \rangle = 0 \right\},$$

and using coordinates $\mathbf{q} = (q_0, \ldots, q_n)$ and $\mathbf{p} = (p_0, \ldots, p_n)$, the standard Liouville form λ_{can} of T^*S^n under this identification becomes the restriction to $TS^n \subset \mathbb{R}^{n+1} \times \mathbb{R}^{n+1}$ of $\sum_{j=0}^n p_j \, dq_j$. Consider the smooth affine variety

$$V = \left\{ (z_0, \ldots, z_n) \in \mathbb{C}^{n+1} \mid z_0^2 + \ldots + z_n^2 = 1 \right\}.$$

(a) Show that

$$\Phi : V \to T^*S^n : \mathbf{x} + i\mathbf{y} \mapsto \left(\frac{\mathbf{x}}{|\mathbf{x}|}, -|\mathbf{x}|\mathbf{y} \right)$$

defines a diffeomorphism such that using coordinates $\mathbf{x} = (x_0, \ldots, x_n)$ and $\mathbf{y} = (y_0, \ldots, y_n)$ for $\mathbf{x} + i\mathbf{y} \in \mathbb{C}^{n+1}$, $\Phi^* \lambda_{\mathrm{can}} = -\sum_{j=0}^n y_j \, dx_j$. In particular, Φ is a symplectomorphism $(V, \omega_{\mathrm{st}}) \to (T^*S^n, \omega_{\mathrm{can}})$.

(b) Show that for $n = 2$, the map $\pi : V \to \mathbb{C} : (z_0, z_1, z_2) \mapsto z_0$ is a Lefschetz fibration with exactly two critical points, at $(\pm 1, 0, 0)$.

(c) Find a suitable compact subdomain $W \subset V$ with boundary and corners so that the map in part (b) restricts to W as an allowable bordered Lefschetz fibration over $\mathbb{D}^2$ with two singular fibers, and regular fibers that are annuli.

By Proposition 9.34, one deduces from this exercise that $(\mathbb{S}T^*S^2, \xi_{\mathrm{can}})$ is supported by a planar open book with cylindrical fibers and monodromy isotopic to the square of a right-handed Dehn twist. It is therefore contactomorphic to the lens space $(L(2, 1), \xi_{\mathrm{st}})$ that appeared in Example 9.29, also known as the standard contact structure on $\mathbb{RP}^3 = S^3/\mathbb{Z}_2$.

Remark 9.39. The smooth variety $V \subset \mathbb{C}^{n+1}$ in the above exercise is an example of a *Brieskorn variety*. These furnish many interesting examples of contact manifolds with exact fillings on which one can explicitly see Lefschetz fibrations and open books; see [KvK16] for a survey.

We have the following analogue of Theorem 3.33 (see [LVW] for details).

Theorem 9.40. *Given a bordered Lefschetz fibration $\Pi : W \to \mathbb{D}^2$, the space of compatible symplectic structures is nonempty and connected. In particular, any bordered Lefschetz fibration determines (uniquely up to symplectic deformation equivalence) a strong symplectic filling (W, ω) of the contact manifold supported by its induced open book at the boundary. Moreover, if the Lefschetz fibration is allowable, then the compatible symplectic structure can be arranged to make (W, ω) an exact filling, and its Liouville deformation class is uniquely determined.[4]*

□

---

[4] A stronger statement is actually true: as shown in [BV15, LVW], allowable Lefschetz fibrations admit Stein structures. In fact, every statement following Theorem 9.40 that involves exact fillings or Liouville deformation is also true for Stein fillings and Stein deformation; see [LVW]. Conversely, Loi-Piergallini [LP01] and Akbulut-Özbağcı [AÖ01b, AÖ01a] proved that every Stein

Note that the allowability condition in the last statement is clearly necessary, as symplectic bordered Lefschetz fibrations that are not allowable contain singular fibers with closed symplectic submanifolds as irreducible components, contradicting exactness. The result is not especially hard to prove—the methods involved are the same as in our proof of Theorem 3.33. What's more surprising is that for *planar* contact manifolds, there is a converse.

Theorem 9.41 (Wendl [Wen10b] and Niederkrüger-Wendl [NW11]). *Suppose (W, ω) is a weak symplectic filling of (M, ξ) and ξ is supported by a planar open book $\pi : M \backslash K \to S^1$. Then (W, ω) admits a symplectic bordered Lefschetz fibration in which each singular fiber has only one critical point, and which restricts to the given open book at the boundary. Moreover, the Lefschetz fibration is allowable if and only if (W, ω) is minimal, and it is uniquely determined up to isotopy by the symplectic deformation class of ω.*

The combination of these two theorems implies that the weak symplectic fillings of a contact manifold with a given planar open book can be classified up to deformation equivalence if one understands topologically all the possible bordered Lefschetz fibrations that can restrict to that open book at the boundary. This is sometimes easy and sometimes very hard, but it is in any case a purely *topological* question; it has nothing intrinsically to do with symplectic or contact geometry. The relationship between minimality and allowability is easy to understand: if $\Pi : W \to \mathbb{D}^2$ is a bordered Lefschetz fibration with genus zero fibers and only one critical point in every singular fiber, then Proposition 3.30 implies that it is allowable if and only if none of its singular fibers contain exceptional spheres. Any non-allowable Lefschetz fibration with genus 0 fibers can thus be blown down to produce one that is allowable, and this changes nothing near the boundary, so we obtain:

Corollary 9.42. *If (M, ξ) is planar and weakly symplectically fillable, then it is also exactly fillable, and Liouville deformation classes of exact fillings are in bijective correspondence to weak symplectic deformation classes of minimal weak fillings. Moreover, all of its weak fillings are deformation equivalent to blowups of exact fillings.* □

Recall that by Proposition 9.34, an open book is not the boundary of *any* Lefschetz fibration unless its monodromy is a composition of right-handed Dehn twists, thus:

Corollary 9.43. *If (M, ξ) is supported by a planar open book whose monodromy is not isotopic to a composition of right-handed Dehn twists, then (M, ξ) is not weakly symplectically fillable.* □

---

filling in dimension four admits an allowable Lefschetz fibration, and more recently, Giroux and Pardon [GP17] extended Donaldson's methods [Don99] to establish this result in all dimensions.

It is not always easy to judge whether a given mapping class on a surface is a product of right-handed Dehn twists, but some progress on this was made by Plamenevskaya and Van Horn-Morris [PV10], who used the above corollary to deduce some new examples of tight but non-fillable contact manifolds. Theorem 9.41 has also been applied by Wand [Wan12] to define new obstructions to planarity of contact structures.

The simplest open books to which one can apply Theorem 9.41 are those with cylindrical pages (Examples 9.27–9.29, also Exercise 9.38): every compactly supported diffeomorphism of the cylinder is isotopic to δ^k for some $k \in \mathbb{Z}$, where δ is a right-handed Dehn twist. It follows that the corresponding contact manifold is fillable if and only if $k \geq 0$, and in this case the number of critical points of any allowable Lefschetz fibration that fills it must be precisely k, thus that Lefschetz fibration is uniquely determined up to diffeomorphism, and all others filling the same open book are blowups of it. This implies:

Corollary 9.44. *The manifolds S^3, $S^1 \times S^2$ and $L(k, k-1)$ with their standard contact structures each have unique weak fillings up to symplectic deformation and blowup.*   □

As we saw in Sect. 9.1, the classification of fillings of S^3 is actually a much older result, and the corresponding result for $S^1 \times S^2$ can be derived from it using an equally old result of Eliashberg [Eli90] about fillings of connected sums. The fillings of lens spaces were classified up to diffeomorphism by Lisca [Lis08], so the above classification up to symplectic deformation is a slight improvement on this. By Exercise 9.38, the special case of $L(k, k-1)$ with $k = 2$ gives the uniqueness of fillings of $(\mathbb{S}T^*S^2, \xi_{can})$: up to diffeomorphism this classification was originally established by McDuff as a corollary of her characterization of rational and ruled surfaces [McD90], and it was later improved to a classification up to Stein deformation equivalence by Hind [Hin00], using punctured holomorphic curve techniques. Some newer applications of Theorem 9.41 to prove uniqueness of symplectic fillings are explained in [PV10].

Remark 9.45. Theorem 9.41 was originally proved for strong fillings in [Wen10b], and the extension to weak fillings was carried out in [NW11] using stable Hamiltonian structures to define suitable cylindrical ends over fillings with non-exact boundary. Unlike the special case of (S^3, ξ_{st}), this is not simply a matter of applying Proposition 8.18 to deform from weak to strong, as every closed oriented 3-manifold admits a planar contact structure, so they are not all rational homology spheres. There is nonetheless a fundamentally cohomological reason why Theorem 9.41, unlike the classification of fillings of $\mathbb{T}^3$ in the previous section, is not limited to strong fillings: given any weak filling (W, ω) of a contact 3-manifold (M, ξ) supported by an open book $\pi : M \backslash K \rightarrow S^1$, ω is necessarily exact near the binding K, and can thus be deformed to look like a strong filling on a neighborhood of K. In [LVW], Theorem 9.41 is extended to so-called *spinal open books*, in which the solid tori $S^1 \times \mathbb{D}^2$ that form neighborhoods of the binding in an open book are replaced by more general domains of the form $S^1 \times \Sigma$ for compact oriented surfaces Σ with boundary; the *spine* of a contact 3-manifold supported by a spinal open book is then a disjoint union of domains of this form such that the circles $S^1 \times \{\text{const}\} \subset S^1 \times \Sigma$ are closed Reeb orbits for a suitable choice of contact form. In general, it turns out that any weak filling that is exact on the spine of a supporting planar spinal open book can be deformed to a strong filling, and it can further be deformed to an exact or Stein filling if it is minimal. In the case of $(\mathbb{T}^3, \xi_k)$ for each $k \in \mathbb{N}$, one can find a planar spinal open book that supports the contact structure, but its spine consists of domains of the form $[a, b] \times S^1 \times S^1 \subset \mathbb{T}^3$, which have nontrivial cohomology. Note that all of the weak fillings of $(\mathbb{T}^3, \xi_k)$ constructed in Exercise 9.21 are non-exact on domains of this form: this is why they cannot be deformed to strong fillings.

We conclude by sketching the proofs of Theorems 9.31 and 9.41, focusing on the strong filling case of the latter result. The connection between holomorphic curves and planar open books that underlies both theorems can be traced back at least as far as a series of papers by Hofer, Wysocki and Zehnder in the mid-1990s, exploring applications of the newly developed theory of finite-energy pseudoholomorphic curves in symplectizations. One particularly compelling application was a 3-dimensional contact analogue of McDuff's characterization of symplectic ruled surfaces:

Theorem 9.46 (Hofer-Wysoki-Zehnder [HWZ95b]). *Suppose (M, ξ) is a closed contact 3-manifold with a contact form α that admits a contractible, simply covered, nondegenerate Reeb orbit $\gamma \subset M$ with the following properties:*

(1) γ has the smallest period among all orbits of R_α;
(2) γ bounds an embedded disk $\mathcal{D} \subset M$ such that $d\alpha|_{T\mathring{\mathcal{D}}} > 0$;
(3) The Conley-Zehnder index of γ with respect to a trivialization of $\xi|_{\mathcal{D}}$ is 3.

Then (M, ξ) is contactomorphic to (S^3, ξ_{st}).

The idea behind this result is roughly that $\mathcal{D}$ gives rise to an embedded finite-energy J-holomorphic plane in $\mathbb{R} \times M$ with index 2 and self-intersection 0, so that combining Proposition 8.51 with a compactness argument, this plane generates a 2-dimensional moduli space of embedded curves in $\mathbb{R} \times M$ whose projections to M form an S^1-family of open disks transverse to the Reeb vector field, foliating $M \setminus \gamma$. These planes projected into M thus form the pages of an open book decomposition $\pi : M \setminus \gamma \to S^1$, with α as a Giroux form, and this decomposition identifies (M, ξ) with the picture of (S^3, ξ_{st}) that we explained in Example 9.26. Variations on this argument have been used in other contexts to characterize particular contact manifolds via generalizations of open books [HT09, HLS15], or to produce global surfaces of section for Reeb flows [HWZ98, HS11, CHP].

The above result is a very special case involving the simplest possible open book, but after the Giroux correspondence arose in 2001, it became natural to ask whether all open books supporting contact structures can similarly be realized as S^1-families of holomorphic curves. One possible statement along these lines is the following result from [Wen10c]; for an alternative version, see [Abb11].

Theorem 9.47. *Suppose $\pi : M \backslash K \to S^1$ is an open book with pages of genus g, supporting a contact structure ξ. Then there exists a Giroux form α and an $\mathbb{R}$-invariant almost complex structure J on $\mathbb{R} \times M$ such that every connected component of K is a Reeb orbit with odd Conley-Zehnder index, and every page $\pi^{-1}(\phi)$ is the image under the projection $\mathbb{R} \times M \to M$ of an embedded J-holomorphic curve in $\mathbb{R} \times M$ with only positive ends and index $2 - 2g$.* □

The proof of this theorem in [Wen10c] requires no special technology; it is a more or less explicit construction.[5] Note that the moduli space of J-holomorphic curves obtained in this statement is actually *two*-dimensional, because one obtains not only an S^1-family corresponding to the pages but also all the $\mathbb{R}$-translations of these curves. On the other hand, the virtual dimension of this moduli space is $2 - 2g$, by a computation very similar to what we saw in the example of $S^2 \times \Sigma_g$ (see Remark 5.2). It follows that the holomorphic pages are well behaved if they have genus zero—in fact in this case they satisfy the requirements for automatic transversality in Sect. 8.3.7 and a nice implicit function theorem (Proposition 8.51)—but for $g > 0$ they are unstable and do not carry any information beyond the very specific setup of $(\mathbb{R} \times M, J)$. This is the reason why Theorems 9.31 and 9.41 require the open book to be planar. It remains unclear whether holomorphic curves can be used to prove the Weinstein conjecture for non-planar contact 3-manifolds, and the analogue of Theorem 9.41 on fillings in the non-planar case is false, as shown by Wand [Wan15].

Remark 9.48. Hofer proposed in [Hof00] a possible solution to the problem with the indices of higher genus embedded holomorphic curves: the idea was to generalize the equation for holomorphic curves in symplectizations by introducing an extra parameter that varies in the $2g$-dimensional space $H^1_{dR}(\Sigma_g)$. When regarded as solutions to this more general problem, the curves in Theorem 9.47 have index 2 regardless of their genus and are stable under perturbations—this is the motivating idea in Abbas's version of the holomorphic open book [Abb11]. On the other hand, the compactness theory of Hofer's generalized equation is still not well understood, and there is no good candidate for a definition of the equation in non-$\mathbb{R}$-invariant settings, so it remains unclear whether this approach will have interesting applications. Taubes's solution [Tau07] to the Weinstein conjecture in 2006 made this question seem a great deal less urgent.

With Theorem 9.47 in mind, the reader may now be able to guess the outline of the proofs of Theorems 9.31 and 9.41. For the strong filling case of Theorem 9.41,

---

[5]We are simplifying the discussion slightly, because the statement of Theorem 9.47 did not say that J belongs to $\mathcal{J}(M, \alpha)$, and in fact this only becomes true after a perturbation; in the general construction, J is adapted to a stable Hamiltonian structure rather than a contact form. Perturbing to contact data and making the J-holomorphic pages survive this perturbation does require some technology, namely the implicit function theorem, and it only works in the $g = 0$ case since that is the only case in which the dimension of the moduli space equals its virtual dimension. See [Wen10c] for details.

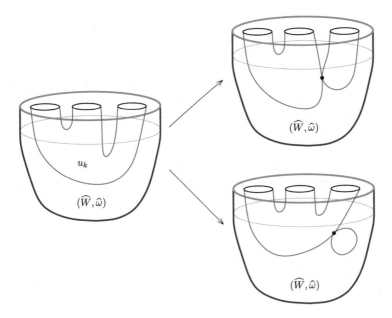

Fig. 9.8 Two examples of possible nodal degenerations for holomorphic curves arising from a planar open book with three binding components. Both scenarios produce Lefschetz singular fibers; the second produces a non-allowable Lefschetz fibration and is only possible if the filling is not minimal

we assume (W, ω) is a given strong filling of (M, ξ), and start by extending (W, ω) to its completion $(\widehat{W}, \widehat{\omega})$. This can be done so that $\widehat{\omega}$ near infinity takes the form $d(e^t \alpha)$, with α chosen to be the Giroux form coming from Theorem 9.47. Then after an $\mathbb{R}$-translation, the holomorphic pages of Theorem 9.47 can be regarded as embedded J-holomorphic curves of index 2 foliating a neighborhood of infinity in $(\widehat{W}, \widehat{\omega})$, and this foliation extends to a Lefschetz fibration of $\widehat{W}$ in much the same way that an embedded J-holomorphic sphere of index 2 in a closed symplectic 4-manifold generates the fibers of a blown-up ruled surface. More precisely, let $\overline{\mathcal{M}}$ denote the compactified moduli space of curves in $\widehat{W}$ arising from the holomorphic pages. Since the asymptotic orbits of these curves are all simply covered, none of them are *bad* orbits in the sense of Definition 8.36, so Corollary 8.38 gives the smooth portion of this moduli space $\mathcal{M}$ an orientation, making $\overline{\mathcal{M}}$ into a compact oriented surface with one boundary component (corresponding to the S^1-family of holomorphic pages). Index counting arguments as in Chap. 4 prove that $\overline{\mathcal{M}}$ contains at most finitely many nodal curves, and one then uses the adjunction formula to show that they look like Lefschetz singular fibers (see Fig. 9.8) and fit together with the smooth curves to form a foliation of $\widehat{W}$, with the nodes as isolated singular points. The natural projection

$$\Pi : \widehat{W} \to \overline{\mathcal{M}}$$

sending every point to the unique (possibly nodal) curve that passes through it can then be viewed as the "completion" of a bordered Lefschetz fibration, which is allowable if (W, ω) is minimal since the nodal curves then can never have any spherical components. Now one uses the structure of the open book to show that $\overline{\mathcal{M}}$ must be diffeomorphic to the disk: in particular, its boundary is contractible. To see this, one can choose a loop $\gamma : S^1 \to M$ forming a meridian on the boundary of a small neighborhood of a binding component. This loop passes exactly once through every page, thus if we regard M as the "ideal boundary" of $\widehat{W}$ at infinity, $\Pi \circ \gamma : S^1 \to \overline{\mathcal{M}}$ parametrizes $\partial \overline{\mathcal{M}}$. But γ, while not contractible in $M \backslash K$, is clearly contractible in $\widehat{W}$: after moving it down into the interior of the cylindrical end, we can fill it with a disk in the region foliated by holomorphic curves, and the image of this disk under Π thus forms a contraction of $\partial \overline{\mathcal{M}}$ in $\overline{\mathcal{M}}$, implying $\overline{\mathcal{M}} \cong \mathbb{D}^2$.

Finally, one needs to show that if the filling is modified by a symplectic deformation $\{\omega_s\}_{s \in [0,1]}$, then the Lefschetz fibration constructed above extends to a family of isotopic ω_s-symplectic Lefschetz fibrations for all $s \in [0, 1]$. This essentially just requires choosing a generic smooth homotopy $\{J_s\}_{s \in [0,1]}$ consisting of ω_s-compatible almost complex structures and replacing the moduli space above by a parametric moduli space dependent on s. The key technical points are as follows:

(1) Since all asymptotic Reeb orbits have odd Conley-Zehnder index and we are only considering curves of genus 0, all curves in the parametric moduli space, including the smooth components of nodal curves, satisfy automatic transversality (Corollary 8.41).

(2) Another consequence of the fact that all asymptotic Reeb orbits have odd Conley-Zehnder index is that all conceivable curves arising in the compactfication of the parametric moduli space have even index (Exercise 8.33). Hence the bound $\operatorname{ind}(u) \geqslant -1$ for simple J_s-holomorphic curves at nongeneric parameter values is improved to $\operatorname{ind}(u) \geqslant 0$, and for this reason, compactness works out the same as in the generic case.

As a consequence, the construction of the J_s-holomorphic Lefschetz fibration works for every parameter $s \in [0, 1]$ and depends smoothly on the parameter. This completes our sketch of the proof of Theorem 9.41.

Remark 9.49. Returning to the proof of Theorem 9.20 on fillings of $(\mathbb{T}^3, \xi_1)$, here is the missing argument to rule out the second nodal degeneration shown in Fig. 9.4. Let $\overline{\mathcal{M}}$ denote either of the two compactified families of holomorphic cylinders that we considered in a (completed) minimal strong filling $(\widehat{W}, \widehat{\omega})$ of the torus. Then a priori, $\overline{\mathcal{M}}$ can contain a finite set of nodal curves, but by the same argument sketched above for Theorem 9.41, they fit together with the smooth curves in $\mathcal{M}$ to foliate $\widehat{W}$ and form the completion of a bordered Lefschetz fibration $\Pi : \widehat{W} \to \overline{\mathcal{M}}$. In contrast to Theorem 9.41, $\overline{\mathcal{M}}$ does not have the topology of the disk; in fact, $\overline{\mathcal{M}}$ has two boundary components, corresponding to the two S^1-families of holomorphic cylinders whose projections to $\mathbb{T}^3$ foliate the complement of a pair of Morse-Bott 2-tori. A variation on the argument used above to prove $\overline{\mathcal{M}} \cong \mathbb{D}^2$ then shows in this case that $\overline{\mathcal{M}} \cong [-1, 1] \times S^1$, so we have a bordered Lefschetz fibration of the filling (W, ω) over the annulus. Instead of an open book, the restriction of such a Lefschetz fibration to the boundary defines a more general decomposition, consisting in this case of two trivial cylinder-fibrations over S^1, which we should think of as the pages of a generalized open book.[6] Just as singular fibers in a Lefschetz fibration over $\mathbb{D}^2$ contribute Dehn twists to the monodromy of the open book at the boundary, the *composition* of the monodromies of these two cylinder-fibrations must be isotopic to a product of right-handed Dehn twists, again one for every singular fiber of the Lefschetz fibration over $[-1, 1] \times S^1$. But the fibers are cylinders and the composition of monodromies is trivial, so the only way to produce this as a product of right-handed Dehn twists is if there are no singular fibers at all, hence no nodal degenerations in the moduli space.

Theorem 9.31 is a slightly easier variation on the above argument, just as Theorems 9.4 and 9.5 on fillings and Reeb orbits for (S^3, ξ_{st}) follow by variations on the same argument. The existence of nodal degenerations which form the singular fibers of the Lefschetz fibration in Theorem 9.41 plays no role in Theorem 9.31: the crucial fact is that since only finitely many such degenerations are possible, there still must be a family of smooth curves that reach to $-\infty$ in any cobordism with a negative end, thus breaking off to produce a building with lower levels asymptotic to Reeb orbits in the concave boundary. One could just as well have stated Theorem 9.31 in a slightly more general way, and we'll conclude with this statement:

Theorem 9.50 (Easy Exercise Based on [ACH05]). *Suppose (M, ξ) is supported by a planar open book whose pages have $k \geqslant 1$ boundary components, and (M', ξ') is another contact manifold admitting an exact symplectic cobordism to (M, ξ). Then for every contact form for ξ', there exists a compact oriented surface Σ, with genus zero and at least 1 but at most k boundary components, admitting a continuous map $\Sigma \to M'$ whose restriction to each component of $\partial\Sigma$ is a closed Reeb orbit.* □

---

[6]This type of decomposition of $(\mathbb{T}^3, \xi_1)$ is called a *summed open book* in [Wen13], as it can be viewed as the result of attaching two open books together via contact fiber sums along binding components. It is also a special case of a *spinal open book* as defined in [LVW]. These notions make it possible to apply the technology of punctured holomorphic spheres to many contact 3-manifolds that are not planar.

Appendix A
Generic Nodes $\Rightarrow$ Lefschetz Critical Points

The purpose of this appendix is to sketch a proof of the folk theorem that a J-holomorphic foliation with an isolated transverse nodal singularity can be identified locally with the neighborhood of a critical point in a Lefschetz fibration. The proof outlined here uses the compactness theory of punctured holomorphic curves in the asymptotically cylindrical setting, see Sect. 8.3.4 and [BEH+03, Bao15]. It is based on conversations with Sam Lisi related to our joint work with Jeremy Van Horn-Morris [LVW].

Assume in the following that (M, ω) is a closed symplectic 4-manifold with a tame almost complex structure J. This is geared toward application to the theorems in Chap. 1, though the statement has a straightforward generalization in the context of SFT compactness, allowing (M, ω) to be e.g. the completion of a compact symplectic manifold with convex boundary, as needed for Theorem 9.41. Recall from Exercise 3.21 that critical points of Lefschetz fibrations in dimension four can be modeled in complex coordinates by the map $\pi(z_1, z_2) = z_1 z_2$.

Proposition A.1. *Suppose* $\overline{\mathcal{M}}^{\mathcal{F}} \subset \overline{\mathcal{M}}_g(J)$ *is a subset of the space of stable nodal holomorphic curves in* (M, J) *with the following properties:*

(i) *$\overline{\mathcal{M}}^{\mathcal{F}}$ is homeomorphic to a 2-dimensional disk;*

(ii) *$\overline{\mathcal{M}}^{\mathcal{F}}$ contains a nodal curve u_0 in its interior, which is embedded except at its nodes, all of which are transverse double points;*

(iii) *$\mathcal{M}^{\mathcal{F}} := \overline{\mathcal{M}}^{\mathcal{F}} \backslash \{u_0\}$ is a smooth 2-dimensional family of smoothly embedded curves;*

(iv) *There exists a point $p \in M$, which is the image of a node of u_0, such that a neighborhood of p in M has a smooth 2-dimensional foliation $\mathcal{F}$, with p as an isolated singularity, where the leaves of $\mathcal{F}$ are pieces of curves in $\overline{\mathcal{M}}^{\mathcal{F}}$.*

© Springer International Publishing AG, part of Springer Nature 2018
C. Wendl, *Holomorphic Curves in Low Dimensions*, Lecture Notes
in Mathematics 2216, https://doi.org/10.1007/978-3-319-91371-1

Then there exists a smooth coordinate chart $\varphi : \mathcal{U} \to \mathbb{C}^2$ *on a neighborhood* $\mathcal{U} \subset M$
of p, sending p to 0, such that

$$(\varphi_t)_* \mathcal{F} \to \mathcal{F}_0 \quad \text{in} \quad C^\infty_{\text{loc}}(\mathbb{C}^2) \text{ as } t \to \infty,$$

where $\varphi_t(x) := t\varphi(x) \in \mathbb{C}^2$ *for* $t > 0$, *and* $\mathcal{F}_0$ *denotes the singular foliation on* $\mathbb{C}^2$
tangent to the fibers of the map $(z_1, z_2) \mapsto z_1 z_2$.

To prove this, we first construct a coordinate chart $\varphi : \mathcal{U} \to \mathbb{C}^2$ near p with a few
special properties. We will require in the first place that $\varphi(p) = 0$ and $(\varphi_* J)(0) = i$.
Using local existence of holomorphic disks through p, one can also arrange so that
φ gives

$$(\dot{M}, \dot{J}) := (M \backslash \{p\}, J)$$

the structure of an almost complex manifold with an *asymptotically* cylindrical
negative end at p, cf. [BEH+03, Example 3.1(2)]. More precisely, this means that if
we rescale the coordinates so as to assume $\varphi(\mathcal{U})$ contains the unit ball $B^4 \subset \mathbb{C}^2$ and
define

$$\Phi : (-\infty, 0] \times S^3 \hookrightarrow \dot{M} : (r, x) \mapsto \varphi^{-1}(e^{2r} x),$$

then the family of almost complex structures $J^s := (f^s)^* \Phi^* J$ on $(-\infty, 0] \times S^3$
defined via $\mathbb{R}$-translation $f^s(r, x) := (r - s, x)$ for $s \geqslant 0$ satisfies

- $J^s \partial_r = R_{\text{st}}$ for all $s \geqslant 0$;
- $J^s \to J^\infty \in \mathcal{J}(S^3, \alpha_{\text{st}})$ as $s \to \infty$,

where α_{st} denotes the standard contact form on $S^3 = \partial\overline{B}^4$ with its induced Reeb
vector field R_{st} whose orbits are the fibers of the Hopf fibration, and J^∞ is the
standard complex structure i under the identification

$$\mathbb{R} \times S^3 \to \mathbb{C}^2 \backslash \{0\} : (r, x) \mapsto e^{2r} x. \tag{A.1}$$

The first condition permits a natural definition of energy in the asymptotically
cylindrical context so that families of curves in (M, J) with uniformly bounded
energy also automatically have bounded energy as curves in $(\dot{M}, \dot{J})$, possibly with
extra punctures approaching the negative end at p. Moreover, the convergence
$J^s \to J^\infty$ is exponentially fast, so that Bao's extension [Bao15] of the standard
asymptotic and compactness results from [BEH+03] is valid in this setting. There
is one more useful condition on $\varphi : \mathcal{U} \to \mathbb{C}^2$ that we are free to impose: the nodal
curve u_0 is assumed to have two components u_0^+ and u_0^- passing through $\mathcal{U}$ that
intersect transversely at p, and we can choose φ so that these two components,
when viewed in coordinates, are tangent at p to $\mathbb{C} \times \{0\}$ and $\{0\} \times \mathbb{C}$ respectively.

Now using Φ to identify a punctured neighborhood of p with $(-\infty, 0] \times S^3$, the
curves u_0^+ and u_0^- appear in $(-\infty, 0] \times S^3$ as $\dot{J}$-holomorphic half-cylinders with

negative ends approaching the Reeb orbits

$$\gamma^+(t) := (e^{2\pi it}, 0), \qquad \gamma^-(t) := (0, e^{2\pi it})$$

in S^3. To prove the proposition, pick any sequence $(r_k, x_k) \in (-\infty, 0] \times S^3$ such that $r_k \to -\infty$, x_k converges to some point $x_\infty \in S^3$ outside the images of γ^+ and γ^-, and (r_k, x_k) for every k is not in the image of u_0^+ or u_0^-. This determines a sequence $u_k \in \mathcal{M}^{\mathcal{F}}$ of curves through (r_k, x_k). The next lemma implies Proposition A.1.

Lemma A.2. *The sequence u_k of J-holomorphic curves in $\dot{M}$ converges in the sense of [BEH+03] to a holomorphic building whose main level is the nodal curve u_0 (regarded as a possibly disconnected curve in $\dot{M}$ with two negative ends approaching p), and it has exactly one lower level, consisting of an embedded J^∞- holomorphic cylinder whose image under the identification of $\mathbb{R} \times S^3$ with $\mathbb{C}^2\backslash\{0\}$ via (A.1) is a fiber of the map $\mathbb{C}^2 \to \mathbb{C} : (z_1, z_2) \mapsto z_1 z_2$.*

Proof. It suffices to show that every subsequence of u_k has a further subsequence converging to a specific building of the stated form. By [BEH+03, Bao15], we can start by extracting a subsequence that converges to *some* building u_∞. Its main level clearly fits the stated description since the nodal curve u_0 is necessarily the limit of any sequence of curves in $\mathcal{M}^{\mathcal{F}}$ that pass through a sequence of points approaching p. The lower levels form a J^∞-holomorphic building in $\mathbb{R} \times S^3$ of arithmetic genus zero with exactly two positive ends, asymptotic to the orbits γ^+ and γ^-. Under the identification $\mathbb{R} \times S^3 = \mathbb{C}^2\backslash\{0\}$, it is not difficult to classify all curves in $\mathbb{R} \times S^3$ that have precisely two positive ends asymptotic to these particular orbits: up to parametrization, they appear in $\mathbb{C}^2$ as pairs of meromorphic functions of the form

$$u(z) = (az + b, 1/z + c)$$

for some constants $a, b, c \in \mathbb{C}$ with $a \neq 0$. Here the parametrization has been chosen so that the curve has positive punctures at 0 and ∞ asymptotic to γ^- and γ^+ respectively. It has a negative puncture at any point $z_0 \in \mathbb{C}\backslash\{0\}$ such that $az_0 + b = 1/z_0 + c = 0$, which means $z_0 = -1/c = -b/a$, so there can be at most one such puncture, existing only if $bc = a$, and it is asymptotic to the orbit

$$\gamma_0(t) = (Ae^{2\pi it}, Be^{2\pi it})$$

where $(A, B) := \frac{(a, -c^2)}{|(a, -c^2)|} \in S^3 = \partial \overline{B}^4$. Note that since $bc = a \neq 0$, A and B are both nonzero, so γ_0 is different from both γ^+ and γ^- and is therefore nontrivially linked with both in S^3. It also has the smallest possible period for closed orbits of R_{st}, hence there must be exactly one additional lower level, which is a plane asymptotic to γ_0, and the nontrivial linking implies that this plane intersects the trivial cylinders over γ^+ and γ^-. But by local positivity of intersections, an intersection of this type implies intersections of u_k with u_0^+ and u_0^- for sufficiently large k; this follows from the fact that $u_0^{\pm}$ have ends end negatively asymptotic to $\gamma^{\pm}$. This contradicts the assumption that the curves in $\overline{\mathcal{M}}^{\mathcal{F}}$ form a foliation near p, so we conclude that the curve $u(z) = (az + b, 1/z + c)$ cannot have a negative puncture, and u_{∞} therefore has only one lower level.

Similarly, u intersects the trivial cylinder over either γ^{+} or γ^{-} if there is any $z \in \mathbb{C}\backslash\{0\}$ with $az + b = 0$ or $1/z + c = 0$; note that we have already ruled out the scenario in which both vanish at the same time. This again leads to a contradiction by positivity of intersections, so we conclude $b = c = 0$ and thus $u(z) = (az, 1/z)$. Note finally that a curve u of this form is uniquely determined up to $\mathbb{R}$-translation by the fact it must pass through $\mathbb{R} \times \{x_{\infty}\}$. □

Corollary A.3. *Under the assumptions of Proposition A.1, a neighborhood of the nodal singularity p admits complex coordinates in which the leaves of $\mathcal{F}$ are the fibers of the map $(z_1, z_2) \mapsto z_1 z_2$.*

Proof. It suffices to show that this is true for the rescaled foliation $\mathcal{F}_t := (\varphi_t)_* \mathcal{F}$ on $B^4 \subset \mathbb{C}^2$ if $t > 0$ is sufficiently large. One can first choose a parametrization of the leaves of $\mathcal{F}_0$ via their transverse intersections with a suitable fixed 2-disk, and use the same disk to parametrize the leaves of $\mathcal{F}_t$ for large t. An ambient isotopy taking $\mathcal{F}_0$ to $\mathcal{F}_t$ for t large can then be constructed by perturbing each leaf of $\mathcal{F}_0$ in the orthogonal direction to produce the corresponding leaf of $\mathcal{F}_t$. □

Bibliography

[Abb11] C. Abbas, Holomorphic open book decompositions. Duke Math. J. **158**(1), 29–82 (2011)

[ACH05] C. Abbas, K. Cieliebak, H. Hofer, The Weinstein conjecture for planar contact structures in dimension three. Comment. Math. Helv. **80**(4), 771–793 (2005)

[AF03] R.A. Adams, J.J.F. Fournier, *Sobolev Spaces*. Pure and Applied Mathematics (Amsterdam), vol. 140, 2nd edn. (Elsevier/Academic, Amsterdam/New York, 2003)

[AÖ01a] S. Akbulut, B. Özbağci, Lefschetz fibrations on compact Stein surfaces. Geom. Topol. **5**, 319–334 (2001)

[AÖ01b] S. Akbulut, B. Özbağci, Erratum: Ą"Lefschetz fibrations on compact Stein surfaces" [Geom. Topol. **5**, 319–334 (2001) (electronic); MR1825664 (2003a:57055)]. Geom. Topol. **5**, 939–945 (2001)

[ABW10] P. Albers, B. Bramham, C. Wendl, On nonseparating contact hypersurfaces in symplectic 4-manifolds. Algebr. Geom. Topol. **10**(2), 697–737 (2010)

[AD14] M. Audin, M. Damian, *Morse Theory and Floer Homology* Universitext (Springer/EDP Sciences, London/Les Ulis, 2014). Translated from the 2010 French original by Reinie Erné

[Bao15] E. Bao, On J-holomorphic curves in almost complex manifolds with asymptotically cylindrical ends. Pac. J. Math. **278**(2), 291–324 (2015)

[BGZ] K. Barth, H. Geiges, K. Zehmisch, The diffeomorphism type of symplectic fillings (2016). Preprint arXiv:1607.03310

[BV15] R.I. Baykur, J. Van Horn-Morris, Families of contact 3-manifolds with arbitrarily large Stein fillings. J. Differ. Geom. **101**(3), 423–465 (2015). With an appendix by S. Lisi and C. Wendl

[BEM15] M.S. Borman, Y. Eliashberg, E. Murphy, Existence and classification of overtwisted contact structures in all dimensions. Acta Math. **215**(2), 281–361 (2015)

[Bou02] F. Bourgeois, A Morse-Bott approach to contact homology. Ph.D. Thesis, Stanford University (2002)

[BM04] F. Bourgeois, K. Mohnke, Coherent orientations in symplectic field theory. Math. Z. **248**(1), 123–146 (2004)

[BEH+03] F. Bourgeois, Y. Eliashberg, H. Hofer, K. Wysocki, E. Zehnder, Compactness results in symplectic field theory. Geom. Topol. **7**, 799–888 (2003)

[Bow12] J. Bowden, Exactly fillable contact structures without Stein fillings. Algebr. Geom. Topol. **12**(3), 1803–1810 (2012)

[Bra15] B. Bramham, Periodic approximations of irrational pseudo-rotations using pseudo-holomorphic curves. Ann. Math. (2) **181**(3), 1033–1086 (2015)

C. Wendl, *Holomorphic Curves in Low Dimensions*, Lecture Notes in Mathematics 2216, https://doi.org/10.1007/978-3-319-91371-1

[Bre93] G.E. Bredon, *Topology and Geometry* (Springer, New York, 1993)

[Che02] Y. Chekanov, Differential algebra of Legendrian links. Invent. Math. **150**(3), 441–483 (2002)

[CE12] K. Cieliebak, Y. Eliashberg, *From Stein to Weinstein and Back: Symplectic Geometry of Affine Complex Manifolds*. American Mathematical Society Colloquium Publications, vol. 59 (American Mathematical Society, Providence, RI, 2012)

[CM07] K. Cieliebak, K. Mohnke, Symplectic hypersurfaces and transversality in Gromov-Witten theory. J. Symplectic Geom. **5**(3), 281–356 (2007)

[CW] A. Cioba, C. Wendl, Unknotted Reeb orbits and nicely embedded holomorphic curves (2016). Preprint arXiv:1609.01660

[Col08] V. Colin, Livres ouverts en géométrie de contact (d'aprés Emmanuel Giroux). Astérisque **317**, Exp. No. 969, vii, 91–117 (2008) (French, with French summary). Séminaire Bourbaki. vols. 2006/2007

[CZ83a] C. Conley, E. Zehnder, An index theory for periodic solutions of a Hamiltonian system, in *Geometric Dynamics (Rio de Janeiro, 1981)*. Lecture Notes in Mathematics, vol. 1007 (Springer, Berlin, 1983), pp. 132–145

[CZ83b] C.C. Conley, E. Zehnder, The Birkhoff-Lewis fixed point theorem and a conjecture of V. I. Arnol'd. Invent. Math. **73**(1), 33–49 (1983)

[CS99] J.H. Conway, N.J.A. Sloane, *Sphere Packings, Lattices and Groups*. Grundlehren der Mathematischen Wissenschaften [Fundamental Principles of Mathematical Sciences], vol. 290, 3rd edn. (Springer, New York, 1999). With additional contributions by E. Bannai, R.E. Borcherds, J. Leech, S.P. Norton, A.M. Odlyzko, R.A. Parker, L. Queen, B.B. Venkov

[CHP] D. Cristofaro-Gardiner, M. Hutchings, D. Pomerleano, Torsion contact forms in three dimensions have two or infinitely many Reeb orbits (2017). Preprint arXiv:1701.02262

[Deb01] O. Debarre, *Higher-Dimensional Algebraic Geometry*. Universitext (Springer, New York, 2001)

[Don99] S.K. Donaldson, Lefschetz pencils on symplectic manifolds. J. Differ. Geom. **53**(2), 205–236 (1999)

[Don02] S.K. Donaldson, *Floer Homology Groups in Yang-Mills Theory*. Cambridge Tracts in Mathematics, vol. 147 (Cambridge University Press, Cambridge, 2002). With the assistance of M. Furuta and D. Kotschick

[DK90] S.K. Donaldson, P.B. Kronheimer, *The Geometry of Four-Manifolds*. Oxford Mathematical Monographs (The Clarendon Press/Oxford University Press, New York, 1990). Oxford Science Publications

[Dra04] D.L. Dragnev, Fredholm theory and transversality for noncompact pseudoholomorphic maps in symplectizations. Commun. Pure Appl. Math. **57**(6), 726–763 (2004)

[Eli89] Y. Eliashberg, Classification of overtwisted contact structures on 3-manifolds. Invent. Math. **98**(3), 623–637 (1989)

[Eli90] Y. Eliashberg, Filling by holomorphic discs and its applications, in *Geometry of Low-Dimensional Manifolds, 2 (Durham, 1989)*. London Mathematical Society Lecture Note Series, vol. 151 (Cambridge University Press, Cambridge, 1990), pp. 45–67

[Eli91] Y. Eliashberg, On symplectic manifolds with some contact properties. J. Differ. Geom. **33**(1), 233–238 (1991)

[Eli96] Y. Eliashberg, Unique holomorphically fillable contact structure on the 3-torus. Int. Math. Res. Not. **2**, 77–82 (1996)

[Eli98] Y. Eliashberg, Invariants in contact topology, in *Proceedings of the International Congress of Mathematicians*, vol. II, Berlin, 1998, pp. 327–338

[Eli04] Y. Eliashberg, A few remarks about symplectic filling. Geom. Topol. **8**, 277–293 (2004)

[EGH00] Y. Eliashberg, A. Givental, H. Hofer, Introduction to symplectic field theory. Geom. Funct. Anal. Special Volume, 560–673 (2000)

[EM02]   Y. Eliashberg, N. Mishachev, *Introduction to the h-Principle*. Graduate Studies in Mathematics, vol. 48 (American Mathematical Society, Providence, RI, 2002)

[Etn04]   J.B. Etnyre, Planar open book decompositions and contact structures. Int. Math. Res. Not. **79**, 4255–4267 (2004)

[Etn06]   J.B. Etnyre, Lectures on open book decompositions and contact structures, in *Floer Homology, Gauge Theory, and Low-Dimensional Topology*. Clay Mathematics Proceedings, vol. 5 (American Mathematical Society, Providence, RI, 2006), pp. 103–141

[EH02]   J.B. Etnyre, K. Honda, On symplectic cobordisms. Math. Ann. **323**(1), 31–39 (2002)

[Eva98]   L.C. Evans, *Partial Differential Equations*. Graduate Studies in Mathematics, vol. 19 (American Mathematical Society, Providence, RI, 1998)

[FS]   J.W. Fish, R. Siefring, Connected sums and finite energy foliations I: contact connected sums (2013). Preprint arXiv:1311.4221

[Flo88a]   A. Floer, The unregularized gradient flow of the symplectic action. Commun. Pure Appl. Math. **41**(6), 775–813 (1988)

[Flo88b]   A. Floer, An instanton-invariant for 3-manifolds. Commun. Math. Phys. **118**(2), 215–240 (1988)

[Gay06]   D.T. Gay, Four-dimensional symplectic cobordisms containing three-handles. Geom. Topol. **10**, 1749–1759 (2006)

[Gei08]   H. Geiges, *An Introduction to Contact Topology*. Cambridge Studies in Advanced Mathematics, vol. 109 (Cambridge University Press, Cambridge, 2008)

[GZ10]   H. Geiges, K. Zehmisch, Eliashberg's proof of Cerf's theorem. J. Topol. Anal. **2**(4), 543–579 (2010)

[GZ13]   H. Geiges, K. Zehmisch, How to recognize a 4-ball when you see one. Münster J. Math. **6**, 525–554 (2013)

[GW17]   C. Gerig, C. Wendl, Generic transversality for unbranched covers of closed pseudo-holomorphic curves. Commun. Pure Appl. Math. **70**(3), 409–443 (2017)

[Ger]   A. Gerstenberger, Geometric transversality in higher genus Gromov-Witten theory (2013). Preprint arXiv:1309.1426

[Ghi05]   P. Ghiggini, Strongly fillable contact 3-manifolds without Stein fillings. Geom. Topol. **9**, 1677–1687 (2005)

[GH]   P. Ghiggini, K. Honda, Giroux torsion and twisted coefficients (2008). Preprint arXiv:0804.1568

[Gir94]   E. Giroux, Une structure de contact, même tendue, est plus ou moins tordue. Ann. Sci. École Norm. Sup. (4) **27**(6), 697–705 (1994) [French, with English summary]

[Gir]   E. Giroux, Links and contact structures. Lecture given at the Georgia International Topology Conference, University of Georgia, May 24, 2001. Notes available at http://www.math.uga.edu/~topology/2001/giroux.pdf

[Gir02]   E. Giroux, Géométrie de contact: de la dimension trois vers les dimensions supérieures, in *Proceedings of the International Congress of Mathematicians*, vol. II, Beijing, 2002, pp. 405–414

[GP17]   E. Giroux, J. Pardon, Existence of Lefschetz fibrations on Stein and Weinstein domains. Geom. Topol. **21**(2), 963–997 (2017)

[Gom95]   R.E. Gompf, A new construction of symplectic manifolds. Ann. Math. (2) **142**(3), 527–595 (1995)

[Gom04a]   R.E. Gompf, Toward a topological characterization of symplectic manifolds. J. Symplectic Geom. **2**(2), 177–206 (2004)

[Gom04b]   R.E. Gompf, Symplectic structures from Lefschetz pencils in high dimensions, in *Proceedings of the Casson Fest, Geometry Topology Monograph*, vol. 7 (Geometry Topology Publication, Coventry, 2004), pp. 267–290 [electronic]

[Gom05]   R.E. Gompf, Locally holomorphic maps yield symplectic structures. Commun. Anal. Geom. **13**(3), 511–525 (2005)

[GS99]   R.E. Gompf, A.I. Stipsicz, *4-Manifolds and Kirby Calculus*. Graduate Studies in Mathematics, vol. 20 (American Mathematical Society, Providence, RI, 1999)

[Gro85] M. Gromov, Pseudoholomorphic curves in symplectic manifolds. Invent. Math. **82**(2), 307–347 (1985)

[Hat02] A. Hatcher, *Algebraic Topology* (Cambridge University Press, Cambridge, 2002)

[Hin00] R. Hind, Holomorphic filling of $\mathbf{RP}^3$. Commun. Contemp. Math. **2**(3), 349–363 (2000)

[Hir94] M.W. Hirsch, *Differential Topology* (Springer, New York, 1994)

[Hof93] H. Hofer, Pseudoholomorphic curves in symplectizations with applications to the Weinstein conjecture in dimension three. Invent. Math. **114**(3), 515–563 (1993)

[Hof00] H. Hofer, Holomorphic curves and real three-dimensional dynamics. Geom. Funct. Anal. Special Volume, 674–704 (2000). GAFA 2000 (Tel Aviv, 1999)

[HLS97] H. Hofer, V. Lizan, J.-C. Sikorav, On genericity for holomorphic curves in four-dimensional almost-complex manifolds. J. Geom. Anal. **7**(1), 149–159 (1997)

[HZ94] H. Hofer, E. Zehnder, *Symplectic Invariants and Hamiltonian Dynamics* (Birkhäuser, Basel, 1994)

[HWZ95a] H. Hofer, K.Wysocki, E. Zehnder, Properties of pseudo-holomorphic curves in symplectisations. II. Embedding controls and algebraic invariants. Geom. Funct. Anal. **5**(2), 270–328 (1995)

[HWZ95b] H. Hofer, K.Wysocki, E. Zehnder, A characterisation of the tight three-sphere. Duke Math. J. **81**(1), 159–226 (1995)

[HWZ96a] H. Hofer, K.Wysocki, E. Zehnder, Properties of pseudoholomorphic curves in symplectisations. I. Asymptotics. Ann. Inst. H. Poincaré Anal. Non Linéaire **13**(3), 337–379 (1996)

[HWZ96b] H. Hofer, K.Wysocki, E. Zehnder, Properties of pseudoholomorphic curves in symplectisations. IV. Asymptotics with degeneracies, in *Contact and Symplectic Geometry* (Cambridge University Press, Cambridge, 1994/1996), pp. 78–117

[HWZ98] H. Hofer, K.Wysocki, E. Zehnder, The dynamics on three-dimensional strictly convex energy surfaces. Ann. Math. (2) **148**(1), 197–289 (1998)

[HWZ99] H. Hofer, K.Wysocki, E. Zehnder, Properties of pseudoholomorphic curves in symplectizations. III. Fredholm theory. Top. Nonlin. Anal. **13**, 381–475 (1999)

[HWZ03] H. Hofer, K.Wysocki, E. Zehnder, Finite energy foliations of tight three-spheres and Hamiltonian dynamics. Ann. Math. (2) **157**(1), 125–255 (2003)

[HWZa] H. Hofer, K.Wysocki, E. Zehnder, Applications of polyfold theory I: the polyfolds of Gromov-Witten theory (2011). Preprint arXiv:1107.2097

[HWZb] H. Hofer, K.Wysocki, E. Zehnder, Polyfold and Fredholm theory (2017). Preprint arXiv:1707.08941

[HS11] U. Hryniewicz, P.A.S. Salomão, On the existence of disk-like global sections for Reeb flows on the tight 3-sphere. Duke Math. J. **160**(3), 415–465 (2011)

[HLS15] U. Hryniewicz, J.E. Licata, P.A.S. Salomão, A dynamical characterization of universally tight lens spaces. Proc. Lond. Math. Soc. (3) **110**(1), 213–269 (2015)

[HMS15] U. Hryniewicz, A. Momin, P.A.S. Salomão, A Poincaré-Birkhoff theorem for tight Reeb flows on S^3. Invent. Math. **199**(2), 333–422 (2015)

[Hut10] M. Hutchings, Embedded contact homology and its applications, in *Proceedings of the International Congress of Mathematicians*, vol. II (Hindustan Book Agency, New Delhi, 2010), pp. 1022–1041

[HT99] M. Hutchings, C.H. Taubes, An introduction to the Seiberg-Witten equations on symplectic manifolds, in *Symplectic Geometry and Topology (Park City, UT, 1997)*. IAS/Park City Mathematical Series, vol. 7 (American Mathematical Society, Providence, RI, 1999), pp. 103–142

[HT09] M. Hutchings, C.H. Taubes, The Weinstein conjecture for stable Hamiltonian structures. Geom. Topol. **13**(2), 901–941 (2009)

[IP99] E.-N. Ionel, T.H. Parker, Gromov invariants and symplectic maps. Math. Ann. **314**(1), 127–158 (1999)

[IS99] S. Ivashkovich, V. Shevchishin, Structure of the moduli space in a neighborhood of a cusp-curve and meromorphic hulls. Invent. Math. **136**(3), 571–602 (1999)

[KM94]   M. Kontsevich, Yu. Manin, Gromov-Witten classes, quantum cohomology, and enumerative geometry. Commun. Math. Phys. **164**(3), 525–562 (1994)

[KM07]   P. Kronheimer, T. Mrowka, *Monopoles and Three-Manifolds*. New Mathematical Monographs, vol. 10 (Cambridge University Press, Cambridge, 2007)

[KvK16] M. Kwon, O. van Koert, Brieskorn manifolds in contact topology. Bull. Lond. Math. Soc. **48**(2), 173–241 (2016)

[LM96a] F. Lalonde, D. McDuff, J-curves and the classification of rational and ruled symplectic 4-manifolds, in *Contact and Symplectic Geometry* (Cambridge, 1994). Publications of the Newton Institute, vol. 8 (Cambridge University Press, Cambridge, 1996), pp. 3–42

[LM96b] F. Lalonde, D. McDuff, The classification of ruled symplectic 4-manifolds. Math. Res. Lett. **3**(6), 769–778 (1996)

[Lan93] S. Lang, *Real and Functional Analysis*, 3rd edn. (Springer, New York, 1993)

[Lan99] S. Lang, *Fundamentals of Differential Geometry* (Springer, New York, 1999)

[Lax07] P.D. Lax, *Linear Algebra and Its Applications*. Pure and Applied Mathematics (Hoboken), 2nd edn. (Wiley-Interscience, Hoboken, NJ, 2007)

[Laz00] L. Lazzarini, Existence of a somewhere injective pseudo-holomorphic disc. Geom. Funct. Anal. **10**(4), 829–862 (2000)

[Lee03] J.M. Lee, *Introduction to Smooth Manifolds*. Graduate Texts in Mathematics, vol. 218 (Springer, New York, 2003)

[Li99]   T.-J. Li, Smoothly embedded spheres in symplectic 4y-manifolds. Proc. Am. Math. Soc. **127**(2), 609–613 (1999)

[LL95]   T.-J. Li, A.-K. Liu, Symplectic structure on ruled surfaces and a generalized adjunction formula. Math. Res. Lett. **2**(4), 453–471 (1995)

[LL99]   T.-J. Li, A.-K. Liu, The equivalence between SW and Gr in the case where $b^+ = 1$. Int. Math. Res. Not. **7**, 335–345 (1999)

[LÖ]   Y. Li, B. Özbağci, Fillings of unit cotangent bundles of nonorientable surfaces (2016). Preprint arXiv:1609.01891

[LMY17] T.-J. Li, C. Y. Mak, K. Yasui, Calabi-Yau caps, uniruled caps and symplectic fillings. Proc. Lond. Math. Soc. (3) **114**(1), 159–187 (2017)

[Lis08] P. Lisca, On symplectic fillings of lens spaces. Trans. Am. Math. Soc. **360**(2), 765–799 (2008)

[LVW]   S. Lisi, J. Van Horn-Morris, C. Wendl, On Symplectic Fillings of spinal open book decompositions (in preparation)

[Liu96] A.-K. Liu, Some new applications of general wall crossing formula, Gompf's conjecture and its applications. Math. Res. Lett. **3**(5), 569–585 (1996)

[LP01]   A. Loi, R. Piergallini, Compact Stein surfaces with boundary as branched covers of B^4. Invent. Math. **143**(2), 325–348 (2001)

[Mas14] P. Massot, Topological methods in 3-dimensional contact geometry, in *Contact and Symplectic Topology*. Bolyai Society Mathematical Studies, vol. 26 (János Bolyai Mathematical Society, Budapest, 2014), pp. 27–83

[MNW13] P. Massot, K. Niederkrüger, C.Wendl, Weak and strong fillability of higher dimensional contact manifolds. Invent. Math. **192**(2), 287–373 (2013)

[McD90] D. McDuff, The structure of rational and ruled symplectic 4-manifolds. J. Am. Math. Soc. **3**(3), 679–712 (1990)

[McD91] D. McDuff, Symplectic manifolds with contact type boundaries. Invent. Math. **103**(3), 651–671 (1991)

[McD92] D. McDuff, Erratum to: The structure of rational and ruled symplectic 4-manifolds. J. Am. Math. Soc. **5**(4), 987–988 (1992)

[McD92] D. McDuff, Immersed spheres in symplectic 4-manifolds. Ann. Inst. Fourier (Grenoble) **42**(1–2), 369–392 (1992) [English, with French summary]

[McD94] D. McDuff, Singularities and positivity of intersections of J-holomorphic curves, *Holomorphic Curves in Symplectic Geometry*. Progress in Mathematics, vol. 117 (Birkhäuser, Basel, 1994), pp. 191–215. With an appendix by Gang Liu

[MS94] D. McDuff, D. Salamon, *J-Holomorphic Curves and Quantum Cohomology*. University Lecture Series, vol. 6 (American Mathematical Society, Providence, RI, 1994)

[MS96] D. McDuff, D. Salamon, A survey of symplectic 4-manifolds with $b^+ = 1$. Turk. J. Math. **20**(1), 47–60 (1996)

[MS12] D. McDuff, D. Salamon, *J-Holomorphic Curves and Symplectic Topology*. American Mathematical Society Colloquium Publications, vol. 52, 2nd edn. (American Mathematical Society, Providence, RI, 2012)

[MS17] D. McDuff, D. Salamon, Introduction to symplectic topology, 3rd edn. (Oxford University Press, Oxford, 2017)

[MW95] M.J. Micallef, B. White, The structure of branch points in minimal surfaces and in pseudo-holomorphic curves. Ann. Math. (2) **141**(1), 35–85 (1995)

[Mil63] J. Milnor, *Morse Theory: Based on Lecture Notes by M. Spivak and R. Wells*. Annals of Mathematics Studies, vol. 51 (Princeton University Press, Princeton, NJ, 1963)

[Mil65] J. Milnor, *Lectures on the H-Cobordism Theorem*, Notes by L. Siebenmann and J. Sondow, (Princeton University Press, Princeton, NJ, 1965)

[Mil97] J.W. Milnor, *Topology from the Differentiable Viewpoint*. Princeton Landmarks in Mathematics (Princeton University Press, Princeton, NJ, 1997). Based on notes by David W. Weaver; Revised reprint of the 1965 original

[MH73] J. Milnor, D. Husemoller, *Symmetric Bilinear Forms*. Ergebnisse der Mathematik und ihrer Grenzgebiete, Band 73 (Springer, New York, 1973)

[MS74] J.W. Milnor, J.D. Stasheff, *Characteristic Classes*. Annals of Mathematics Studies, vol. 76 (Princeton University Press, Princeton, NJ, 1974)

[Mom11] A. Momin, Contact homology of orbit complements and implied existence. J. Modern Dyn. **5**(3), 409–472 (2011)

[Nel15] J. Nelson, Automatic transversality in contact homology I: regularity. Abh. Math. Semin. Univ. Hambg. **85**(2), 125–179 (2015)

[Nie06] K. Niederkrüger, The plastikstufe–a generalization of the overtwisted disk to higher dimensions. Algebr. Geom. Topol. **6**, 2473–2508 (2006)

[NW11] K. Niederkrüger, C. Wendl, Weak symplectic fillings and holomorphic curves. Ann. Sci. ẠL Ecole Norm. Sup. (4) **44**(5), 801–853 (2011)

[OV12] A. Oancea, C. Viterbo, On the topology of fillings of contact manifolds and applications. Comment. Math. Helv. **87**(1), 41–69 (2012)

[OO05] H. Ohta, K. Ono, Simple singularities and symplectic fillings. J. Differ. Geom. **69**(1), 1–42 (2005)

[ÖS04a] B. Özbağci, A.I. Stipsicz, *Surgery on Contact 3-Manifolds and Stein Surfaces*. Bolyai Society Mathematical Studies, vol. 13 (Springer, Berlin, 2004)

[OS04b] P. Ozsváth, Z. Szabó, Holomorphic disks and topological invariants for closed three-manifolds. Ann. Math. (2) **159**(3), 1027–1158 (2004)

[Pal66] R.S. Palais, Homotopy theory of infinite dimensional manifolds. Topology **5**, 1–16 (1966)

[Par16] J. Pardon, An algebraic approach to virtual fundamental cycles on moduli spaces of pseudo-holomorphic curves. Geom. Topol. **20**(2), 779–1034 (2016)

[PV10] O. Plamenevskaya, J. Van Horn-Morris, Planar open books, monodromy factorizations and Stein fillings. Geom. Topol. **14**, 2077–2101 (2010)

[Rab78] P.H. Rabinowitz, Periodic solutions of Hamiltonian systems. Commun. Pure Appl. Math. **31**(2), 157–184 (1978)

[RS93] J. Robbin, D. Salamon, The Maslov index for paths. Topology **32**(4), 827–844 (1993)

[Rua94] Y. Ruan, Symplectic topology on algebraic 3-folds. J. Differ. Geom. **39**(1), 215–227 (1994)

[Rua96] Y. Ruan, Topological sigma model and Donaldson-type invariants in Gromov theory. Duke Math. J. **83**(2), 461–500 (1996)

[RT97] Y. Ruan, G. Tian, Higher genus symplectic invariants and sigma models coupled with gravity. Invent. Math. **130**, 455–516 (1997)

[Sal99]   D. Salamon, Lectures on Floer homology, in *Symplectic Geometry and Topology (Park City, UT, 1997)*. IAS/Park City Mathematical Series, vol. 7 (American Mathematical Society, Providence, RI, 1999), pp. 143–229

[Sch95]   M. Schwarz, Cohomology operations from S^1-cobordisms in Floer homology. Ph.D. Thesis, ETH Zürich (1995)

[Sei08a] P. Seidel, A biased view of symplectic cohomology, in *Current Developments in Mathematics, 2006* (International Press, Somerville, MA, 2008), pp. 211–253

[Sei08b] P. Seidel, *Fukaya Categories and Picard-Lefschetz Theory*. Zurich Lectures in Advanced Mathematics (European Mathematical Society (EMS), Zürich, 2008)

[Sie08]   R. Siefring, Relative asymptotic behavior of pseudoholomorphic half-cylinders. Commun. Pure Appl. Math. **61**(12), 1631–1684 (2008)

[Sie11]   R. Siefring, Intersection theory of punctured pseudoholomorphic curves. Geom. Topol. **15**, 2351–2457 (2011)

[Sie17]   R. Siefring, Finite-energy pseudoholomorphic planes with multiple asymptotic limits. Math. Ann. **368**(1–2), 367–390 (2017)

[Sma59]   S. Smale, Diffeomorphisms of the 2-sphere. Proc. Am. Math. Soc. **10**, 621–626 (1959)

[Sma65]   S. Smale, An infinite dimensional version of Sard's theorem. Am. J. Math. **87**, 861–866 (1965)

[SV17]   S. Sivek, J. Van Horn-Morris, Fillings of unit cotangent bundles. Math. Ann. **368**(3–4), 1063–1080 (2017)

[Sti02]   A.I. Stipsicz, Gauge theory and Stein fillings of certain 3-manifolds. Turk. J. Math. **26**(1), 115–130 (2002)

[Tau95]   C.H. Taubes, The Seiberg-Witten and Gromov invariants. Math. Res. Lett **2**(2), 221–238 (1995)

[Tau96a] C.H. Taubes, Counting pseudo-holomorphic submanifolds in dimension 4. J. Differ. Geom. **44**(4), 818–893 (1996)

[Tau96b] C.H. Taubes, SW ⇒ Gr: from the Seiberg-Witten equations to pseudo-holomorphic curves. J. Am. Math. Soc. **9**(3), 845–918 (1996)

[Tau00]   C.H. Taubes, *Seiberg Witten and Gromov Invariants for Symplectic 4-Manifolds*, ed. by R. Wentworth. First International Press Lecture Series, vol. 2 (International Press, Somerville, MA, 2000)

[Tau07]   C.H. Taubes, The Seiberg-Witten equations and the Weinstein conjecture. Geom. Topol. **11**, 2117–2202 (2007)

[Tho54]   R. Thom, Quelques propriétés globales des variétés différentiables. Comment. Math. Helv. **28**, 17–86 (1954) [French]

[Thu76]   W.P. Thurston, Some simple examples of symplectic manifolds. Proc. Am. Math. Soc. **55**(2), 467–468 (1976)

[TW75]   W.P. Thurston, H.E. Winkelnkemper, On the existence of contact forms. Proc. Am. Math. Soc. **52**, 345–347 (1975)

[vK]   O. van Koert, Lecture notes on stabilization of contact open books (2011). Preprint arXiv:1012.4359

[Wan12]   A. Wand, Mapping class group relations, Stein fillings, and planar open book decompositions. J. Topol. **5**(1), 1–14 (2012)

[Wan15]   A. Wand, Factorizations of diffeomorphisms of compact surfaces with boundary. Geom. Topol. **19**(5), 2407–2464 (2015)

[Wei78]   A. Weinstein, Periodic orbits for convex Hamiltonian systems. Ann. Math. (2) **108**(3), 507–518 (1978)

[Wen08]   C. Wendl, Finite energy foliations on overtwisted contact manifolds. Geom. Topol. **12**, 531–616 (2008)

[Wen10a] C. Wendl, Automatic transversality and orbifolds of punctured holomorphic curves in dimension four. Comment. Math. Helv. **85**(2), 347–407 (2010)

[Wen10b] C. Wendl, Strongly fillable contact manifolds and J-holomorphic foliations. Duke Math. J. **151**(3), 337–384 (2010)

[Wen10c] C. Wendl, Open book decompositions and stable Hamiltonian structures. Expo. Math. **28**(2), 187–199 (2010)

[Wen13] C. Wendl, A hierarchy of local symplectic filling obstructions for contact 3-manifolds. Duke Math. J. **162**(12), 2197–2283 (2013)

[Wena] C. Wendl, Generic transversality in symplectizations. Two-part blog post, available at https://symplecticfieldtheorist.wordpress.com/2014/11/27/

[Wenb] C. Wendl, Signs (or how to annoy a symplectic topologist). Blog post, available at https://symplecticfieldtheorist.wordpress.com/2015/08/23/

[Wenc] C. Wendl, Lectures on holomorphic curves in symplectic and contact geometry (2010). Preprint arXiv:1011.1690v2

[Wend] C. Wendl, Lectures on Symplectic Field Theory. EMS Series of Lectures in Mathematics (to appear). Preprint arXiv:1612.01009

[Wene] C. Wendl, Transversality and super-rigidity for multiply covered holomorphic curves (2016). Preprint arXiv:1609.09867v3

[Wenf] C. Wendl, Contact 3-manifolds, holomorphic curves and intersection theory (2017). Preprint arXiv:1706.05540

[Zeh03] K. Zehmisch, The Eliashberg-Gromov tightness theorem. Diploma Thesis, Universität Leipzig (2003)

[Zeh15] K. Zehmisch, Holomorphic jets in symplectic manifolds. J. Fixed Point Theory Appl. **17**(2), 379–402 (2015)

[Zin08] A. Zinger, Pseudocycles and integral homology. Trans. Am. Math. Soc. **360**(5), 2741–2765 (2008)

[Zin16] A. Zinger, The determinant line bundle for Fredholm operators: construction, properties, and classification. Math. Scand. **118**(2), 203–268 (2016)

Index

abstract open book, 260
adjunction formula
    for closed holomorphic curves, 64
    for punctured holomorphic curves, 238
allowable Lefschetz fibration, 263
almost complex manifold, 21
    with asymptotically cylindrical ends, 275
    with cylindrical ends, 212
    with pseudoconvex boundary, 243
almost complex structure, 21
    adapted to a contact form, 210
    asymptotically cylindrical, 275
    callibrated, *see* almost complex structure,
        compatible with a symplectic form
    compatible with a symplectic form, 4, 22
    tamed by a symplectic form, 22
arithmetic genus, 46, 78
Arnol'd conjecture, 193
aspherical symplectic manifolds, 194
asymptotic defect, 237
asymptotic eigenfunctions, 232, 238
asymptotic marker, 225
asymptotic trivializations, 219, 236
asymptotic winding number, 232
asymptotically cylindrical almost complex
    structure, 275
asymptotically cylindrical map, 220, 238
automorphism group of a holomorphic curve,
    29

bad Reeb orbit, 229
Baire set, 30
base point of a Lefschetz pencil, 80
Betti numbers, 174

biholomorphic, 22, 24
binding of an open book decomposition, 259
birational equivalence, 10, 136
blowdown
    complex, 69
    smooth or almost complex, 69, 75
    symplectic, 6, 73, 74
blowdown map, 68, 69, 75
blowup
    complex, 67
    smooth or almost complex, 69, 75
    symplectic, 6, 71, 74
bordered Lefschetz fibration, 263
    allowable, 263
    horizontal boundary of, 263
    symplectic, 265
    vertical boundary of, 263
bordism between pseudocycles, 148
branch point of a holomorphic map, 24
branched cover, 24
breaking orbit, 223, 225

canonical 1-form, *see* standard Liouville form
    on cotangent bundles
canonical class of a symplectic 4-manifold, 14
canonical line bundle, 14
Cauchy-Riemann type operator, 54
characteristic line field, 197
circle compactification of a punctured Riemann
    surface, 215
coherent orientations, 228
comeager, 30
compactification
    for closed holomorphic curves, 46

LECTURE NOTES IN MATHEMATICS   ⌁ Springer

Editors in Chief: J.-M. Morel, B. Teissier;

Editorial Policy

1. Lecture Notes aim to report new developments in all areas of mathematics and their applications – quickly, informally and at a high level. Mathematical texts analysing new developments in modelling and numerical simulation are welcome.

   Manuscripts should be reasonably self-contained and rounded off. Thus they may, and often will, present not only results of the author but also related work by other people. They may be based on specialised lecture courses. Furthermore, the manuscripts should provide sufficient motivation, examples and applications. This clearly distinguishes Lecture Notes from journal articles or technical reports which normally are very concise. Articles intended for a journal but too long to be accepted by most journals, usually do not have this "lecture notes" character. For similar reasons it is unusual for doctoral theses to be accepted for the Lecture Notes series, though habilitation theses may be appropriate.

2. Besides monographs, multi-author manuscripts resulting from SUMMER SCHOOLS or similar INTENSIVE COURSES are welcome, provided their objective was held to present an active mathematical topic to an audience at the beginning or intermediate graduate level (a list of participants should be provided).

   The resulting manuscript should not be just a collection of course notes, but should require advance planning and coordination among the main lecturers. The subject matter should dictate the structure of the book. This structure should be motivated and explained in a scientific introduction, and the notation, references, index and formulation of results should be, if possible, unified by the editors. Each contribution should have an abstract and an introduction referring to the other contributions. In other words, more preparatory work must go into a multi-authored volume than simply assembling a disparate collection of papers, communicated at the event.

3. Manuscripts should be submitted either online at www.editorialmanager.com/lnm to Springer's mathematics editorial in Heidelberg, or electronically to one of the series editors. Authors should be aware that incomplete or insufficiently close-to-final manuscripts almost always result in longer refereeing times and nevertheless unclear referees' recommendations, making further refereeing of a final draft necessary. The strict minimum amount of material that will be considered should include a detailed outline describing the planned contents of each chapter, a bibliography and several sample chapters. Parallel submission of a manuscript to another publisher while under consideration for LNM is not acceptable and can lead to rejection.

4. In general, **monographs** will be sent out to at least 2 external referees for evaluation.

   A final decision to publish can be made only on the basis of the complete manuscript, however a refereeing process leading to a preliminary decision can be based on a pre-final or incomplete manuscript.

   Volume Editors of **multi-author works** are expected to arrange for the refereeing, to the usual scientific standards, of the individual contributions. If the resulting reports can be forwarded to the LNM Editorial Board, this is very helpful. If no reports are forwarded or if other questions remain unclear in respect of homogeneity etc, the series editors may wish to consult external referees for an overall evaluation of the volume.

5. Manuscripts should in general be submitted in English. Final manuscripts should contain at least 100 pages of mathematical text and should always include

   – a table of contents;
   – an informative introduction, with adequate motivation and perhaps some historical remarks: it should be accessible to a reader not intimately familiar with the topic treated;
   – a subject index: as a rule this is genuinely helpful for the reader.
   – For evaluation purposes, manuscripts should be submitted as pdf files.

6. Careful preparation of the manuscripts will help keep production time short besides ensuring satisfactory appearance of the finished book in print and online. After acceptance of the manuscript authors will be asked to prepare the final LaTeX source files (see LaTeX templates online: https://www.springer.com/gb/authors-editors/book-authors-editors/manuscriptpreparation/5636) plus the corresponding pdf- or zipped ps-file. The LaTeX source files are essential for producing the full-text online version of the book, see http://link.springer.com/bookseries/304 for the existing online volumes of LNM). The technical production of a Lecture Notes volume takes approximately 12 weeks. Additional instructions, if necessary, are available on request from lnm@springer.com.

7. Authors receive a total of 30 free copies of their volume and free access to their book on SpringerLink, but no royalties. They are entitled to a discount of 33.3 % on the price of Springer books purchased for their personal use, if ordering directly from Springer.

8. Commitment to publish is made by a *Publishing Agreement*; contributing authors of multiauthor books are requested to sign a *Consent to Publish form*. Springer-Verlag registers the copyright for each volume. Authors are free to reuse material contained in their LNM volumes in later publications: a brief written (or e-mail) request for formal permission is sufficient.

Addresses:

Professor Jean-Michel Morel, CMLA, École Normale Supérieure de Cachan, France
E-mail: moreljeanmichel@gmail.com

Professor Bernard Teissier, Equipe Géométrie et Dynamique,
Institut de Mathématiques de Jussieu – Paris Rive Gauche, Paris, France
E-mail: bernard.teissier@imj-prg.fr

Springer: Ute McCrory, Mathematics, Heidelberg, Germany,
E-mail: lnm@springer.com

Printed in the United States
By Bookmasters